T0338258

m-HEALTH

m-HEALTH

Fundamentals and Applications

ROBERT S. H. ISTEPANIAN
BRYAN WOODWARD

 IEEE Engineering in Medicine
and Biology Society, *Sponsor*

 IEEE Press Series in Biomedical Engineering
Metin Akay, *Series Editor*

For general information on our other products and services or for technical support, please contact our Customer Care Department within the United States at (800) 762-2974, outside the United States at (317) 572-3993 or fax (317) 572-4002.

Wiley also publishes its books in a variety of electronic formats. Some content that appears in print may not be available in electronic formats. For more information about Wiley products, visit our web site at www.wiley.com.

Library of Congress Cataloging-in-Publication Data is available.

ISBN: 978-1-118-49698-5

Printed in the United States of America

10 9 8 7 6 5 4 3 2 1

*'Behold, I will bring to it health
and healing, and I will heal them and reveal to them abundance of
prosperity and security'*

Jeremiah 33:6

'Mespot,' to the day when the palm trees will smile again

and

To the centenary of 'Forget-Me-Not.'

Robert S. H. Istepanian

CONTENTS

ABOUT THE AUTHORS

ROBERT S. H. ISTEPANIAN

Robert Istepanian is recognized as one of the leading authorities and pioneers of mobile healthcare and the first scientist to have coined and defined the concept of "m-Health." He holds a Ph.D. in Electronic and Electrical Engineering from Loughborough University, UK, and he has held several academic and research posts in the UK and Canada. These included visiting professor at the Department of Electrical and Electronic Engineering, Imperial College, London; professor of Data Communications for Healthcare and founding Director of the Medical Information and Network Technologies Research Centre at Kingston University, London; senior lectureships in the Universities of Portsmouth and Brunel University, UK; and associate professor at Ryerson University, Toronto with adjunct professorship at the University of West Ontario, Canada. He has also held a visiting professorship at St. George's Medical School, University of London, and was the Leverhulme distinguished visiting fellow at the Centre for Global e-Health Innovation, University of Toronto.

Professor Istepanian was awarded the 2009 IEEE award for the best and most cited paper by the IEEE Engineering in Medicine and Biology Society for his seminal paper on mobile healthcare (m-Health) published in 2004. He was also the recipient of the IEE Heaviside Award in 1999 from the Institution of Electrical Engineering, UK. He has led numerous funded multidisciplinary research projects on m-Health, e-Health, and telehealth funded by the UK Engineering and Physical Research Council, the European Commission, the British Council, the Royal Society, the Royal Academy of Engineering, and the Leverhulme Trust, in addition to sponsored projects and clinical trials funded by global telecom and mobile industries.

He has served as the Vice-Chair of the International Telecommunication Union focus group on standardization of machine-to-machine (M2M) communications.

He has also served as an expert on numerous assessment and peer evaluation panels on healthcare technology innovations, well-being, m-Health, and e-Health, including the Dutch–Philips partnership program on "Healthy Life Style", the Science Foundation Ireland Strategic Research Cluster Grants program, the Finnish Strategic Centers of Science, Technology and Innovation, and the Canada Foundation for Innovation. In addition, he has been a peer reviewer for the following UK Funding bodies: EPSRC, BBSRC, Wellcome Trust, Department of Health, Service Delivery Organisation, Health Innovation Challenge Fund, National Institute of Health Research, BUPA Foundation, and Diabetes-UK. Further, he has served on the editorial board of *IEEE Transactions on Information Technology in Biomedicine*, *IEEE Transactions on NanoBioScience*, *IEEE Transactions on Mobile Computing*, *International Journal of Telemedicine and Applications*, *Journal of Mobile Multimedia*, and *Journal of World Medical & Health Policy*, and as guest editor of the first two of these journals.

Professor Istepanian has served on numerous IEEE committees and chaired organizing and technical committees of national and international conferences in the United Kingdom, the United States and elsewhere, including the Telemed Conferences at the Royal Society of Medicine, London, the IET, London, the 2000 World Medical Congress, Chicago, and the successive IEEE Engineering in Medicine and Biology International Annual Conferences. He has been invited to present numerous keynote lectures at international conferences and meetings in the UK, Europe, the US, Canada, and other countries. His publications exceed 200 peer-reviewed papers and books on mobile communications for healthcare, m-Health, control systems, and biomedical signal processing.

BRYAN WOODWARD

Bryan Woodward holds two UK doctorates, a Ph.D. in physics from the University of London (Imperial College) and a D.Sc. in electronic engineering from Loughborough University. He has held positions with the UK Atomic Energy Authority, the Royal Australian Navy, Guy's Hospital Medical School, the Australian Atomic Energy Commission, and Loughborough University, where he was Head of the Department of Electronic and Electrical Engineering and a professor with the department's Centre for Mobile Communications Research.

Professor Woodward has been an external examiner for higher degrees at universities in the United Kingdom, France, India, and Australia; a referee for professorial appointments at 12 universities; an invited lecturer in Australia, Burma, China, India, France, Poland, and the United Kingdom; and an expert assessor for peer review research panels for the Australian, Canadian, and Spanish governments, for the UK Engineering and Physical Sciences Research Council (EPSRC), and for the European Commission's 5th and 6th Framework Programmes. Furthermore, he has been a chief examiner and moderator for the UK Engineering Council examinations and a consultant to over 20 companies. He has published over 60 academic journal papers and 120 international conference papers, as well as many articles for professional and popular magazines, and he has also done over 30 radio interviews. Finally, he has been a publications referee and book reviewer for *Electronics Letters*;

IEEE Communications Magazine; *IEEE Journal of Information Technology in Biomedicine* (as Associate Editor and editorial board member); *International Journal of Electronic Healthcare*; *International Journal of Telemedicine and Applications; Journal of Mobile Multimedia*; *Medical Engineering and Physics*; *Optics and Lasers in Engineering*; *Proceedings of the IEE (Circuits, Devices and Systems)*; *Proceedings of the IEE (Communications)*; and *Ultrasonics*.

Professor Woodward has participated in or led 10 multinational research projects funded by the European Commission and others funded by the EPSRC, the UK Department of the Environment, the Indian Department of Science and Technology, and industrial companies. He has also co-ordinated a major m-Health project funded by the British Council's UK–India Education and Research Initiative (UKIERI), with the aim of using mobile communications to improve the monitoring of heart disease and diabetes, which are prevalent in both developed and developing countries. The UK partners were Loughborough University and Kingston University; while the Indian partners were the Indian Institute of Technology Delhi, the All-India Institute of Medical Sciences, and Aligarh Muslim University.

Having retired, Bryan Woodward is now an Emeritus Professor of Loughborough University.

FOREWORD

The prominence of mobile health technologies as a driver for national and international healthcare strategies will undoubtedly grow as modern medicine advances into the 21st century. With smartphone penetration nearly ubiquitous in both the developed and developing world, the global potential to enable high-quality, cost-effective healthcare services - and meaningful patient engagement with patients and the public - is enormous. However, there are unique challenges in tailoring these m-Health strategies to make them accessible in the developing world and to an ageing population burdened with chronic disease. We must also address the important issues of public trust in data sharing, security, consent and privacy that will enable the profound benefits of digital, connected healthcare systems or, which, as likely, could inhibit progress if not tackled head-on.

m-Health: Fundamentals and Applications is a wonderfully comprehensive introduction to the subject of m-Health with valuable examples of studies and successful applications of this rapidly emerging innovation in healthcare. It highlights the crucial work that needs doing if we are to close the gap between what we know—in terms of the clinical evidence supporting m-Health innovation—and the challenges of consumer acceptability that may prevent wider adoption and diffusion of this exciting technological platform.

Professor the Lord Ara Darzi of Denham OM KBE PC FRS

Director, Institute of Global Health Innovation, Imperial College London

PREFACE

Mobile health (m-Health): Is it one of the greatest technological breakthroughs of our time or just another much-hyped smart healthcare technology bubble that could burst soon? Such a paradoxical view is perhaps an accurate reflection of the current status of m-Health. This important, if not essential, healthcare technology is known today to millions of people, both medical and nonexpert alike, as a powerful and transformative concept much needed for twenty-first century healthcare services.

This book has been written to continue the story of m-Health and its development since 2003. Over a decade ago, when m-Health was first introduced and defined, there was no indication then that it would be transformed into today's global multibillion dollar industry, albeit viewed critically and cautiously by the medical and healthcare communities.

M-Health was first defined as *mobile computing, medical sensor, and communications technologies for health care*. This simple yet powerful interpretation of m-Health as a scientific and technological concept has been driven to successful implementation by enthusiastic stakeholders and by rapid developments of these three enabling pillars. Unsustainable healthcare costs and ever-increasing demands for better access and quality of care make m-Health an important technology concept. Unfortunately, m-Health has been distorted and undermined by misleading interpretations, leading to the current spectrum of contradictions and paradoxical views. The collision of the end objectives, requirements, and evidence from opposing business and medical targets is fuelling this *status quo* and inhibiting the as yet unseen potential of m-Health. As an example of this scenario, we all see today major industrial power houses from global telecom, mobile phone, pharmaceutical, health, and insurance companies, and other health-related industries, all vigorously advocating different "consumer m-Health" products and services in a variety of standards and formats.

They range from smart consumer well-being trackers and health monitors, smart health watches, and various targeted healthcare and mobile disease management tools. These and other consumer-based m-Health monitoring devices are becoming increasingly popular and widely used in spite of the absence of large-scale clinical evidence of their healthcare outcomes and improved patient care. The proponents of this consumer's face of m-Health argue that this represents the best realistic path for future predictive healthcare and well-being, and that it potentially alleviates the current burdens of the symptomatic healthcare costs.

At the opposite end of the spectrum, we witness an increasing level of interest in the academic and medical research communities, which target cutting-edge research conducted in different areas of mobile healthcare, leading to many publications, reports, and articles that reflect the clinical outcomes of these studies. Mobile health is also being increasingly taught in related medical and health information training courses. However, regardless of the clinical outcomes of m-Health, there is an increasing trend by some healthcare providers to voice a cautionary note concerning the hype of m-Health, with nonconviction as to the real benefits, questioning its clinical effectiveness and efficacy. These are increasingly justified by the lack of global evidence of large-scale endorsements and acceptance of m-Health by healthcare providers and services. This picture, however, detracts from the clear global health benefits of m-Health.

Increasingly, experts and nonexperts alike are also confused by the plethora of alternative terms and abbreviations being used, such as *connected health*, *smart health*, and even *digital health*, which perhaps reflect this conundrum. These terms are being increasingly used to either replace or justify a new beginning or even shy away from m-Health for one reason or another. Perhaps these newer terms might also reflect the answer to the key question that everyone has been asking for years: *Is m-Health dead or has it just moved address?*

The answer clearly lies in the powerful market forces and economic benefits already mentioned, in addition to the daily supplement of hundreds of m-Health-related documents published in research and NGO reports, academic papers, books, market analysis documents, and online blogs and articles, as well as annual conferences and summits organized globally, reflecting a decade-long evolution of this healthcare technology concept.

Consideration for brevity and the desire to avoid wearying the well-informed by cataloguing what they will regard as obvious has led us to omit from these pages lengthy explanations of certain broad technical issues as much as possible. These issues might be unknown to that large group of general readers who look perplexed when the name "m-Health" arises in conversations, and who only brighten up when it is explained to them that in its most simplistic form it is the use of smartphones for healthcare! This book may, however, serve to bring before the wider spectrum of interested readers some clarification of such a "black hole" and outline something of the infinite variety of the concept. Furthermore, we hope that it will help both expert and lay readers to understand the complexity of m-Health. For this reason we have omitted mathematical equations from the text, but have referenced more detailed papers and books where appropriate.

Chapter 1 charts the evolution of m-Health more than a decade ago and how it was transformed from a mere academic concept to a global, albeit controversial, healthcare technology phenomenon.

Chapters 2–4 describe in detail the basics of the three enabling scientific technological elements of m-Health (sensors, computing, and communications). We describe how each of these key ingredients has evolved and matured over the last decade. We describe, for example, the rapid evolution of m-Health in parallel with the maturing process of its enabling technologies from biowearable sensors to the wireless and mobile communication technologies of 4G and 5G systems and beyond. We also detail in these chapters the impact of new computing and Internet paradigms from the Internet of things (IoT) to Web 2.0 and Health 2.0 on m-Health. We also discuss the role of the current m-Health Apps phenomenon and their clinical efficacy and design challenges, together with other issues such as the role of social networking and healthy data mining concepts on the future advances of "m-Health 2.0."

Chapter 5 illustrates some of the relevant medical aspects and clinical applications of m-Health. We endeavor to clarify some of the concerns and varying views that are being discussed and advocated by the medical community, particularly on the clinical efficacies and effectiveness of some of these smartphone-centric m-Health interventions and applications. These applications are supplemented by clinical examples and current studies, particularly in acute and chronic disease management, and in other important medical conditions. The studies provide clear clinical outcomes in some areas as well as ambiguous and unclear evidence in others.

Chapter 6 presents one of the most rewarding and successful areas of m-Health, which is the endorsement of the success of mobile health as a global health phenomenon. In this chapter, we describe successful applications and deployments of m-Health in various global health settings, particularly in developing countries. We also describe some examples of m-Health in postconflict regions in the world. These examples represent ample proof of the success of m-Health as a transformative concept for better and more effective healthcare delivery, especially in those areas where it is most needed, and where its clinical evidence is clear and its economic impact is justified.

Chapter 7 discusses m-Health markets, business and ecosystem models, and policy-related issues. This illustrates how consumer-led "m-Health" markets are, and will continue to be, one of the driving forces behind the global proliferation of m-Health markets, especially in specific areas of wellness and health monitoring, regardless of the healthcare outcomes and medical efficacy objectives and the pros and cons of markets.

In the last chapter, Chapter 8, we discuss the future of m-Health and we present a vision for its future direction and how this concept can potentially shape and transform healthcare services in the coming decades of the twenty-first century.

Finally, although it is not an easy task to write a book on m-Health and at the same time cover all the important aspects in one volume, we have attempted to include the most relevant issues. This book is mainly written to increase the general awareness and importance of m-Health, not only to interested stakeholders, such as clinicians, healthcare providers, patients, consumers, telecommunications and mobile phone industries, and health insurers, but also to interested lay readers. The aim is to describe the initial philosophy of m-Health, its evolution, and current state of the art, where it is heading and, most importantly, how it can transform some the current healthcare services to better, more efficient, and affordable means of personalized care delivery.

ROBERT S.H. ISTEPANIAN
London, UK

BRYAN WOODWARD
Loughborough, UK

ACKNOWLEDGMENTS

The authors would like to express their deep gratitude to Lord Darzi of Denham of Imperial College London for his very gracious and generous foreword for this book.

Robert S. H. Istepanian would like to acknowledge the support of the late Professor Swamy Laxminarayan, founding Editor-in-Chief of *IEEE Transactions on Information Technology in Biomedicine* (now *IEEE Journal of Biomedical and Health Informatics*), for his vision and leadership in publishing one of the first papers on m-Health in the Transactions.

He would also like to thank all his clinical, academic, and industrial colleagues with whom he collaborated over the last two decades. Special thanks are due to Jose Lacal (Stryker MAKO), Kunle Ibidun (formerly with Orange, France Telecom), Yuan Ting Zhang (Chinese University of Hong Kong), Emil Jovanov (University of Alabama, Huntsville, AL), Costas Pattichis (University of Cyprus), Aura Ganz (University of Massachusetts, Amherst, MA), Nada Philip, Ala Sungoor, Bee Tang, and Barbara Pierscionek (Kingston University, London, UK), Nazar Amso and John Gregory (Cardiff University Medical School, UK), Ken Earle (St. George's Medical School and NHS Trust, London, UK), Tony Constantinides (Imperial College, London, UK), Garik Markarian (Lancaster University, UK), Adel Sharif (Surrey University, UK), Hamed Al-Raweshidy (Brunel University, UK), Alex Jadad, Joseph Cafazo, and Tony Easty (Centre of Global e-Health Innovations, University of Toronto, Ontario), Kaamran Raahemifar (Ryerson University, Toronto, Ontario), and others I may have inadvertently omitted.

Special Acknowledgement: To Bryan, what can I say? Fate brought us together one autumn day in October 1990 when I stood for the first time at your office door at Loughborough University as your new Ph.D. student. Perhaps now you wish you had the Star Trek "Tricoder" to "energize" me away to another Galaxy! Many thanks for your wonderful friendship and English sense of humor, and most of all for all the years of support that I will not forget.

Bryan Woodward would like to thank former colleagues, research students, and final-year students of the Department of Electronic and Electrical Engineering at Loughborough University, particularly David Mulvaney, Sekharjit Datta, Paul Harvey, Omar Farooq (now with Aligarh Muslim University, India), Fadlee Rasid (now with University of Putra Malaysia, Malaysia), Anoop Vyas (now with Indian Institute of Technology Delhi, India), and Bhaskar Thakkar (now with G H Patel College of Engineering and Technology, Gujarat, India).

Special Acknowledgement: My 40-year career at Imperial College London, Guy's Hospital Medical School, the Australian Atomic Energy Commission, and Loughborough University would never have come to fruition but for my good fortune to have met a great teacher when I was 15 years old. The most influential person in my life was the late Harry Morgan, who taught me the power and beauty of the English language and whose inspirational teaching during a difficult period I will remember all my life.

Most of the contracts and grants for our research on m-Health has been awarded by the Engineering and Physical Sciences Research Council, the European Commission's IST, FP7 and Marie Curie Programmes, industrial sponsorships (Motorola USA, Orange, and France Telecom), The Leverhulme Trust, The Royal Society, The Royal Academy of Engineering, The British Council's United Kingdom–India Education and Research Initiative, and the Indian Department of Science and Technology.

We are also particularly indebted to Mr. Harry Istepanian for his excellent work and support in preparing all the graphics and figures in the book, with the assistance of Mr. Dilip Romesh Aravinda (figures graphic design) and Ms. Barbara Lauger (proof reading).

Many thanks are also due to Ms. Mary Hatcher at John Wiley-IEEE Press for offering us the opportunity to publish this work and also for her patience during the much delayed writing process. We would also like to thank Mr. Brady A. Chin at John Wiley-IEEE Press, Danielle Lacourciere (Wiley) and Shikha Pahuja (Thomson Digital) for their editorial assistance in the final preparation of this book.

Finally, we acknowledge our families for their unfailing support and encouragement during the years, and their unrecorded kindness that has rendered our work less difficult.

ACRONYMS

AAA	Authentication, Authorization, and Accounting
AAL	Ambient Assisted Living
ACA	Affordable Care Act
AECOPD	Acute Exacerbation of Chronic Obstructive Pulmonary Disease
AED	Academy for Educational Development
AHIMA	American Health Information Management Association
AI	Artificial Intelligence, Adherence Index
API	Application Programming Interface
ART	Anti-Retroviral Therapy
ASHA	Accredited Social Health Activists
ATM	Asynchronous Transfer Mode
BAN	Body Area Network
BANN	Body Area Nano-Network
BASN	Body Area Sensor Network
BG	Blood Glucose
BLE	Bluetooth Low Energy
BMI	Body Mass Index
BPM (bpm)	Beats Per Minute
BPSK	Binary Phase Shift Keying
BRICS	Brazil, Russian Federation, India, China, and South Africa
BSN	Body Sensor Network
BVP	Blood Volume Pulse
BWL	Behavioral Weight Loss
CCM	Chronic Care Model
CDISC	Clinical Data Interchange Standards Consortium

CDMA	Code-Division Multiple Access
CGM	Continuous Glucose Monitor
CHA	Continua Health Alliance
CHD	Coronary Heart Disease
CHW	Community Healthcare Worker
COPD	Chronic Obstructive Pulmonary Disease
CPS	Cyber-Physical System
CRED	Center for Research on the Epidemiology of Disasters
CRM	Cardiac Rhythm Management
CVD	Cardio Vascular Disease
D-AMPS	Digital Advanced Mobile Phone Access
DICOM	Digital Imaging and Communications in Medicine
DID	Device IDentification
DoS	Denial of Service
DPWS	Devices Profile for Web Services
DSCDMA	Direct Sequence Code-Division Multiple Access
ECG	Electro Cardio Gram
EDGE	Enhanced Data Rates for GSM Evolution
EEG	Electro Encephalo Gram
EHR	Electronic Health Record
EMA	Ecological Momentary Assessment
EMG	Electro Myo Gram
EMR	Electronic Medical Records
EPC	Evolved Packet Core or Electronic Product Code
EPR	Electronic Personal Record
ETSI	European Telecommunications Standard Institute
EU	European Union
EV-DO	Evolution-Data Optimized
FC	Frequency Channel, Frequency Control
FCC	Federal Communications Commission
FDA	Food and Drug Administration
FDD	Frequency-Division Duplex
GB	gigabyte
GDM	Gestational Diabetes Mellitus
GFSK	Gaussian Frequency Shift Keying
GHS	Ghana Health Services
GOLD	Global Initiative for Chronic Obstructive Lung Disease
GPRS	General Packet Radio Service
GSM	Global System for Mobile Communications, Group Spécial Mobile
GSMA	Group Spéciale Mobile Association
HA	Home Agent
HARQ	Hybrid Automatic Repeat Request
HBC	Human Body Communications
HbA1c	glycated hemoglobin (A1c)
HDFS	Hadoop Distributed File System

HetNet	Heterogeneous Networks
HIMSS	Healthcare Information and Management Systems Society
HIPAA	Health Insurance Portability Accountability Act
HITECH	Health Information Technology for Economic and Clinical Health
HIV	Human Immunodeficiency Virus
HR	Heart Rate
HRV	Heart Rate Variability
HSDPA	High Speed Downlink Packet Access
HSPA	High Speed Packet Access
HSPA+	Evolved High Speed Packet Access
HSUPA	High Speed Uplink Packet Access
HTTP	Hypertext Transfer Protocol
IaaS	Infrastructure as a Service
ICT	Information and Communication Technology
IDRC	International Development Research Council
IEEE	Institution of Electrical and Electronics Engineers (USA)
IETF	Internet Engineering Task Force
IHD	Ischemic Heart Disease
IHE	Integrating the Healthcare Enterprise
IHTSDO	International Health Terminology Standardization Committee
IHTT	Institute of Health Technology Transformation
IMS	Information Management System
IMT	International Mobile Telecommunications
IoE	Internet of Everything
IoT	Internet of Things
IP	Internet Protocol
IrDA	Infrared Data Association
ISM	Industrial, Scientific, and Medical (band)
ISO	International Organization for Standardization
IT	Information Technology
ITS	Intelligent Transport System
ITU	International Telecommunications Union
IVR	Interactive Voice Response
IWBAN	Implantable Wireless Body Area Network
IWG	Innovation Working Group
JIC	Joint Initiative Council
KM	Knowledge Mobilization
LEARNS	LEprosy Alert and Response Network and Surveillance System
LED	Light-Emitting Diode
LOS	Line-Of-Sight
LTE	Long Term Evolution
LTE-A	Long Term Evolution Advanced
LoWPAN	Low-Power Wireless Personal Area Network
MAC	Media Access Control

MARP	Most At-Risk Populations
MBAN	Medical Body Area Network
MBOFDM	Multiband Orthogonal Frequency-Division Multiplexing
MCC	Mobile Cloud Computing
MC-CDMA	Multi-Carrier Code-Division Multiple Access
MCOT	Mobile Cardiac Outpatient Telemetry
MDDS	Medical Device Data Systems
MDG	Millennium Development Goal
MEC	Mobile Edge Computing
MENA	Middle East and North Africa Region
MGMP	Mobile Gateway/Mobile Patient
MGSP	Mobile Gateway/Static Patient
MHRA	Medicine and Health Care Products Regulatory Agency
MICS	Medical Implant Communications Service
MIMO	Multiple-Input Multiple-Output
MMA	Mobile Medical Apps
MMC	Massive Machine Communication
MMS	Multimedia Messaging Service
MoH	Ministry of Health
MNO	Mobile Network Operator
MOS	Mean Opinion Score
m-QoE	Medical Quality of Experience
m-QoS	Medical Quality of Service
MTD	Machine-Type Device
M2M	Machine-to-Machine
M4RH	Mobile for Reproductive Health
NB	narrowband
NCD	Non Communicable Disease
NFC	Near-Field Communications
NGN	Next-Generation Networks
NGO	Nongovernment Organization
NHS	National Health Service
NICE	National Institute for Health and Care Excellence
N-LOS	Non-Line-Of-Sight
OFDMA	Orthogonal Frequency-Division Multiple Access
OLAP	Online Analytical Processing
OMI	Operational Medicine Institute
OQPSK	Offset Quadrature Phase-Shift Keying
PaaS	Platform as a Service
PAN	Personal Area Network
PANACeA	Pan-Asian Collaboration for e-Health Adoption and Application
PCC	Patient-Centered Care
PDA	Personal Digital Assistant

PGHD	Patient-Generated Health Data
PHC	Primary Health Centre
PHD	Personal Health Device
PHR	Public Health Record
PHY	PHYsical layer
PIN	Personal Identification Number
POC	Point Of Care
POTS	Plain Old Telephone Service
PPG	Photo Plethysmo Graphy
PQRST	Refers to Specific Points on an Electrocardiogram
PwC	PricewaterhouseCoopers
P2P	Peer-to-Peer
QCI	Quality of Service Class Identifier
QoE	Quality of Experience
QoS	Quality of Service
RCT	Randomized Control Trial
R&D	Research and Development
RFID	Radio Frequency Identification
RHM	Remote Health Monitoring
RMNCH	Reproductive, Maternal, Newborn, and Child Health
ROI	Return Of Investment
ROM	Read-Only Memory
RR	Respiratory Rate
SaaS	Software as a Service
SBA/FD	Skilled Birth Attendance and Facility Delivery
SC-FDMA	Single-Carrier Frequency-Division Multiple Access
SCII	Subcutaneous Insulin Infusion
SCL	Service Capabilities Layer
SD	Standard Deviation
SDN	Software-Defined Networking
SDO	Standards Development Organizations
SGSP	Static Gateway/Static Patient
SGMP	Static Gateway/Mobile Patient
SHARP	Strengthening HIV/AIDS Response Partnerships
S-ICD	Subcutaneous Implantable Cardiac Defibrillator
SIM	Subscriber Identity Module
SMAC	Social Networking, Mobile, Analytics and Cloud
SMBG	Self-Monitoring Blood Glucose
SME	Small-to-Medium-Sized Enterprise
SMS	Short Message Service
SNMP	Simple Network Management Protocol
SOC	System-On-Chip
SO-FDMA	Scalable Orthogonal Frequency-Division Multiple Access
SpO_2	Blood Oxygen Saturation

STI	Sexually Transmitted Infections
TB	Tuberculosis
TCP/IP	Internet Protocol
TDD	Time-Division Duplex
TDM	Time-Division Multiplexing
TDMA	Time-Division Multiple Access
TTC	Text to Change
T1D	Type 1 Diabetes
T2D	Type 2 Diabetes
UID	User Identification
UKIERI	United Kingdom–India Education and Research Initiative
UMTS	Universal Mobile Telecommunications System
UNHCR	United Nations High Commissioner for Refugees
USAID	United States Agency for International Development
UWB	Ultra-Wide Band
VLAN	Virtual Local Area Network
WAN	Wide Area Network
WBAN	Wireless Body Area Network
W-CDMA	Wideband Code-Division Multiple Access
WHO	World Health Organization
WIBSN	Wearable and Implantable Body Sensor Network
Wi-Fi	Wireless Fidelity
WiMAX	Worldwide Interoperability for Microwave Access
WISE	Wireless Intelligent Sensors
WLAN (and Wi-Fi)	Wireless Local Area Network
WMAN	Wireless Metropolitan Area Network
WMTS	Wireless Medical Telemetry Services
WPAN	Wireless Personal Area Network
WSN	Wireless Sensor Network
WWAN	Wireless Wide Area Network
WWBAN	Wearable Wireless Body Area Network
1G	First Generation of mobile phones
2G	Second Generation of mobile phones
2.5G	"Two and a half G," midstage between 2G and 3G
3G	Third Generation of mobile phones
3GPP-LTE	Third Generation Partnership Program Long Term Evolution
3.5G	"three and a half G," midstage between 3G and 4G
3.9G	pre-4G Generation of mobile phones
4G	Fourth Generation of mobile phones
4PSK	Quadrature Phase Shift Keying
5G	Fifth Generation of mobile phones
5GPP	Fifth Generation Public–Private Partnership
8PSK	8 Phase Shift Keying

Units

kbps	kilobits per second
Gbps	gigabits per second
Mbps	megabits per second
mmHg	millimeters of mercury
mg/dl	milligram per deciliter (measurement of blood glucose level)
mmol/l	millimole per liter (measurement of blood glucose level)

1

INTRODUCTION TO m-HEALTH

We have a seemingly insatiable desire to communicate, search, find, share, watch, listen, learn, experience, download and upload information. Year on year the demand grows exponentially.

Peter Cochrane, Technology Futurologist.

1.1 INTRODUCTION

When the concept of mobile health (m-Health) was first introduced and defined in 2003, there was no indication a decade on that it would become the fourth ICT for healthcare pillar after telemedicine, telehealth, and e-Health. Since then, numerous papers and books have appeared in the literature on this topic, together with a multi-billion dollar healthcare delivery-related industry embracing m-Health as its main service. While m-Health is rapidly expanding in the commercial world and has effectively evolved into a separate applied scientific discipline, it is still "under the radar" in general and probably some way from being fully adopted by the medical world, although this gap is narrowing.

The introduction of smartphone technologies has played a major influential role in the evolution of m-Health, albeit in a focused way, particularly in well-being and health-monitoring applications. This role is both a blessing and a curse: the blessing is that technological breakthroughs in mobile and Internet communications are putting m-Health on the global radar, while the curse is the false notion among many people

m-Health: Fundamentals and Applications, First Edition. Robert S. H. Istepanian and Bryan Woodward.
© 2017 by The Institute of Electrical and Electronics Engineers, Inc. Published 2017 by John Wiley & Sons, Inc.

that m-Health is merely another "App" (application). This book is an attempt to present the full picture.

Nowadays, if you type the word "m-Health" into a search engine, the result will be an astounding 100 million or more hits. The 2009 inaugural m-Health Summit, a partnership between the Foundation for the National Institutes of Health and the m-Health Alliance, attracted 800 people; this attendance increased to over 4000 participants from 56 countries at the Fifth Summit in 2013 (Slabodkin, 2013). This is the level of interest that m-Health has generated globally among healthcare providers, academia, and the telecommunications, biomedical, and pharmaceutical commercial sectors.

In this chapter, first we look back briefly at where it all started, then we consider the digital world and the impact it has had on the concept of m-Health. Finally we discuss the correlation between mobile communications and Internet technologies, their role in the future of m-Health, and consequently their social impact on future healthcare services; this is currently being termed "smart health" or "digital health."

We start by defining the various labels that constitute the "Information and Communication Technology (ICT) for Health" domains and how they have been adopted in relation to m-Health, sometimes incorrectly. The notion of m-Health (which we define later) implies an underlying digital ingredient, and the ubiquity of the digital world is exploited in all healthcare applications. The main components that constitute the "building blocks" of the m-Health concept, that is, wireless medical sensors, mobile communications, network connectivity, and the Internet, are themselves invariably digital systems. This combination comprises a powerful trilogy that has spawned an impressive number of applications, but in the years ahead, now with the availability of 4G connectivity, cloud services, Internet of Things (IoT) and Web 2.0, and doubtless future technologies previously undreamed of, these advances will transform the world of medicine and healthcare in a truly spectacular way.

1.2   THE CONCEPT OF m-HEALTH: THE BEGINNINGS

The parable of inventing the wheels on travel luggage is perhaps the best way to describe the beginnings of m-Health. On August 16, 2012, the *USA Today Travel* newspaper published a special report in their travel section (Clark, 2012) entitled "The innovation that revolutionized the traveler's world had humble beginnings," which highlighted how a simple idea, the invention of travel luggage with wheels, can cause millions of people to say: "why didn't I think of that?" The originator of the idea was Robert Plath, a Northwest Airlines pilot who noticed, while queuing behind passengers at an airport security check point in 1987, how many of them were struggling to detach their bags from bulky metal luggage trolleys. Nowadays, almost every sizeable travel bag and case has in-built wheels, which has changed everyone's travel experience for the better. But no one today has heard of Bob Plath, the originator of an innovation used daily by millions of travelers!

The relevance of this story is that it has a parallel with the inception of m-Health. Few know that the Eureka moment on m-Health came in 1996 when the lead author of this book was reading a magazine article at San Francisco Airport about using the then new cellular phones to remind elderly patients to take their medications. From that moment, the concept of using a mobile phone for other medical applications emerged, followed by the publication of the first paper on the modeling of a wireless telemedicine system (Istepanian, 1998). Around 1999–2000, several landmark papers mentioned the principle behind m-Health and presented the basic concept but did not define it clearly as "m-Health" (Istepanian, 1999, 2000; Istepanian and Laxminarayan, 2000).

What is m-Health? Literally, it means "mobile health," which in the context of this book means the use, for health-related applications, of sensors, mobile devices such as "smartphones," tablets, and laptops, and the huge infrastructure of communications networks already in place. New language abounds in m-Health, not least confusing expressions such as "mobile health eco-system," which defines the scope of mobile health in terms of patient, clinician, healthcare provider, service provider, mobile device, and applications (GSMA, 2011).

To the best of our knowledge, the first citation of the word "m-Health" was mentioned in 2003 (Istepanian and Lacal, 2003), followed by the first definition of the concept as *mobile computing, medical sensor and communications technologies for healthcare* (Istepanian et al., 2004). These three basic building blocks are now widely adopted globally for m-Health as one of the main domains of ICT for health. It is also well known that there were no "m-Health" or "mobile health" terms mentioned anywhere prior to these seminal papers. These facts on the beginning of the concept and its original definition have never been documented *prima facie*.

In 2011, the World Health Organization (WHO) defined m-Health as covering "medical and public health practice supported by mobile devices, such as mobile phones, patient monitoring devices, personal digital assistants (PDAs), and other wireless devices" (WHO, 2011). This and other confusing definitions have contributed to the status quo and to ambiguity over the concept's original definition.

Why is m-Health important? One reason is to reverse the traditional event-driven situation in which a patient consults a doctor only when he or she is ill. Instead, recent advances in m-Health allow the patient to be continuously monitored (which may be resisted by some people and is possibly controversial) by a smart sensor network to detect any indication of illness before it occurs. This "well-being" monitoring model can prevent emergencies and empowers patients to be proactive. It therefore allows the realization of a patient-led (or patient-centric) healthcare model rather than a doctor-led model, something that governments are keen to support as a potential means of reducing health expenditure. Some 6 billion mobile phone users globally who are concerned about their health can now exploit the unprecedented evolution of m-Health, with phones that are increasingly equipped with touch screens and Apps that control sensors, Global Positioning System (GPS), camera, and many other features, linked globally by a variety of communications networks. The latest smartphones are even equipped with embedded heart rate and well-being sensors that can calculate real-time heart rates and exercise levels with a touch of the screen. This does not imply that m-Health is merely a smartphone health delivery system as is widely recognized. We will discuss this issue further throughout the book.

There is another reason why m-Health is important; it is that advances in its three main building blocks have enabled the parallel development of innovative mobile healthcare applications and services. To cite an example of its importance, a search of "m-Health" on YouTube results in more than 10 million hits. Another example where the concept entered a major yet controversial leap in its evolutionary process was the introduction of the first smartphone (the iPhone) in 2007; nowadays, tens of thousands of medical Apps are downloaded and used for different medical, well-being, and healthcare delivery applications worldwide on Apple Store or Google Play. This relates to a further reason, and perhaps the most important one, which is the market potential of m-Health, particularly for developing new healthcare delivery models and services that can be more efficient and cost-effective.

In a study by the IMS Institute of Health Informatics, it was estimated that by October 2013 there were 43,689 Apps under the categories of "healthcare and fitness" or "medical," out of which only 23,682 were considered genuine healthcare categories (IMS online).

Table 1.1 shows the results of a survey indicating an increasing trend by physicians and healthcare providers to perform different healthcare activities on mobile devices between 2010 and 2014 (PwC, 2014). Although these results relate to the United States and other high-income countries, similar trends are becoming increasingly evident in many global healthcare service delivery settings.

This evolution would not have been feasible without rapid advances in wireless communications and network technologies from about 2000, when progression from the Global System for Mobile Communications, or Group Spécial Mobile (GSM), to the General Packet Radio Service (GPRS) made it possible to send packet data over networks linked to the Internet. This, for example, enabled for the first time medical "vital signs" such as electrocardiogram (ECG), blood pressure, body temperature, and photoplethysmography (PPG) signals to be sent to a remote server using cellular

TABLE 1.1 Survey of the Increased Trend of Healthcare Activities Performed by Physicians on Mobile Devices 2010–2014

Healthcare Activities Performed by Physicians on Mobile Devices	% in 2010	% in 2014
Access Electronic Health Records (EHR)	12	45
Prescribe medications	14	41
Review images	7	32
Communicate with patients	21	31
Receive data from medical devices	11	20
Initiate and track a referral	6	17
Conduct clinical consultation from different location than the patient's	5	12
Monitor patients who are hospitalized	N/A	17
Receive data from a mobile App	N/A	14
Patient tracks data		

Adapted from PwC HRI (2014).

networks (Istepanian, 1998). This synthesis of mobility and healthcare that forms the basis of m-Health has been quickly adopted by telecommunications, and mobile and medical devices industries, reflecting the huge potential markets envisaged; this has already been realized, albeit with mostly consumer-driven rather that clinically driven outcomes. For example, one study estimated the opportunities in the global mobile healthcare market to be worth between $50 and $60 billions (McKinsey, 2014). In 2014, there were more than 60 industry-sponsored and academic conferences and events relating to m-Health (m-Health Insight, 2014). These activities reflect the global interest and commercial opportunities in m-Health.

The concept is still evolving in its second decade, and as sensing, computing, and networking technologies become ever more refined, advanced, and accessible, there will be a growing demand for all that m-Health has to offer in the future.

1.3 TAXONOMY OF TELEMEDICINE, TELEHEALTH, e-HEALTH, AND m-HEALTH

Before we present details of the m-Health domain, it is imperative to clarify some of the different terminologies used in the ICT for health domains. The increasing number of terms has become so confusing, to experts as well as to lay people, that attempts to classify them have been made, including developing different "taxonomies" (Bashshur et al., 2011). These aim to establish a clear classification of how telemedicine fits with other related domains, notably telehealth, e-Health, and m-Health. These labels have led to the spawning of several international journals since the early 1990s, notably *Telemedicine and e-Health Journal* and *Journal of Telemedicine and Telehealth*, and numerous books, reports, and literature reviews (Bashshur et al., 1997; Yang and Hui, 2008; Olla and Tan, 2009; Currell et al., 2010).

A practical reason for this classification is to attempt to shed light on the true effect of these domains in terms of the cost, quality of service and access to care, or for treatment or monitoring. Further considerations are the place for medical research, health policy, and reimbursement of network charges. For many years it has been asserted that the main benefit of remote medical access is to save time and money, especially for patients, who then do not need to travel from home to a hospital or clinic for a consultation with a doctor. This has still to be verified but it is a worthy goal; for example, if new m-Health procedures could save just 1% of the UK's National Health Service budget, this would amount to more than $1.5 billion annually.

The presumed advantage of establishing the taxonomy—as with flora and fauna—is to bring order out of confusion, to try to fit the various domains and components into a pattern. By its nature, this is not an exact science and numerous iterations may be expected before some definitive blueprint is widely accepted.

In has been proposed that ICT for health domains, or classes, may be thought of as consisting of four key "domains of care" (Bashshur et al., 2011): telemedicine, telehealth, e-Health, and m-Health, as illustrated in Fig. 1.1. The premise is that these terms are not interchangeable because they represent different concepts and related activities, and this becomes evident if we consider their origins.

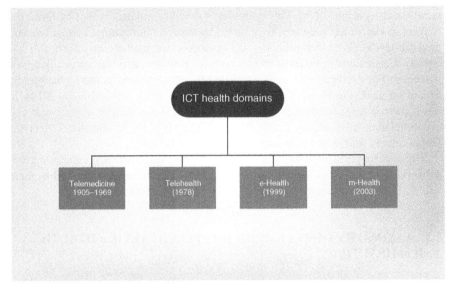

FIG. 1.1   Key domains of ICT for health. (Adapted from Bashshur et al. (2011).)

Telemedicine has been in use since about 1969 (Currell et al., 2010). Since then, telemedicine has been extensively adopted as an umbrella term, and it is appropriate to quote some of key definitions as a starting point: "telemedicine involves the use of modern information technology, especially two-way-interactive audio/visual tele-communications, computers, and telemetry, to deliver health services to patients and to facilitate information exchange between primary care physicians and specialists at some distances from each other" (Bashshur et al., 1997). Another broad definition defines telemedicine as "the use of telecommunications for medical diagnosis and patient care. It involves the use of telecommunications technology as a medium for the provision of medical services to sites that are at a distance from the provider. The concept encompasses everything from the use of standard telephone services through high speed, wide bandwidth transmission of digitized signals in conjunction with computers, fiber optics, satellites and other sophisticated peripheral equipment and software" (Scannell et al., 1995). More concisely, telemedicine may also be defined as the use of communications to exchange medical diagnostic and therapeutic informa-tion, usually between a doctor and a patient who are in different places (Woodward et al., 2001).

The simplest application of telemedicine is, of course, a telephone call or text message between a doctor and a patient. Far from trivializing the subject, this is an example of the literal meaning of telemedicine: *tele*, meaning *far* or *distant*, and *medicine*, meaning *to treat* or *to cure*. A recent review outlined the empirical evidence of telemedicine interventions for diabetes as an example of the effectiveness of telemedicine in some important clinical specialisms and chronic diseases (Bashshur, 2013; Bashshur et al., 2015).

Telehealth came along in 1978 to mean a broadening of the scope of telemedicine to include public health, health education, health services, environmental and industrial health, among others. In terms of taxonomy, telehealth may be seen as a separate domain but there is no hard and fast rule about this. In general, telehealth (sometimes assumed to come under the telemedicine umbrella) deals with remote monitoring by health professionals of a patient's physiological data for diagnosis and disease management. There is also the term *telecare*, which is generally defined as the use of different sensors and alarms to assist people to live independently. This again leads confusingly to the erroneous notion of m-Health as being another telehealth application that uses wireless technologies for remote monitoring using mobile phones.

e-Health has an origin that is still debatable, but there is a general consensus that it started in about 1999–2000 with the start of "dot com bubble." There is now a profusion of words with the "e-" prefix, such as e-book, e-ticket, e-business, e-commerce, and e-government. It seems, then as now, that anyone can make up such a word! In 2005, a study on this topic revealed 51 definitions of "what is e-Health?" (Hans et al., 2005). There is no doubt that more of these definitions have been introduced since then.

e-Health was defined by the WHO as "the cost-effective and secure use of ICT in support of health and health-related fields, including health-care services, health surveillance, health literature, and health education, knowledge and research." The key common ground for e-Health is the use of technology, electronic processing, and communication networks for different healthcare services.

A proposal for the components in the e-Health domain is shown in Fig. 1.2; here, e-Health comprises electronic health records, health information, clinical decision

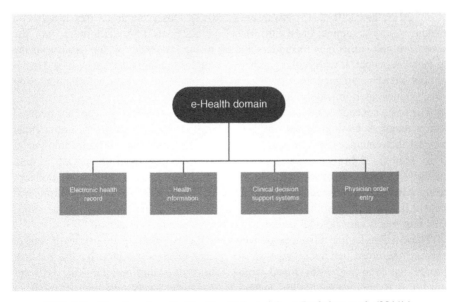

FIG. 1.2   Key domains of e-Health. (Adapted from Bashshur et al. (2011).)

support systems, and physician order entry (Bashshur et al. 2011). This concept has also led some global institutions to categorize m-Health as an offshoot of e-Health with an added wireless element.

The WHO, in their Global Observatory for e-Health service report, asserted that there is so far no standardized definition of m-Health, citing it as "medical and public health practice supported by mobile devices, such as mobile phones, patient monitoring devices, PDAs, and other wireless devices" (WHO, 2011). However, some of the terminologies and interpretation used in this definition (e.g., the now obsolete PDA) are strongly debatable and add to the confusion surrounding the nature of the m-Health concept. There is also the obvious view that the inclusion of m-Health as a subdomain of e-Health reflects the current paradox of the differentiation between the two concepts.

m-Health, as mentioned earlier, was first coined and defined in 2003 as *mobile computing, medical sensor and communications technologies for healthcare* (Istepanian and Lacal, 2003; Istepanian et al., 2004). This basic definition is still valid today, as it simply represents the essence of what the m-Health concept is all about. It also reflects the adoption of this domain as a triumph of digital ubiquity and technical evolution that we witness today, from technologies such as smartphones, 4G cellular networks and beyond, cloud computing, smart wearable medical sensors linked with the IoT, and machine-to-machine (M2M) communications. Telecommunication industries were the first to embrace mobile health as part of their future business and revenue models, with Vodafone being the earliest in 2006. As already mentioned, this domain is defined as much in terms of associated technology as with health concepts (Istepanian et al., 2006).

Several derivative definitions of m-Health have been cited that erroneously categorize mobile health as part of e-Health, telehealth, or telemedicine (Free et al., 2013; Fiordelli, 2013; Kwan et al., 2013). Observers argue that it does not matter how you define m-Health. However, such arbitrary definitions fuel the confusion and sometimes misunderstanding to the lay reader on the essence of the m-Health concept. It widens the gap between understanding the science of m-Health as opposed to viewing it as the provision for healthcare services and information using smartphones, as it is widely understood.

The basic concept of m-Health is illustrated in Fig. 1.3. This concept has evolved into a major technological and applied scientific domain with global commercial interest that embraces advances in the healthcare, computing, and telecommunications sectors. It brings together academic researchers, medical specialists, and business experts worldwide to achieve innovative solutions in healthcare delivery using advances in these technologies. Further developments in the near future will benefit from ultrafast mobile broadband connectivity with new Internet architectures, enabling much wider global access to healthcare services on demand.

These technological advances are well known to be spiraling almost out of control and are therefore inevitably way ahead of any changes adopted by the medical profession, even in the developed world. While something may be achieved technically, it is another matter for doctors to introduce a new treatment or mode of working or clinical procedure into everyday use. This is a challenge for the future.

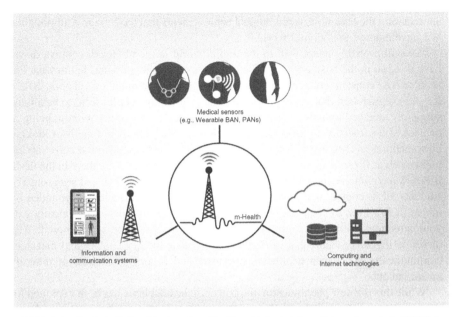

FIG. 1.3 Basic building blocks of m-Health. (Adapted from Istepanian et al. (2004).)

1.4 m-HEALTH AND DIGITAL UBIQUITY

In simple terms, m-Health exemplifies our digital world from the healthcare perspective, so it is worth taking a brief sojourn into this world to consider how it might affect us when truly personalized health systems are available to everyone. Some have found alternative terms for m-Health, for example, "connected health" or "digital health," to reflect the fact that most of what is happening in the new digital communications era is based on one form or another of mobile devices and terminals (Williams, 2013).

The "digital ubiquity" terminology is of vital importance in any aspect of science or engineering to do with "signals," or how we represent information. For the lay reader, an analog signal is continuous, while a digital signal represents discrete samples of the analog original. Hundreds of textbooks describe them and how one can be converted into the other. Although we will not repeat the process, it is at the heart of many practical engineering applications, including biomedical and other related disciplines and applications that we will consider in this book. In practice, "digital" represents progress, as manifested by the demise of now obsolete analog television sets in favor of digital sets.

Now that we are so dependent on mobile devices—smartphones, smart tablets, Kindles, tablets, laptops, and the like—we have unconsciously adopted a lexicon of technical language without necessarily understanding its meaning. We send e-mails and texts, we use Facebook, we tweet using Twitter, and Skype (and use ghastly new verbs like these!), and feel completely at ease with this new jargon without appreciating the enormous technological achievements that have made it all possible. We have entered the "digital world."

We will probably adapt well to the digital world in the m-Health context, as we have done so in the business world, because this digital ubiquity has led to what has been called "constant connectivity" or "continual digital stimulation" (Topol, 2012). It is no exaggeration that many people, particularly young people, seem to be almost surgically attached to their smartphones. They cannot go 5 minutes without being in contact with someone else out there in the ether or checking their e-mails or looking up something on the Internet. People walk along the street tapping out texts or talking to themselves (or so it appears), often ignoring their friends who are there in the flesh. It is now a common experience to walk into a university dining hall and see groups of students all in contact with someone else who is not present, while surrounded by others. It is a form of social exclusion: someone who is in touch electronically gets your attention, while someone who is there in your presence gets ignored. It is therefore highly probable that m-Health will become the key enabler and means of communications between healthcare providers and future generations of digitally aware patients.

While this is a new phenomenon, the power of the telephone has been exploited for decades. One of the authors (B.W.), when working as a research associate at Guy's Hospital Medical School, London in the early 1970s, found that it was often difficult to get an appointment with the Head of the Anatomy Department, who had a zealously protective secretary. The way to get to talk to him was simply to phone him; without realizing it, he somehow felt compelled to answer the ringing menace on his desk!

As Topol puts it, the digital revolution has introduced a new syndrome of "constant connectivity" with a tsunami of digital data—always checking e-mails, making phone calls, sending texts, taking photographs and sending them instantly, to such an extent that this form of anti-social behavior, to any non-addict, has now become "normal." While it is not entirely acceptable, it is difficult to get young people to break the habit, even temporarily, for example, during a meeting, dinner in a restaurant, or even a theatre performance.

What drives us to this kind of behavior? Many suggestions have been put forward to explain this always-on lifestyle, such as a "digitally induced attention deficit disorder" (Richtel, 2010), which induces smartphone devotees, the new "digerati" (digitally literate), to seek "information without knowledge" and subject themselves to "data diarrhœa" and a "deluge of digital drivel" (Topol, 2012). The "Google effect" and the addiction to screen watching is all too self-evident in many people; it is the preserve of a new species called *Homo distractus* and if we are guilty of joining this club, we may be said to be "outsourcing our brains to the cloud" (Keller, 2011).

So, many of us—again mainly young people and the digital savvy generation—are effectively preconditioned to take on the technology of m-Health in the near future, as evidenced by the increased use of smartphones, social networking, and downloading Apps. The key challenge for m-Health is to convert this digital behavior to realize positive and better health outcomes. This is one of the main research targets in the development process of new mobile health systems and their clinical impact and benefit issues.

With a smartphone, we are half way there. Only the body sensors need to be added to the mix. Some of these medical sensors, such as those for heart rate monitoring and for exercise and fitness applications, are already embedded in some of the newer smartphone models on the market. One of the many concerns with social networking in the healthcare domain is that people with medical problems, real or imagined, are seeking diagnoses and potential cures from the Internet rather than from their doctor. There are also thousands of Apps that purport to monitor and diagnose a vast number of health parameters. It is a worrying trend because it is one thing to obtain the data but quite another to make sense of it. The machine cannot make a reliable diagnosis; that requires the intervention of a doctor, whom many people are reluctant to consult unless they are actually feeling unwell. Ethical and privacy issues, along with concerns of using downloaded Apps, will inevitably be subject to scrutiny by the medical world in the coming years.

It is estimated that the use of smartphone technology can create more than 5 billion points of contact between consumers, healthcare workers, health system administrators, and firms in supply chains for health commodities (World Bank, 2012). Ironically, while smartphones may be seen as a potential cure-all, with reservations, there has been concern for many years that their over-use may be a health problem in itself, notably brain tissue heating, although most studies have yielded negative or inconclusive results (Patrick et al., 2008). This issue has been largely underestimated, especially with the massive increase and sustainable use of mobile phones globally. These issues were studied extensively with yet unclear long-term outcomes on human health (WHO, 2006, 2013) and warrant further research within the development process of future mobile health systems.

With major technological advances in smartphones and cellular network structures, commensurate with wider frequency spectra and operational modalities of the forthcoming 5G-based smartphones, tablets, and the IoT connecting billions of everyday devices worldwide, this issue needs careful revisiting. From the healthcare perspective, these and other factors, such as cybersecurity, privacy, and ethical concerns, constitute the "dark side" of m-Health that warrants further attention and research.

The current debate is also raging on the privacy and security of so-called "big medical data" obtained from patients, because the huge amount of data that can be stored is virtually unlimited. In 1969, NASA put Armstrong and Aldrin on the moon using a computer with the computing power and memory capacity of a pocket calculator. Their systems were triplicated, so that if one of them failed they still had two backups, and if a second one failed they had a final backup. Think how much easier the moon landing would have been with a modern tablet or a smartphone, with their sophisticated software and huge memories!

The volume of the "big health data" and what to do with it is another consequence of this digital ubiquity. Nowadays, thousands of servers are out there, each with a vast memory. The label that has been adopted—"cloud"—reflects the impression of some immense computer but is arguably an understatement given the galactic scale of its memory. The amount of memory available in the 1980s, when the first laptops came on the market, was stated in kilobytes (1000 bytes, or 1000×8-bit words) and later in megabytes (a million bytes); to put this in context, a digital photographic image needs typically 500 kB to 1.5 MB of memory so an early laptop could not store even one digital photograph. Nowadays, memory is stated in gigabytes (1000 MB), terabytes (1000 GB), petabytes (1000 terabytes), or even exabytes (1000 PB). These figures are beyond the normal human scale of appreciation. Making sense of this, if a photograph uses 1 MB, then an exabyte of memory could store a trillion photographs, that is, a million million. A new scientific area of "big mobile health data" is now born out of these developments.

The deluge of digital information from a "sea of sensors," including embedded sensors in smartphones, tablets, laptops, and other gadgets allows us to sense every parameter imaginable. The trend in the next decade is likely to be an exponential increase of data from these devices (The Economist Group, 2010). In the context of this book we have the capability to digitize almost anything related to biology and physiology, and although there is huge interest world-wide in doing this, with resources and effort from researchers and from industry, the medical profession remains—perhaps not without reason—cautious to some extent to such change. Medical ethics and privacy-related issues will be a major subject in the near future once these medical sensing and communications advances become more widely accessible and applicable. Technologists can show doctors what is possible, but for doctors to accept their suggestions is still a distant objective. Inevitably, and regardless of these discussions, it will take some time, perhaps a generation, before we have a truly "patient-led" or "patient-centered" society, as decreed by some Western governments, which are always striving, usually controversially, to increase "efficiency."

1.5 THE PARADIGM SHIFT OF MOBILE CONNECTIVITY AND m-HEALTH SERVICES

Figure 1.4 shows some of the key technological advances between the advent of the mobile phone in 1973 and the evolution of m-Health since 2003. We have come a long way in terms of technological progress from the field telephone, as used by the military in the early years of the twentieth century, but it was still a mobile system, albeit a hefty box encumbered with wires. More than half a century later, there were portable phones the size of a toaster, then in 1971 came what must have been a real breakthrough with the introduction of the Radiotelephone Chopper Bicycle, which was demonstrated at the Ideal Home Exhibition in London. In 1973, the first mobile phone call was made from a truly handheld device, as seen in one of the James Bond films when our hero used a phone the size of a house brick, now long since consigned to obsolescence. In the era up until about the 1990s, these budding mobile phone technologies are generally looked upon as the first generation (1G), although the term was not widely used at the time.

From about 1990, the first "normal-sized" lightweight mobile phones came on the market and in 1992 the first text was sent using the newly introduced short message service (SMS). There was little inkling then that "texting" (to use a new verb in the

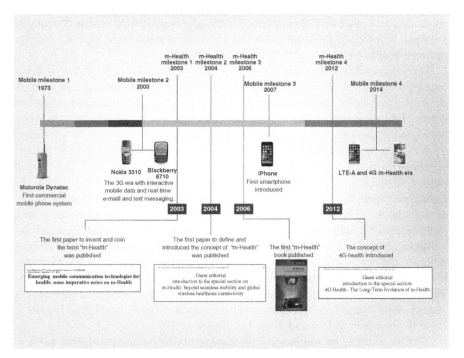

FIG. 1.4 A historical perspective of the evolution of mobile phones and m-Health.

English lexicon) would be so incredibly popular, probably superseding voice calls by an order of magnitude because of its cost-effectiveness. This second generation (2G) had limitations in terms of both bit rates (9.6 kbps) and operational coverage, which was much lower than for the Plain Old Telephone Service (POTS) of fixed lines. This provided limited capability for real-time data transmission but with a comparatively narrow spectrum available for medical applications.

A decade later, by about 2000, people had a taste for global communications and wanted more than a simple phone that they could carry around: They wanted a device of the same size that could do a lot more, such as take photographs and take on the role of a small computer. We were now into the third generation (3G) era of the now defunct PDA and the pre-smartphone era; in 2003, the Blackberry was born, then in 2007 Steve Jobs introduced the iPhone, heralding the beginning of the smartphone era that has had a profound effect on m-Health. A further milestone came in 2008 when Google brought in their Android software, which has since become a standard software feature on their smartphones and tablets.

In general, two major achievements have affected the evolution of m-Health since the early 2000s. The first was the launch of 3G, which brought a further advancement of cellular networks and data services. It allowed for the first time simultaneous voice, data, multimedia, wireless messaging, and mobile computing services, with improved quality of services for most of the first m-Health applications, for example, vital sign monitoring and real-time diagnostic applications. The concept of including high-speed data and other services integrated with voice transmission emerged as one of the main innovations in future telecommunications systems, especially for patient-centered healthcare.

The second achievement was the introduction of the iPhone in 2007. This ushered a new yet controversial era for m-Health in which smart healthcare applications were developed in their thousands and could be downloaded by patients with a touch of the screen. This phenomenon moved m-Health from the mere research base to the wider world and made it the major mobile health application industry that we see today.

We are now, since 2012, in era of the fourth generation (4G), which has the capability of five times the data rate of 3G, with "theoretical" download rates of around 20 Mbps and upload rate of 22 Mbps (in practical and usability terms, the smartphone users have much lower Mbps download and upload rates than these). As of today, most 4G licenses from telecom operators are being rolled out globally, with higher speeds that will have a major impact on the mobile business and potentially on various mobile phone-centric m-Health services. A major advantage of using 4G is its ubiquity and personalization, with Internet access at the core of its connectivity and related services. It is expected that post-2020, the fifth generation (5G) will be launched, with smart wireless capabilities and data rates approaching the gigabyte per second range. These will open unprecedented opportunities for m-Health. Details of these technologies will be explained in Chapter 4.

This connectivity paradigm will have a major impact on future m-Health services and applications, especially for developing countries. This is highly significant for m-Health because it supports mobility, for both patients and doctors, and also allows recent developments in cloud computing and storage to be integrated with m-Health services, thereby obviating the need to store patient data on individual devices or servers and thus ushering the new era of "Big m-Health Data."

New cellular technologies from 4G, 5G, and beyond will also support the use of M2M connectivity and IoT paradigms as transformative communication protocols that will have a profound impact on future mobile health systems. At present, the only roadblock in this context is the lack of interoperability of the "machines"; these are sensors, body area networks, handheld devices (used as "gateways" between patients and the outside world), computers, their connections, and so on. For health monitoring, handheld devices, such as smartphones or tablets, will probably become redundant because future smart body area networks will be able to transmit directly to more conventional networks, as used for telecommunications. The urgent need for interoperability is probably the biggest challenge to manufacturers of these systems, who have to collaborate, make compromises, and therefore evolve business models that are beneficial to all, including the manufacturers themselves. The closest parallel scenario today is in advances in the homes automation sector, in which home security applications and utilities monitoring can be controlled remotely via smartphone connectivity.

Many of today's smart mobile phones, even the latest and greatest, will eventually become outdated, so it is pointless to promote particular brands. It will be interesting to reflect in, say, 10 years after their launch, how many of them have stood the test of time. Will the Apple iPhone 6, Samsung Galaxy S6, Motorola MotoX, and so on still be around? As we read these pages, in few years from now possibly all these models will be obsolete. It is fair to predict that they will be superseded by more appealing versions with ever greater numbers of Apps, but it is likely that these gadgets will have inbuilt intelligent sensors that can be individually calibrated. The Apps themselves would not need to be hosted on individual phones because they would be sited in a "cloud" and downloaded when needed (Davis, 2013).

Perhaps "the need for speed" sums up the current 4G systems as well as any phrase, and in the m-Health context its use by the London Air Ambulance is a good example (Phelan, 2013). The major advantages of 4G here are the added speed of carrying out navigation to pinpoint the site where an accident has occurred, or where a patient is waiting, and the capability of transmitting high-resolution video images to a trauma specialist in a hospital.

Present and future technologies, from 4G to 5G and beyond, will support the development of effective medical care delivery well into the twenty-first century. The new wireless technologies will allow both doctors and patients to roam freely, while maintaining access to critical medical information (Istepanian and Zhang, 2012). It was predicted more than a decade ago that increasing data traffic and demands from different medical applications will be compatible with the data rates of 4G systems for true mobile applications. However, it is the organization and bureaucracy that will be slow to catch up and the following oft-quoted "equation" is worthy of mention (Istepanian and Lacal, 2003):

Current organization + new technology = expensive current organization

Hence, the expectations are for new generations of mobile technologies to be acceptable for challenges that include the following:

- *Aging Society:* There are increasing numbers of older adults with concomitant health problems and fewer young people, so that to sustain the economy, those older people will have to be persuaded to continue working longer. To be able to do this, a greater emphasis on the health of the elderly will mean an increase in demand for healthcare. At the moment, an obstacle to the implementation of m-Health is that commercial organizations do not regard the aging health economy as large enough to invest time and research in. We are probably at the cusp of a big change, because the growing demand for healthcare services and the reduced supply of service providers and caregivers will mean that m-Health suddenly acquires a heightened importance.
- *Fragmentation:* There is a fragmentation of care caused by the need to push down costs while still keeping abreast of rapid technological and medical advances. Such cooperation in healthcare will probably be achieved more by "patient power" than by government directives.
- *Expectation:* There is increased patient expectation because of easier access to information, which will mean that the preeminence of doctors will be challenged. Lifestyle changes will mean that affluent patients will tend to demand consultation and treatment at any time and wherever they are, although patients

at the lower end of the socioeconomic scale may have to settle for lower expectations.

- *Personalization:* Increased complexity of assessment, diagnosis, investigations, and treatment will mean a knowledge explosion and m-Health may serve a useful function of rapid dissemination of the skills and knowledge, especially in the genomic age. The increasing interdisciplinary area of personalized health will become a fundamental part of what future m-Health systems will look like and how they will function, as will be shown in later chapters.

- *Data and Security:* The availability of health-related data is particularly important and crucial for the development process of future m-Health systems. There is increasing influx of health-related data generated by smartphone users globally. This vast amount of data will require robust security and privacy measures and processing mechanisms that need to tackle the increasing worries and potential threats of when and where these data can be accessed and used, how these can be accessed by clinicians and healthcare providers, and so on.

1.6   IMPACT OF m-HEALTH ON CULTURAL, COMMERCIAL, AND OPERATIONAL CHANGES

A nation's health service is fashioned by its economy, demography, culture, and medical tradition, among other factors. This identity poses a challenge to m-Health, which can improve this service. In addition, there is the problem of "component management," which derives from the situation that the providers and payers of healthcare view health challenges only through the specific window of care for which they are responsible.

One of the main incentives to progress in healthcare is reimbursement and revenue models, which is basic to the cost of healthcare. Thus, telecommunications and Internet service provider companies are increasingly forced to organize their packages into reimbursable models. In the current economic climate, this issue is paramount for the future use of m-Health services.

Any task that falls outside these packages tends to be overlooked or receives low priority. From the mobile health perspective, component management systems serve patients poorly. The current healthcare service emphasis is on treatment rather than prevention, and there is a lack of incentive for providers to treat the entire disease process, which leads to an uncoordinated delivery system. However, there is a new wave of change advocated by mobile health providers to shift symptomatic care models to preventative care models, and future mobile health systems can be the enablers.

Some other key factors that may accelerate the diffusion of m-Health with social and medical impact are as follows:

- From a management perspective, planning for future implementation of m-Health services needs to favor mobile solutions to replace the old fixed computer-based Internet connectivity services.

- From an economic perspective, m-Health costs and savings need to appear on the same account to identify any benefits.
- From a government perspective, there is a need to fund promising m-Health initiatives that may be integrated into the healthcare system, and to make comprehensive assessments of m-Health advantages and benefits.
- From an education perspective, there is a need to inform doctors and healthcare providers of the latest technology developments and what can be achieved by adopting future systems based on them.
- From a legal perspective, ethical issues of mobile healthcare services need to be addressed and decided.

In the social context, constant connectivity may be instant but medical treatment of an individual patient is only really effective if it is specific to that patient rather than to the population as a whole. For example, a patient is given the correct dose of the most appropriate drug, taken at specific times and for a particular duration. This is a simplification but it gives the general idea.

Other m-Health challenges and limitations that are still under discussion may be summarized as follows:

- *Lack of Interoperability and Standards:* This will be discussed at length in a later chapter, but essentially it means an incompatibility between the building elements that comprise present mobile technology hardware, medical sensor connectivity, and web access protocols. Furthermore, there is a lack of integration between various medical services that use m-Health to enable access to health records, appointments, drug compliance and administration, referrals, images, and so on. What is needed is a truly integrated "m-Health on demand" universal mode of operation. This is currently being debated by international organizations such as the WHO and the International Telecommunications Union (ITU). Particular effort concerns the impact of M2M communications, for which standards for m-Health services are being finalized.
- *Security and Privacy:* There is increasing demand to develop robust and secure approaches for improved security and privacy of medical data transmitted in m-Health connectivity channels. This is a challenging issue that is currently being addressed by major research and industry institutions.
- *Medical Working Practice and Acceptability:* It is generally difficult to implement any major *modus operandi* in traditional medical practices and hospitals, especially the adoption of new and unproven technologies. This is not necessarily due to staff obstructiveness, or their reluctance to accept changes in working practices, but more because of the layers of complex bureaucracy and organization, together with pressurized schedules, lack of incentive, and a lack of appreciation of any benefits, including cost-saving and better use of time.
- *Business Models and Ecosystems:* There are to date no clear business models for m-Health but only sporadic models. It is difficult terrain to navigate as there is clearly no overarching evidence that a single m-Health business model can work

for different applications and services. Payment and revenue models change from one case to another. This is a challenge that has been addressed extensively by the different telecommunications and service providers that have embraced m-Health systems. However, there are success stories of business models that are working, otherwise how can the huge level of investment, estimated in billions of dollars in this domain, be justified?

There are other factors relating to the technology hype cycle, and the current position of m-Health within the innovation cycle is open to further debate and discussion. These and other issues and barriers for larger deployment of m-Health services globally will be considered in more detail in later chapters.

1.7 SUMMARY

Since the introduction of the concept over a decade ago, m-Health has become one of the four domains of ICT for healthcare. Some would argue that it is the key domain due to its fundamental link with our digital and mobile world. Clearly, mobile health is now not only a mere ICT for health domain but also a science in itself. It represents how technology and related advances in mobile communications, Internet technologies, and medical sensors can better serve medical and healthcare provisions. However, there is ongoing debate and opinions about the exact meaning of m-Health and its role in future healthcare services and well-being applications.

The role of the current 4G and future 5G communications and network technologies, and other wireless technologies and their connectivity with smart wearable sensors and cloud computing platforms, will play an increasingly pivotal role in m-Health systems that will be part of everyday life for many people in the near future. Understanding people's social and behavioral response toward the acceptance of this generation of smart mobile computing and network systems is an important advance in this evolutionary process. Another advance will be the future interface architectures of smart M2M communications and IoT for m-Health that will potentially lead to the development of a new generation of social networking and intelligent computing connectivity care models. The era of a successor, m-Health 2.0, is evolving from these developments. However, all challenges will not be fully realized without bringing together the multidisciplinary areas of medicine, psychology, sociology, economics, engineering, and computer science to achieve the full potential and impact of m-Health that we alluded to in the beginning of this chapter.

REFERENCES

Bashshur RL (2013) Guest Editorial: compelling issues in telemedicine. *Telemedicine and e-Health* 19(5):1–3.

Bashshur R, Sanders JH, William G, and Shannon GW (Eds.) (1997) *Telemedicine: Theory and Practice*, Springfield, IL: Charles C Thomas Publisher Ltd.

Bashshur R, Shannon G, Krupinski E, and Grigsby J. (2011) Policy: the taxonomy of telemedicine. *Telemedicine and e-Health* 17(6):484–494.

Bashshur RL, Shannon GW, Smith BR, and Woodward MA (2015) The empirical evidence for the telemedicine intervention in diabetes management. *Telemedicine and e-Health* 21(5):321–354

Clark J (2012) *Rollaboard luggage celebrates a wheelie big birthday: the innovation that revolutionized the traveler's world had humble beginnings*, USA Today Travel, August 16. Available at http://travel.usatoday.com/news/story/2012-08-16/Rollaboard-luggage-celebrates-a-wheelie-big-birthday/57104830/1 (accessed October 2013).

Currell R, Urquhart C, Wainwright P, and Lewis R (2010) Telemedicine versus face to face patient care: effects on professional practice and health care outcomes. *Cochrane Database of Systematic Reviews* 2, CD002098.

Davis A (2013) 4G allows you to do things you couldn't do before. *The Daily Telegraph: The Age of Mobile*, April 20. Available at http://www.telegraph.co.uk/sponsored/technology/4g-mobile/10001468/4g-speed-benefits.html (accessed November 20, 2013).

The Economist Group (2010) All too much: monstrous amount of data. *The Economist*, February 25. Available at http://www.economist.com/node/15557421 (accessed November 2013).

Fiordelli M (2013) Mapping m-health research: a decade of evolution. *Journal of Medical Internet Research* 15(5):e95.

Free C, Phillips G, Watson L, Galli L, Felix L, Edwards P, Patel V, and Haines A (2013) The effectiveness of mobile-health technologies to improve health care service delivery processes: a systematic review and meta-analysis. *PLoS Medicine* 10(1):e1001363.

GSMA (2011) *Connected mobile health devices: a reference architecture*, White Paper, Version 1.

Hans OH, Rizo C, Enkin M, and Jadad A. (2005) What is e-Health? A systematic review of published definitions. *Journal of Medical Internet Research* 7(1):e1.

IMS Institute of Health Informatics *Patient Apps for improved healthcare: from novelty to mains stream*. Available at http://www.theimsinstitute.org, (accessed January 2016).

Istepanian R (1998) Modelling of GSM-based mobile telemedical system. *Proceedings of the 20th Annual International Conference of the IEEE Engineering in Medicine and Biology Society*, Vol. 3, pp. 1166–1169.

Istepanian RS (1999) Telemedicine in the United Kingdom: current status and future prospects. *IEEE Transactions on Information Technology in Biomedicine* 3(1):158–159.

Istepanian RSH (2000) Guest Editorial: special issue on mobile telemedicine and telehealth systems. *IEEE Transactions on Information Technology in Biomedicine* 4(3):194.

Istepanian RSH and Lacal JC (2003) Emerging mobile communication technologies for health: some imperative notes on m-Health. *Proceedings of the 25th Annual International Conference of the IEEE Engineering in Medicine and Biology Society*, Vol.2, pp. 1414–1416.

Istepanian RSH and Laxminaryan S (2000) Unwired e-med: the next generation of wireless and internet telemedicine systems. *IEEE Transactions on Information Technology in Biomedicine* 4(3):189–194.

Istepanian RSH and Zhang YT (2012) Guest Editorial: 4G-Health: the long-term evolution of m-Health. *IEEE Transactions on Information Technology in Biomedicine* 16(1):1–5.

Istepanian, RSH, Jovanov E, and Zhang YT (2004) m-Health: beyond seamless mobility for global wireless healthcare connectivity—Editorial Paper. *IEEE Transactions on Information Technology in Biomedicine* 8(4):405–414.

Istepanian RSH, Laxminarayan S, and Pattichis CS (2006) *M-Health: Emerging Mobile Health Systems.* New York: Springer.

Keller W. (2011) *Are we 'outsourcing our brains to the cloud?'* Available at www.bookhaven .standard.edu (accessed November 2013; also The Twitter Trap, *New York Times*, May 18, 2013).

Kwan A, Mechael P, and Kaonga NN (2013) State of behaviour change initiatives and how mobile phones are transforming it. In: Donner J and Mechael P. (Eds.), *M-Health in Practice: Mobile Technology for Health Promotion in the Developing World.* Vol. 2, Bloomsbury, pp. 15–31.

McKinsey (2014) *M-Health.* Available at http://www.mckinsey.com/mhealth (accessed February 2014).

mhealthInsight.com (2014) mHealth events for 2014, http://mhealthinsight.com/2013/12/04/ mhealth-events-for-2014/. (accessed October 2014).

Olla P and Tan J (2009) *Mobile Health Solutions for Biomedical Applications.* New York: Medical Information Science Reference.

Patrick K, Griswold WG, Raab F, and Intille SS (2008) Health and the mobile phone. *American Journal of Preventative Medicine* 35(2):177–181.

Phelan D (2013) *When the need for speed is a matter of life and death.* The Daily Telegraph: The Age of Mobile IV, April 20.

Price Water House Coopers (PwC) (2014) *Healthcare delivery of the future: how digital technology can bridge time and distance between clinicians and consumers.* Available at http://www.pwc.com/en_US/us/health-industries/top-health-industry-issues/assets/pwc-healthcare-delivery-of-the-future.pdf (accessed May 2015).

PwC: Health Research Institute (2014) *Healthcare delivery of the future: how digital technology can bridge the gap of time and distance between clinicians and consumers.* Available at http://www.pwc.com/us/en/press-releases/2014/pwc-hri-report-consumer-friendly-technology.html (accessed January 2015).

Richtel M. (2010) Attached to technology and paying a price. *New York Times*, June 6, 2010. Available at www.nytimes.com/2010/06/07/technology/07brain.html (accessed February 2013).

Scannell K, Perednia DA, and Kissman H (1995) *Telemedicine: Past, Present, Future. Current Bibliographies in Medicine.* Maryland: National Library of Medicine.

Slabodkin E (2013) *mHealth Summit 2013: a preview*, FierceHealthcare Online. Available at http://www.fiercemobilehealthcare.com/story/mhealth-summit-2014-preview/2013-12-03. (accessed October 2014).

Topol E (2012) *The Creative Destruction of Medicine: How the Digital Revolution Will Create Better Health Care.* Basic Books.

Williams D (2013) *Keeping the UK at the forefront of the new digital age.* The Daily Telegraph: The Age of Mobile III, April 20.

World Health Organisation (WHO) (2006) *What effects do mobile phones have on people's health?* Report, WHO Regional Office for Europe's Health Evidence Network (HEN). Available at http://www.euro.who.int/data/assets/pdf_file/0006/74463/E89486.pdf (accessed October 2014).

World Health Organisation (WHO) (2011) *m-Health new horizons for health through mobile technologies.* Global Observatory for eHealth Series, Vol. 3, pp. 5–6, WHO (accessed October 2013).

World Health Organisation (WHO) (2013) *What are the health risks associated with mobile phones and their base stations?* Available at http://www.who.int/features/qa/30/en (accessed October 2014).

WHO Fifty-Eighth World Health Assembly (2013) Agenda item 13.17, Geneva: World Health Organization, 2005. Available at http://www.who.int/gb/ebwha/pdf_files/WHA58/WHA58_28-en.pdf. (accessed November 2013).

Woodward B, Richards CI, and Istepanian RSH (2001) Design of a telemedicine system using a mobile telephone. *IEEE Transactions on Information Technology in Biomedicine* (Special Issue on Mobile Telemedicine and Telehealth Systems, Part II) 5(1):13–15.

Qiang CZ, Yamamichi M, Hausman V, Miller R, Altman D, and World Bank (2012) *Mobile Applications for the Health Sector*, ICT Sector report. Available at http://siteresources.worldbank.org/INFORMATIONANDCOMMUNICATIONANDTECHNOLOGIES/Resources/mHealth_report_%28Apr_2012%29.pdf (accessed October 2014).

Yang X and Hui C (2008) *Mobile Telemedicine: A Computing and Network Perspective.* Boca Raton, FL: CRC Press.

2

SMART m-HEALTH SENSING

Declare the past, diagnose the present, and foretell the future.

Hippocrates

2.1   INTRODUCTION

One of the major challenges for m-Health is the design and development of intelligent sensors. These need to be compatible with current and envisaged future wireless and computing technologies. While there have been technical leaps forward in mobile communications and Internet technologies in the past two decades, medical sensing has remained largely unchanged until recently. Although today there is a full range of commercial medical sensors, from wireless, wearable sensors linked to smartphones, and smart wristwatches to pocket wireless ultrasound scanners, we have yet to see the expected acceptability and validation by patients and doctors of these new sensor technologies that match the ubiquitous use of smartphones and Internet technologies.

We are all familiar with the medical profession's trademark accessory, the stethoscope, which is often worn around the neck as a badge that says, "I am a doctor." While the traditional stethoscope is useful for having a quick listen to the heart, it is useless as a means of sourcing medical data that can be sent to a remote location in the scenarios we are considering here. There is now a so-called "superior stethoscope," which comprises a smart adapter attached to a conventional stethoscope. This obviates the need for expensive echocardiograms by streaming heartbeat

m-Health: Fundamentals and Applications, First Edition. Robert S. H. Istepanian and Bryan Woodward.
© 2017 by The Institute of Electrical and Electronics Engineers, Inc. Published 2017 by John Wiley & Sons, Inc.

data to the cloud so that a cardiologist can download it to a smartphone for remote analysis (Park, 2015).

In the age of smart m-Health, we see advances in sophisticated and reliable electronic devices to sense the body's "vital signs," and indeed many other parameters, in a seamless, simple, and ubiquitous way. More than that, we also need "smart sensors," which have a degree of built-in decision-making capability. We are witnessing ongoing and extensive discussions in the academic and medical research world about these sensor technologies and the approaching era where the smartphone will replace the stethoscope. Again, this is yet to be seen and most importantly clinically validated and widely used.

In this chapter we present a new taxonomy of m-Health sensing. We also chart the evolution of medical sensing from the early remote monitoring principles to recent advances in true m-Health monitoring with wearable technologies, body area networks (BAN), wireless implantable bimolecular sensors, and many more devices.

2.2   FUNDAMENTALS OF m-HEALTH SENSING AND A NEW TAXONOMY

Before presenting the fundamental issues of m-Health sensors, we need to introduce an overall medical sensing taxonomy for m-Health. This is a complex process, which is the reason for the absence of any such taxonomy or classification to date. The problem lies with a general misunderstanding of the concept of m-Health, and therefore of the nature of m-Health sensing. The popular, and dominant, view is that m-Health is a wireless monitoring and health status concept using "mobile technology" (including smartphones). This is only a partial view of the overall notion of m-Health. To illustrate this view in terms of sensors, an earlier taxonomy of "wireless health monitoring" was based on the device mode of operation, that is, whether a static or mobile "gateway," and whether the patient was stationary or moving. This taxonomy comprised four categories (Raskovic et al., 2007):

- Static Gateway/Static Patient (SGSP)
- Static Gateway/Mobile Patient (SGMP)
- Mobile Gateway/Static Patient (MGSP)
- Mobile Gateway/Mobile Patient (MGMP)

This original taxonomy of wireless health monitoring emphasizes the need for a new classification that embraces all manner of sensing, an overall sensing taxonomy for m-Health, not just the wireless monitoring aspects. Figure 2.1 shows a new and generalized taxonomy that reflects the earlier issues and the basics of m-Health. This taxonomy is based on an understanding of the fundamental difference between health, illness, wellness, and well-being, and their relevance to the m-Health continuum and applications. These cover wellness to illness and disease management, diagnosis, and treatment of specific diseases, and assistive and prognostic care. This new m-Health

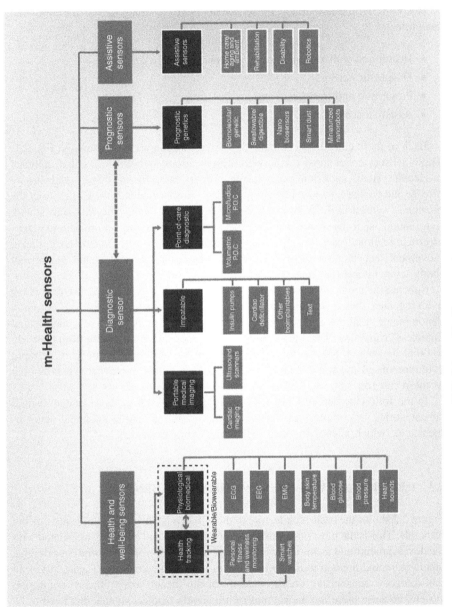

FIG. 2.1 m-Health sensors taxonomy.

taxonomy reflects the different types of sensors in terms of their proximity, direct contract, and invasiveness with the human body; these range from noninvasive, through minimally invasive, to invasive.

Accordingly, the new m-Health sensor taxonomy is composed of the following categories:

- Health and wellness monitoring sensors
- Diagnostic sensors
- Prognostic and treatment sensors
- Assistive sensors

Each of these categories can have their own subcategories, as shown in Fig. 2.1. These sensors cover most of the health applications and categories that embrace m-Health systems, such as in public health, primary care, emergency care, and chronic disease and wellness management applications. These usually use one or more of the biosignals sensed in different m-Health systems, such as *bioacoustic* (heart sound, lung sound, speech), *biochemical* (concentrations, substance compositions), *bio-electric and biomagnetic* (electric potential, ion currents), *biomechanical* (size, shape, movement, acceleration, flow), *biooptical* (color, luminescence), and *biothermal* (body temperature) (Hoffmann and Solzbacher, 2011).

Subclassifications of these main categories are not exhaustive, as this is an evolving area due to ongoing developments in m-Health sensing. The common denominator in all these sensor classes from a technology perspective is their added "wireless" capability. This means how they communicate information from the body to other devices or systems for a particular purpose or action. Also, with the rapid evolution of sensing and mobile technologies, there will be an inevitable overlapping of these categories.

In the following sections, we will describe briefly each of these categories and subcategories for completeness. However, this is an ongoing research and development area, which is evolving as a separate scientific domain.

2.3   HEALTH AND WELLNESS MONITORING SENSORS

Figure 2.2 shows the basic architecture of the wireless health and wellness monitoring principle. This is the most popular model for m-Health, and is closely associated with the developments and technological advances in wireless sensing. Some consider this architecture and domain to be the basis of an m-Health ecosystem (Rajan, 2013). In this category, sensors are the "front end" of any health and wellness monitoring system, because these are the sources of a patient's medical signals that need to be measured and monitored (Tarassenko, 2014).

It is usually convenient to transmit the signals via one of the many available low-power technologies, for example, low-energy Bluetooth or ZigBee, to a smartphone, many of which embrace their own embedded sensors. This allows, for example,

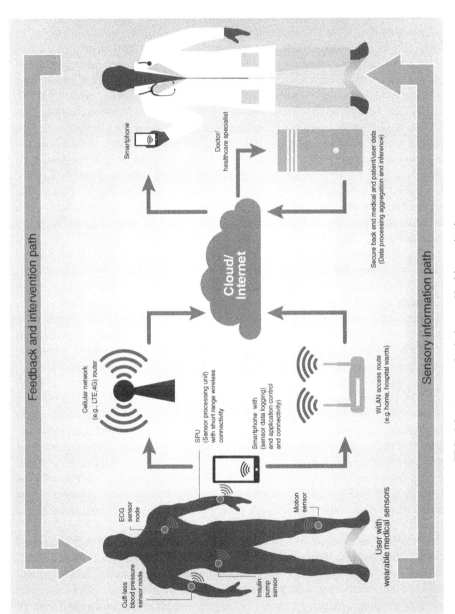

FIG. 2.2 A general wireless m-Health monitoring system.

27

logging and on-the-spot checks by a nurse or paramedic in emergency cases, or at least a preliminary diagnosis by a doctor. In principle, the medical data are usually transmitted either over short or longer distances depending on the nature of the data and the transmissions required. This wireless connectivity then enables specialist caregivers or doctors to access medical data using their own smartphone, tablet, or laptop as web access tools. If necessary, patients can also access their own data, as directed by a doctor or nurse, so that they can examine their logged medical data and monitor and manage their own condition. In the well-being scenario, new smart wearable systems and their applications (Apps) enable patients to log their own health activity data and synchronize these with a smartphone to access the necessary advice.

From the well-being perspective, we are all interested in our own health, even if we manage it badly by consuming too much food, alcohol, and drugs, or by taking too little exercise. Some of us are unlucky, being victims of disease or accident through no fault of our own. As we are now living longer than ever before in developed countries, the cost of our survival is rising year after year to astronomical proportions. One proposed way to offset the cost has been to adopt a patient-centric approach to health and wellness, even though the cost-benefit model has yet to be verified objectively (Bashshur et al., 2013; Free et al., 2013).

Monitoring systems include everything from wearable devices or embedded phone sensors to implantable sensors that are configured into a smart wireless network to obtain information about the body, how it is functioning, or what is going wrong (Kumar et al., 2013).

From a historical perspective, earlier telemedicine applications enabled the adoption of medical sensors for ambulatory and desktop computer-based health monitoring. With advances in m-Health, sensor monitoring has entered a new era, with wearable and nanosensors increasingly linked to smartphone applications.

The simplest and most familiar wearable medical devices are heart rate (HR) monitors, comprising electrodes in a chest band with a short-range wireless link to a wrist-worn display. These are generally used, for example, by athletes to display their heart rate and trigger an audible alert at or above some predetermined maximum rate. The use of such devices reflects a good sense of wellness and an interest in personal health and fitness (Lymberis and Gatzoulis, 2011). As shown in Table 2.1, there can be other physiological signals from many different types of sensors, such as cuff-less blood pressure (BP) meters, thermometers, glucose meters, peak expiratory flow meters, activity and wellness monitors, pulse oximetry sensors, inertial sensors, and countless other devices.

Monitoring the output of sensors requires high technical and environmental specifications because they are used on or in the human body, which moves errati-cally, sweats profusely, and is exposed to extremes of temperature, humidity, and shock. At the very least, sensors need to be physically small and lightweight, dissipate low power, have a long life cycle, be shockproof and waterproof, and operate wirelessly for long periods. In effect, they need to be people-proof.

A wearable network of health-monitoring sensors can be particularly important for observing changes in a patient's "vital signs," such as heart rate, electrocardiogram (ECG), and blood pressure, which allows a doctor to make a diagnosis leading to

TABLE 2.1 Summary of the Main Biomedical and Physiological Parameters Used in Wireless Monitoring

Parameter/BioSignal	Wireless Sensor Example	Basic Function
Electrocardiogram (ECG)	Wearable skin patch or chest electrode	Electrical activity of the heart (continuous waveform of contraction and relaxation phases of cardiac cycles)
Electroencephalogram (EEG)	Wearable scalp electrodes	Measurement of electrical brain activity and other brain potentials
Electromyogram (EMG)	Wearable skin electrodes	Electrical activity of the skeletal muscles that characterize the neuromuscular system
Heart rate	Pulse oximeter or wearable chest vest	Frequency of the cardiac cycle
Heart sounds	Wearable vest with phonocardiograph sensors	Recording of the heart sounds
Body and/or skin temperature	Wearable skin temperature sensor patch	Body or skin temperature
Blood pressure (systolic, diastolic)	Arm cuff-based or cuff-less blood monitoring device	Force of circulating blood on the walls of blood vessels and arteries
Blood glucose levels	Strip-based glucose monitoring devices Wearable patches with embedded microneedles	Measurement of glucose (sugar levels in the blood)
Respiration rate	Wearable piezoelectric or piezoresistive sensors	Breathing rate or movements that indicate inspiration and expiration rate per unit time
Blood oxygen saturation	Pulse oximeter sensor	The concentrations level of oxygen in the blood

improved health, or to be alerted to a life-threatening condition. In the long term, a patient may benefit from continuous, long-term monitoring as part of a diagnostic or wellness regime. Monitoring can enable the optimal treatment of a chronic disease, or simply allow observation during the recovery period following trauma or after surgery. Furthermore, long-term monitoring can capture diurnal and circadian rhythms and cycles in physiological signals, or detect irregularities, which may be used as indicators of cardiac recovery in patients who have experienced myocardial infarction (Mark et al., 2009). Although it may be construed as "Big Brother" watching over a patient (albeit in their own interest), long-term monitoring also allows a doctor to observe whether that patient is adhering to treatment, such as drug therapy, physical rehabilitation after hip or knee surgery, stroke rehabilitation, or brain trauma rehabilitation.

Nowadays, different wireless biosensors in many shapes and sizes are available commercially, with some sensors embedded in smart mobile phones, and tailored with global positioning capabilities that can even pin-point the geographical location of the user during health assessments. Smart wearable sensors in the fitness and wellness commercial sector are the most successful, with a wide range of products in this market (IHS Electronics and Media, 2013). Some examples of smart wearable wrist bands and watches include *Apple's* Watch (Apple, 2016), *Fitbit* Force (Fitbit, 2016), *Jawbone* Up (Jawbone, 2016), and Spree *SmartCap* (Spree, 2016).

In order to provide a better understanding of this category of sensing tools, we describe below the basics of m-Health monitoring principles.

2.4 WHO IS MONITORED?

In principle, anyone can be monitored for a wide variety of medical, physiological, or physical conditions, but of particular importance are elderly patients, who by virtue of their advanced age tend to have more health problems than younger people (Hirt and Scheffler, 2008). Ironically, it is because of better healthcare over the last half century, as well as improved knowledge of healthy eating and the avoidance of bad habits such as smoking, that we are all living longer, and with longevity comes disease and infirmity.

The main reason for monitoring is to improve the quality of life. Typically, and for obvious clinical reasons, these are used for patient monitoring in posttrauma care, or postoperative care. While for serious conditions this may be done in a hospital with static equipment, for patients who are free to move around but who need to be monitored from time to time, at home, at leisure, or at work, there is a need for a mobile sensing system. The challenge here is to obtain measurements that are of comparable quality and reliability to those that can be obtained in a hospital. This may reasonably be questioned by doctors, justifiably so in some specialist areas. For example, most cardiologists would no doubt say that the data from a 3-lead ECG system, such as a body-worn network would normally produce, does not show the amount of clinical and diagnostic details that is available with data from a 12-lead system that would be used in a cardiology department to diagnose a large number of heart conditions (Yanowitz, 2012). In some cases patients may need to be supervised during monitoring; this would probably be unnecessary if the design and implementation of the sensing system is sufficiently advanced, a topic that we consider elsewhere.

In the wellness domain, athletes can be monitored in real time while they are training, for example, during running, cycling, or swimming. This example may not be as important for health for the population at large, but it presents challenges to obtain biomedical data, especially in a sport like swimming, where the transmission of signals is a problem. Another challenge is to monitor a diver in a recompression chamber. Inside the chamber, a system called D-MAS HyperSat measures the diver's clinical vital signs, usually heart rhythm, blood pressure, oxygen saturation, and core temperature, and also allows digital imaging; outside, a doctor provides clinical interpretation and support (D-MAS HyperSat, 2016).

Many people are conscientious or curious about their health, and want to monitor themselves. There are many devices on the open market that allow them to measure heart rate, blood pressure, temperature, and so on, and also a vast number of Apps available on smartphones, but they are generally not designed to provide advice that an experienced doctor can offer. So your blood pressure is high, now what? It is straightforward enough to obtain basic medical data with these Apps, but they must not be considered to provide definitive health diagnoses; it is potentially a dangerous notion to think of them other than as indicators of health status. This is particularly the case with the "fitness and wellness" wristband trackers such as Fitbit, Nike+, Fuelband, Garmin, and Misfit Shine, which represent one of the commercial successes of smart m-Health monitoring. These gadgets count steps, measure sleep, and the fancier models can also tabulate some vital physiological metrics such as heart rate, blood oxygen levels, skin temperature, and body weight by linking these to smartphone applications for further viewing and education. There is ongoing debate on the role of these commercial trackers and their capabilities to narrow the engagement gap via the consumer m-Health route toward preventative and new collaborative care models. However, to date there has been no large, peer-reviewed study to verify the clinical accuracy of these gadgets and to validate their efficacy. Consumers are buying them in their millions, thereby fuelling the phenomenon of *quantified self*, which describes people who are obsessed with monitoring their own health. The widespread adoption of these devices opens new challenges for the future of m-Health.

2.5 WHAT IS MONITORED?

There are a number of universal physiological and physical parameters for bio-sensing and wireless monitoring. Among these are the electrocardiogram, electro-encephalogram (EEG), electromyogram (EMG), blood pressure (systolic and dia-stolic), body or skin temperature, respiratory rate (RR), oxygen saturation, heart rate, perspiration (sweating or skin) conductivity, heart sounds, blood glucose level, and body movements (Jones, 2010).

Table 2.1 shows a summary of the main parameters that constitute the widely used biomedical measurements required for most health and well-being monitoring purposes. Non-biomedical parameters, such as motion and activity monitoring, are also used in most of the wearable and health-monitoring devices. Some of these parameters can be ubiquitously measured using embedded sensors in newer models of 4G smartphones (e.g., heart rate, ECG activity), and also in commercially available wearable wellness devices (e.g., wrist bands). These devices are wirelessly linked to a smartphone App for further analysis and feedback. It is expected that the number of health-related sensors to be integrated in smartphones will increase rapidly in the next few years.

Next, we explain the basics of these biomedical parameters for completeness. The communication and wireless requirements of these will be explained in Chapter 4. Other details on the biosignaling and biomedical principles of these sensors are beyond the scope of this book and may be consulted elsewhere (Hoffmann and Solzbacher, 2011).

2.5.1 Electrocardiogram and Heart Rate

In general, measurement of the heart rate is a familiar and basic means of determining a person's well-being (Hirt and Scheffler, 2008). Resting heart rate varies considerably with age and fitness; generally, the fitter the person, the lower the resting heart rate. An athlete at age 20 might have a resting heart rate of 35–40 bpm (beats per minute), while an active person at age 50–60 is more likely to have a resting heart rate of 50 bpm or more.

The well-known rule of thumb for working out "normal" maximum heart rate is *220 minus age*; so for someone aged 20, this would be 200 bpm, while for someone aged 60 it would be 160 bpm. The formula is seriously flawed, however, as it is based only on age, and takes no account of medical history, weight, general health, and other factors. So although this formula can be a very rough guide, it should not be taken as a scientific measurement, and exceeding the value for your age during vigorous exercise does not necessarily presage an imminent heart attack. One useful indicator of fitness is to recover from maximum heart rate during exercise to resting heart rate within 3 min. This test is often done during a medical examination, using chest electrodes to obtain an ECG.

During ECG monitoring, many heart conditions can be recognized, especially excessively high and low heart rates (tachycardia and bradycardia, respectively), and other flags such as arrhythmia, which would indicate underlying cardiovascular problems. Irregularities can show up, and many measurements are obtainable, such as heart rate variability (HRV), which a cardiologist can use to make a diagnosis. Some conditions, such as premature ventricular contraction or extra-systole, which a person may not be aware of, can easily be missed, and may only show up during a long ECG monitoring session. Readers interested in the complexities of cardiology are referred to a comprehensive tutorial paper on ECG interpretation (Yanowitz, 2012). This paper describes characteristics of the normal ECG, measurement abnormalities, rhythm abnormalities, conduction abnormalities, atrial enlargement, ventricular hypertrophy, myocardial infarction, ST segment abnormalities, T-wave abnormalities, and U-wave abnormalities. The sampling rate to obtain a good-quality ECG is typically around 1 kHz. Generally, heart rate can be acquired effortlessly using ECG data, and in some applications blood volume pulse (BVP) can also be seamlessly captured from the wrist and used to estimate energy values (Firstbeat Technologies, 2012).

Developments in cardiac sensing are moving at a fast pace, with commercial systems ranging from embedded ECG sensors on the back of a smartphone that can detect cardiac arrhythmias in real time by touching the phone to the chest to wearable wrist HR monitoring and tracking sensors linked to a smartphone for exercise and well-being purposes. Other recent developments include, for example, a wireless subcutaneous implantable cardiac defibrillator (S-ICD) that is inserted under the skin beneath the armpit and replaces the leads of traditional defibrillators to detect and regulate any irregular arrhythmia wirelessly (US FDA, 2014).

2.5.2 Blood Pressure

Blood pressure, or more correctly arterial blood pressure, is the pressure exerted on the walls of arteries by the heart's pumping action. Anyone who has had a medical examination has experienced a blood pressure measurement, usually by having an inflatable pressure cuff placed around the upper arm to measure the pressure in the brachial artery. The "normal" maximum (systolic) and minimum (diastolic) values should be close to 120/80 mmHg (millimeters of mercury), so a doctor can easily spot if something is wrong, or suggest underlying causes, such as stress, obesity, smoking, alcohol, physical inactivity, or poor diet, which can precipitate a heart attack or a stroke. Low pressure (hypotension) is generally considered to be 90/60 or lower, and high pressure (hypertension) to be 140/90 or higher. Long-term monitoring of blood pressure using a pressure cuff is not convenient because of the bulkiness and obtrusiveness of the equipment. There are many commercially available wireless cuff-based blood pressure monitoring devices that are linked to a smartphone using Bluetooth connectivity, which allows diabetic or hypertensive patients to measure their own blood pressure at home. The measurements can be used as indicators of hypotension, prehypertension, or hypertension.

2.5.3 Blood Glucose Levels

A sensor for monitoring blood glucose levels is generally part of a mobile diabetes monitoring and management system or a mobile chronic disease management system. It is well known that high levels of blood glucose indicate a complication of diabetes, while low levels can result in fainting or unconsciousness (Topol, 2012). Diabetes patients are all too familiar with the process of pricking a finger to obtain a small blood sample that is used to measure glucose concentration and to take steps to regulate it. This tedious, painful, and inconvenient chore has to be done two to four times a day, but it still gives an intermittent picture because the glucose level can vary considerably, depending on the amount of food and drink consumed, exercise done, and any medicines taken, such as insulin.

For non-diabetics, normal blood glucose level is about 4 mmol/l, or 72 mg/dl, and after a meal the level may increase to 7.8 mmol/l. For diabetics, the level before meals is 4–7 mmol/l for types 1 and 2 diabetes, and after meals it is 9 mmol/l for type 1 and 8.5 mmol/l for type 2 (Diabetes UK, 2013). This topic is described in more detail in Chapter 5.

An improved technique, which is expensive, used a continuous glucose monitor (CGM), in which a microsensor is injected into the interstitial fluid under the skin of the abdomen where it measures the glucose level (Topol, 2012). This is an example of a smart sensor, which transmits every few minutes to a small receiver carried by the patient, giving a near-continuous measurement of blood glucose. Numerous commercial systems and Apps have been developed to measure blood glucose levels wirelessly, using off-the-shelf glucometers linking these measurements to a smartphone and to remote servers for follow-up by a specialist nurse or doctor. Other commercial systems tailored for type 1 diabetes are also on the market, including wirelessly operated insulin pumps. The major challenge from the sensing perspective is how to obtain accurate blood sugar level measurements that are clinically acceptable without the current invasive method. This is an ongoing research topic with many proposals, from microneedles that accurately monitors glucose levels and administers insulin treatment when required (Yoon et al., 2013) to electrochemical biosensor methods (US FDA, 2013), that have yet to see wider patient acceptability, clinical reliability, and efficacy compared to the traditional method of the capillary (finger pricking) blood acquisition approach. More recent advances in wearable sweat biosensors being tested for health tracking and blood glucose measurements is an interesting development in this area (Heikenfeld, 2014).

2.5.4 Blood Oxygen Saturation (SpO$_2$)

The amount of oxygen in the blood, usually referred to as arterial oxygen saturation, is typically 95–100%. A value of 98% means that each blood cell comprises 98% oxygenated hemoglobin and 2% nonoxygenated. A value below about 80% (hypoxemia) indicates an alarming number of possible medical complications, including anemia, emphysema, chronic bronchitis, shock, sleep apnœa, pneumonia, lung cancer, carbon monoxide poisoning, and chronic obstructive pulmonary disease (COPD). It is usually measured with a clip-on probe on the forefinger or ear lobe, using light that is transmitted to and reflected from a blood vessel to give a measurement. Some finger pulse oximeters simply display oxygen saturation as a number, while other also measure and display heart rate and show a photoplethysmography (PPG) waveform; this can be obtained with compact PPG sensors that are Bluetooth-enabled, and wirelessly linked to a smartphone. Technical details of PPG sensors are available elsewhere (Tamura et al., 2014).

While monitoring physiological and physical properties is of primary interest and benefit, there are other less well-known devices, such as wearable and implantable chemical sensors that can be used for real-time daily monitoring of bodily fluids such as tears, sweat, urine, and blood. Such sensors have a clear application in monitoring glucose levels in diabetes patients, but the widespread adoption of chemical sensors has been complicated by the difficulty of collecting samples, where necessary, as well as sensor calibration (which may require a sample), wearability and data security (Jovanov et al., 2003; Diamond et al., 2008).

Readers interested in studying a variety of more specialized applications are referred to a treatise on biomedical sensors (Jones, 2010). The topics in this paper include temperature sensor technology, liquid flow sensors, respiratory gas flow sensors, biomedical sensors for ionizing and nonionizing radiation, medical ultrasound sensors, chemical sensors, and biosensors for medical thermography, and infrared radiation measurements.

2.5.5 Body or Skin Temperature

Body or skin temperature is constant for long periods, with a "normal" value of around 37°C, or 98.6 °F. This value has been quoted since the nineteenth century, but it can be misleading because the temperature varies for different parts of the body. A useful table is available, showing results of temperature measurements for men and women for four sites—oral, rectal, tympanic (ear canal), and axillary (armpit) (Sund-Levander et al., 2002). A departure from the "normal" temperature by even 1° or 2° can signal a serious condition. A high temperature (hyperthermia) usually means a level of feverishness, while a low temperature (hypothermia) means a slowing down of the body's functioning, typically due to exposure to cold weather or immersion in cold water. Temperature is not the easiest of parameters to measure on an ambulatory patient wearing a sensor system, because of the problem of placing the thermometer, but it is arguably the least useful of the vital signs. The sampling rate for monitoring temperature is low, say 1 Hz or lower, because of its invariability. Current wireless thermometer monitoring is also used to indicate human body status (e.g., stressed or normal).

2.5.6 Electroencephalogram

An electroencephalogram shows the electrical activity of the brain by monitoring an array of sensors attached to the head. Its importance lies in the detection of seizures and the diagnosis of epilepsy and mental health disorders. It can also enable clinicians to check for loss of consciousness or dementia, study sleep disorders, monitor a patient during brain surgery, and diagnose injuries to the brain and spinal cord. Further research has included studies of personal traits and sensory inputs, brain signal generation and acquisition, brain signal analysis, and feedback generation (Mihajlović et al., 2014).

Monitoring the EEG has conventionally been a highly complex, static, hospital-based procedure, but there is now interest in home and portable monitoring (Casson and Rodriguez-Villegas, 2013). This requires advances in circuits and algorithms to obtain high-quality EEG recordings for online analysis in noncontrolled environments. The concept of in-ear EEG monitoring using electrodes in the ear canal as part of a standard hearing aid has been proposed; the electrodes are noninvasive, socially acceptable, and securely held in place. To be practical for everyday use, any mobile system needs to be intelligent, wireless, unobtrusive, and comfortable to wear.

Wearable technology can be applied for sensory and gesture control, voice recognition, augmented reality, health monitoring, biometrics, and improving sleep quality and concentration. A further, more futuristic, application is brainwave sensing to enable a person's "thoughts" to "control' equipment, including creating "life compelling experiences" such as inviting a crowd to wear brainwave controllers to use their thoughts to control the lights on Niagara Falls! (Luke, 2014). This technology has been introduced by a Canadian company, InteraXon, which markets the Muse EEG brain-sensing headband, used in conjunction with a smartphone App.

As driver drowsiness is a major cause of road traffic accidents, one suggestion has been to monitor the driver's EEG signal (Li et al., 2015). This has been achieved in a driving simulation experiment with a Bluetooth-enabled headband and a smart watch to evaluate a driver drowsiness detection model in real time. The main aim was to estimate the relative severity of driver drowsiness.

In addition to the above list of widely used medical parameters, there have been many other more specialized and disease-focused sensing applications, such as the wireless monitoring of the intracavitary pressure of the bladder (Coosemans and Puers, 2005), and neural prosthetic devices (Mohseni et al., 2005; Neihart and Harrison, 2005).

2.6   WEARABLE SENSORS FOR m-HEALTH MONITORING

We cite two examples that reflect the importance of the fast evolving area of wearable sensors on m-Health monitoring. The first is from *Scientific American*, which reported the news of a 15-year-old teenager winning the "Science of Action" award for his invention of a wearable pressure sensor for Alzheimer's patients. The sensor can be attached to a foot or a sock and notifies caregivers via their smartphones if a patient who should be sleeping gets out of bed (Sneed and Scheer, 2014).

The second example was reported the same year by Samsung (Galaxy Gear) and Apple (Apple Watch), both of which released their first smart wristwatches incorporating wearable sensors aimed at wellness and fitness monitoring (Schumacher, 2014). These devices use new interfaces that combine a touch screen with physical buttons to translate the experience of using a smartphone on the wrist. Besides telling the time, most of the smart watches on the market today can send messages, give directions, track fitness, and make wireless payments. The additional use of these smart watches for more specific m-Health applications, such as cardiac and other chronic diseases monitoring, could be incorporated sooner rather than later.

These and other innovations reflect the successful outcome of a decade of technological progress in the wearable domain of integrating sensors into smart m-Health monitoring. In general, wearable systems can offer many clinical benefits for patients and healthcare providers:

- The ability to monitor health in different environments, which is usually not possible with wired sensing, thus assisting doctors to better diagnose their patients' health status.
- Patient empowerment by providing education and control of their own health and well-being.
- Improved healthcare management and clinical outcomes, and potentially to assist in medical training and education.
- Potential reduction in healthcare costs.
- The creation of big health data to be used and harnessed for personalized treatments and also for further medical studies.

Although the concept of wearable technology dates back to the 1990s (Pentland, 1996), developments have impacted profoundly on the integration of wearable sensing devices for wireless health monitoring (Sazonov and Neuman, 2014). The key technologies for designing wearable devices include miniaturization, intelligence, networking, digitization, and standardization (Lin and Tang, 2000; Weber et al., 2005; Poon and Zhang, 2008). Wearable technologies, in general, can be realized for m-Health monitoring with either wireless body sensor networks or wearable garments with embedded sensors.

In wearable wireless networks, the sensors measure and transmit any or a multiple of medical parameters listed in Table 1.1, and connect these to an intelligent hub or a smartphone for feedback and advice. A recent example is a simple, wearable "intuitive" thermostat that can measure the body's temperature and inform the wearer why the temperature is high (e.g., exercise, fever) and adjust it accordingly. Another example is a micro-wearable sensor dot on the arm that can gauge intelligently the wearer's diet by tracking sweat, and recommend healthy lifestyle changes (Time, 2014). From a clinical perspective, there is continued debate and ambiguity on the efficacy and long-term benefits of wearable devices. One study indicated that only 15% of U.S. physicians surveyed discuss wearables or health Apps with their patients (Pennic, 2015). These and other similar clinical opinions reflect the concern that as long as consumer-driven m-Health markets dominate this area, these wearable products will remain a fad from the clinical perspective at least for the foreseeable future.

Historically, the challenge has been to design and implement very small, low-power sensors that can stand alone as intelligent devices. Each of these has the capability of sensing and processing medical data, and then communicating with each other over short ranges to form a network (Benhaddou et al., 2008). Each sensor in this network has the capability of detecting vital medical signals that we have already introduced, typically ECG, PPG, HR, BP, and temperature (Joel et al., 2011).

We present here a range of these wireless technologies designed for monitoring patients continuously or intermittently, with both wearable and implantable networks (Darwish and Hassanien, 2011; Bazaka and Jacob, 2013). Further details are discussed in Chapter 4.

There are many confusing labels in use for intelligent sensor networks, so we will first spell out some relevant definitions. In the early days of personal monitoring with electronic sensors, the type of network used was referred to, logically, as a personal area network (PAN), which may have been wired or wireless, and was used for intrabody communications (Jovanov et al., 2000). From about 2000, the term body area network has become more in fashion, and has since been used in the context of monitoring wearable sensors (on the body), and implantable sensors (inside the body) (Zimmerman, 1996). Such an arrangement is also referred to as a network of Wireless Intelligent Sensors (WISE), a wireless sensor network (WSN), a body sensor network (BSN) (Jovanov et al., 2009), a Wearable and Implantable Body Sensor Network (WIBSN), or an Implantable Wireless Body Area Network (IWBAN) (Khan et al., 2009). To add to the confusion, there are yet more terms in use, including Medical Body Area Network (MBAN) and Body Area Sensor Network (BASN).

The distinctive feature of these networks is that they are capable of processing signals, and are interconnected by an extremely low-power, wireless protocol. The individual sensors are examples of system-on-chip (SOC) technology.

A further label used is a Wearable Wireless Body Area Network (WWBAN) (Filho et al., 2013). Without affecting the meaning, this is often referred to more concisely as a Wireless Body Area Network (WBAN). In simple terms, a WBAN represent a wireless communication network linking a patient and a smart node or mobile device (usually smartphone or tablet) through wearable body sensors. Purists may offer fine distinctions between BANs and WBANs, and their technology and formal technical specifications, as defined by IEEE standards, may vary considerably in detail (especially for implantable sensors). However, all these terms and acronyms essentially amount to the same thing, and they constitute the "front end" of any m-Health system.

In general, WBANs represent a tremendous emerging opportunity for m-Health. Their "wearable networks" are special-purpose wireless sensor networks that enable remote health monitoring, originally for e-Health (Lymberis and de Rossi, 2004), now in many m-Health applications, such as well-being and fitness. However, WBANs present a number of technical challenges, including their architecture, density, data rate, latency, and mobility (Cao and Leung, 2009). Generally, a WBAN consists of interconnected sensors that continuously monitor data and send it to a "coordinator," which in turn sends it to a healthcare center through a conventional communications network. Figure 2.3 shows the basic connectivity architecture for a

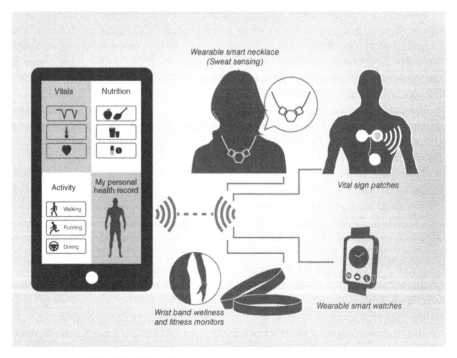

FIG. 2.3   Wearable body area network monitoring system.

typical wearable WBAN. This incorporates a wearable sensor embedded with a short-range communication module (such as low-energy Bluetooth and ZigBee) connected to the Internet via a smartphone App linked with a remote server, or "cloud," that provides user feedback, monitoring, further analysis, and data aggregation, as shown earlier in Fig. 2.2.

This configuration is increasingly used in smartphone-based m-Health monitoring applications that dominate the wearable market. According to the IEEE 802.15.6 TM-2012 standard, a WBAN is a low-power smart device working "on, in, or around the human body" to carry out a variety of medical and nonmedical applications such as personal entertainment (Kim and Cho, 2009). A WBAN operates at typical data rates of 10 kbps to 10 Mbps over a range of up to 1 m, which defines the extent of the domain over which the individual sensors are connected. By the 802.15.6 Task Group's definition, a WBAN is distinguished by short-range, low-power, and reliable wireless transmission. In general, the IEEE 802.15.6 WBAN standard defines narrowband (NB) and ultra-wide band (UWB), as appropriate to any particular application in human body communications (HBC).

To put this into the context of other wider area networks brings us back to the confusion of terms: A wireless personal area network (WPAN), not to be confused with the earlier usage of PAN already referred to above, operates up to about 10 m around the human body or personal vicinity, and is normally associated with nonmedical communications, such as wellness or fitness applications.

In general, a WBAN uses the Medical Device Radiocommunications Service (MedRadio); the frequency bands are now, since 2009, 401–406, 413–419, 426–432, 438–444, and 451–457 MHz (FCC, 2011). A similar service is the Wireless Medical Telemetry Service (WMTS). The WMTS band is also used by other technologies, such as wireless Local Area Networks (WLAN), also known as Wi-Fi (IEEE 802.11/a/b/g/n); Bluetooth (IEEE 802.15.1); UWB (IEEE 802.15.3); and ZigBee (IEEE 802.15.4).

The licensed Medical Implant Communications Service (MICS) is used for data transmission in the 402–405 MHz band; it is dedicated to bidirectional communication with a pacemaker or other implanted electronic device. Many other applications adopt the unlicensed industrial, scientific, and medical (ISM) band (2.4–2.4835 GHz).

The typical transmission bandwidth requirement of a WBAN is 1.2 MHz (Bluetooth, 2014; Cao and Leung, 2009; Kraemer and Katz, 2009). The selection of any band depends on the specific application. This standard also defines the media access control (MAC) protocol, which controls access to a channel or simply of moving data packets to and from one network to another across a shared radio channel.

Wearable network configurations of sensors are more than a random arrangement of sensing devices stuck on the body. The important distinction is that they communicate with one another, and therefore form a truly integrated, intelligent network. Such a "smart wearable network" acts as an early warning system because, in effect, it "listens" to what the body is saying.

An individual intelligent biosensor might be a sophisticated electronic system comprising a sensing element, microelectronic processor circuit, and memory, all powered by a tiny battery. This is often referred to as a *node*, which processes signals representing a particular biomedical parameter and communicates with other nodes in the network (Schenker et al., 2004; Quwaider and Biswas, 2008). The architecture of a typical sensor node in a WBAN system is shown in Fig. 2.4, which comprises a radio module, sensor module, and memory module controlled by a microprocessor module (Cao and Leung, 2009). The sensor module includes a dedicated sensor, a filter unit, and an analog-to-digital converter. Thus, the analog signal from the sensor is bandpass-filtered, and then digitized before transmission. The memory module consists of conventional random access memory, read-only memory, and flash memory. Most of the units involved can now be manufactured on a compact, miniaturized single SOC.

The usual configuration is to have a hierarchical network of smart sensors comprising a *master node* and one or more *slave nodes*. The slave nodes process signals locally, as appropriate for a given parameter, such as heart rate or blood oxygen saturation (Intanagonwiwat et al., 2000), before transmitting data to the master node. The master node controls the transmission of a data stream to a gateway; this is normally a mobile device such as a laptop or smartphone.

Data acquisition can be *point-to-point* or *multipoint-to-point*, depending on the application. For example, monitoring vital signs would require attached or implanted sensors to route data multipoint-to-point, or from slave nodes to a master node. The data are then transmitted wirelessly to a mobile device. In a more unusual application, the distributed detection of an athlete's posture, point-to-point data sharing across a network of on-body sensors, is appropriate (Chen and Bassett, 2005). Recent advances are also focusing on the development of near-zero power consumption sensor nodes tailored for the next generation of wearable devices and biosensor networks (Thotahewa et al., 2014).

2.6.1 Applications of Wireless Body Area Sensors

Medical applications of WBANs are numerous, and the scope is wide and varied, but the common aim is to transmit the wellness, biomedical, and physiological data from a person to a remote health center, hospital, or clinic by using smart, unobtrusive monitoring (Aleksandar et al., 2006; Otto et al., 2006; Braem et al., 2006; Latre et al., 2007; Alemdar and Ersoy, 2010). Most of these applications follow the same wireless monitoring principles discussed earlier and illustrated in Fig. 2.3. Here are a few example applications:

- Patient monitoring of various physiological parameters, monitoring mentally and physically disabled people, and tracking doctors or patients in a hospital (Akyildiz et al., 2002; Noury et al., 2000).
- Epileptic seizure warning, glucose monitoring, and cancer detection (Khan et al., 2009).
- Smart homes for monitoring a person's well-being, or to diagnose a medical problem; for elderly people, this can be used to support their self-care at home (Dervisoglu et al., 2007; Time, 2014).

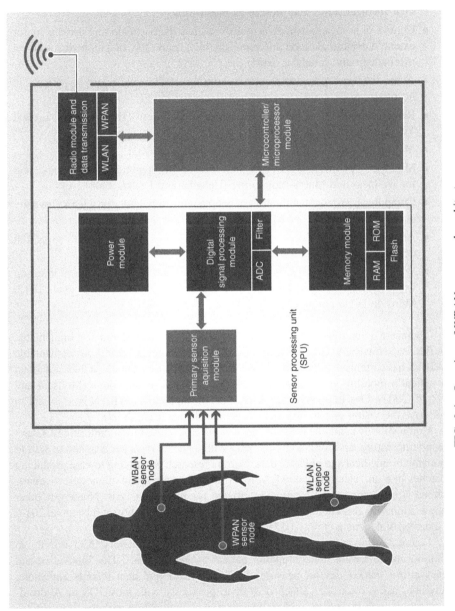

FIG. 2.4 Overview of WBAN sensor node architecture.

41

- Smart medicine and improving drug administration in hospitals, by attaching microsensors to medicine packages to minimize the possibility of incorrectly prescribing drugs to patients, especially if patients wear sensors that identify their medical status and prescribed medication.

- Control of home appliances, access to medical data, and alerting patients to, for example, abnormal blood pressure, very high heart rates, or a high glucose level (McFadden and Indulska, 2004).

- Acute care and post-acute care monitoring of the vital signs of patients in hospitals.

- Remotely monitoring patients with chronic conditions (Herzog, 2004; Lo and Yang, 2005; Loew et al., 2007).

- Monitoring casualties in accidents (Malan et al., 2004).

- Monitoring people at work or leisure for early warning of disease or illness, and for wellness and fitness monitoring (Habetha and Reiter, 2008).

- Computer-assisted rehabilitation of people with ambulatory problems (Jovanov et al., 2005).

- Assisted living of elderly people living at home, including the use of fall detection sensors (Eklund et al., 2005).

- Wearable gait sensors, which have led to commercial systems for monitoring walking and running patterns using smartphone applications. (Tao et al., 2012; Muro-de-la-Herran et al., 2014).

There are numerous other systems that have been considered for implantable WBANs (Schwiebert et al., 2001; Tang et al., 2005; Aziz et al., 2006). An implantable WBAN has certain advantages over a wearable WBAN; for example, it does not affect a patient's mobility and it does not cause skin rashes or infections (Mohseni and Najafi, 2005). One impressive application of an implantable WBAN is as an aid to improve the vision of partially sighted people (Khan et al., 2009).

Commercially available wearable devices have been used for monitoring ECG and skin temperature, and for activity tracking with a sticker or patch attached to skin to transmit biomedical and physical data. Current research is evolving toward producing patches like an "electronic skin." Most of these are wirelessly connected to smartphone Apps equipped with tailored software for further analysis. Numerous other applications are also cited in recent review publications (Sazonov and Neuman 2014; González-Valenzuela et al., 2012).

Figure 2.5 shows an example of a smartphone-based ambulatory ECG arrhythmia monitoring device for noncontinuous, patient-triggered events. This innovative and easy-to-use market device measures ECG and heart rate data records anywhere, anytime, using electrodes attached to or in proximity with most iOS or Android-compatible mobile devices, with good accuracy in conjunction with the "AliveECG" App for further monitoring and diagnosis (AliveCor, 2016).

Similar ambulatory ECG devices are also available on the market and are cited in specialist m-Health sites shown in the Appendix. However, as noted in several

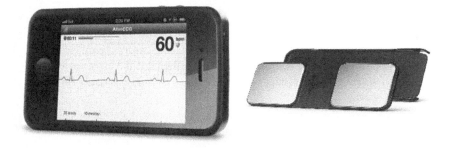

FIG. 2.5   A mobile ECG heart monitor. (Photo courtesy of AliveCor.)

extensive review studies of ambulatory arrhythmia monitoring devices, most of the event monitors (which have been around for some time) have their own limitations as mobile ECG monitoring devices. For example, they will not usually catch the initiation of an arrhythmia accurately, which has diagnostic value, and they can also miss short arrhythmias (Zimetbaum and Goldman, 2010).

To further illustrate the diversity of the health wearables market, Fig. 2.6 shows a wearable-intensive nerve stimulator for managing chronic pain; it is designed particularly for people with painful diabetic neuropathy, fibromyalgia, sciatica, and osteoarthritis, among other conditions (Quellrelief, 2016).

FIG. 2.6   Wearable device for pain management and relief. (Photo courtesy of Quellrelief.)

In order to better utilize and exploit new WBAN features and characteristics, several issues need to be addressed for future use:

- Design and development of very small, low-cost sensors with their associated electronics and wireless capabilities; these will enable the large-scale and cost-effective use and wider deployment of these systems, especially in the developing world.
- Prioritization of the medical and physiological data being monitored, so that the overall system design and functionalities can be tailored and adapted appropriately for different clinical scenarios and environments.
- Development of robust WBAN security and privacy protocols that can be applied in different clinical scenarios and operating conditions.
- Node-based medical data processing and analysis, with capabilities that can provide accurate analysis and alerts in many practical and emergency health scenarios, such as detecting life-threatening signs or symptoms.
- Developments of global models and standards for machine-to-machine (M2M)-based WBAN communications systems for their wireless connectivity with other consumer objects.
- Better harnessing and understanding of the "big wearable health data" generated from these WBAN systems for tailoring personalized patient care and improved therapeutics.
- Large clinical trials to validate the clinical efficacy and effectiveness of wearable devices in key patient m-Health monitoring applications.

2.6.2   Smart Clothing and Textiles

This is another category of wearable technology. In general, a smart clothing item is a garment, usually like a vest or close-fitting shirt, with in-built smart sensors that form a BAN once a patient puts it on (McCann and Bryson, 2009). Sometimes termed also as e-textiles, smart clothing can monitor biometrics such as heart rate or ECG in the same way as any other BAN, but, in principle, in a more convenient way for the wearer (Langenhove, 2007). This is the case for most reasonably able-bodied patients, and is quicker than attaching individual sensors, which may be disposable and therefore temporary. For elderly, frail patients, the effort of putting on smart clothing may be less appealing or practical.

There has been considerable interest in the design and manufacture of smart clothing in recent years because of its obvious market potential, especially in the fashion industry. However, currently, it has far from everyday acceptance or widespread use for ambulatory patients, but it is one of the promising areas of m-Health monitoring, especially with advances in the nanotechnologies. Some early examples of the development of smart clothing include the following:

- An earlier system developed by NASA comprising a wearable patch to control heart rate, blood pressure, and other physiological parameters for astronauts (Lin and Tang, 2000).

- The "Smart Shirt" developed by Sensatex (2016) as a wearable health monitoring system that monitors heart rate, body temperature, and motion of the trunk (Park and Jayaraman, 2004).
- WEALTHY and My Heart projects funded by the European Union, which used cotton shirts with sensors to measure respiratory activity, electrocardiograms, electromyograms, and body posture (Paradiso et al., 2004, 2005; Lymberis and Paradiso, 2008).
- Life Shirt, developed by Vivo Metrics (Halin et al., 2005), a body monitoring system developed by Body Media, and the Nike-Apple iPod Sports kit (Diamond et al., 2008).
- Smart Vest, to monitor various physiological parameters such as electrocardiograms, photoplethysmographs, heart rate, blood pressure, body temperature, and galvanic skin response (Pandian et al., 2007).

Further reviews of wearable systems developed for different physiological monitoring and rehabilitation applications are cited elsewhere (Mundt et al., 2005; Gopalsamy et al., 2005; Pandian et al., 2007, 2008; Patel et al., 2012).

2.7   WEARABLE FITNESS AND HEALTH-TRACKING DEVICES

The commercial appeal of wearable technologies has been successfully adopted by major consumer industries, particularly for sport fitness and health-tracking applications. A recent research market analysis on the current status of wearable devices reported that more than 90 million units were shipped in 2014. Of these, more than 42 million were for sport fitness and health monitoring, with a prediction of more than 57 million in 2015 (Pai, 2014). These statistics reflect the commercial success and consumer appeal of these devices and their potential role in future m-Health monitoring.

Table 2.2 shows examples of wearable fitness and health-tracking devices on the market. This list represents only a sample of the much crowded market of similar devices, with prices ranging from $99 to more than $200. Most of these devices offer identical functionalities for health tracking and wellness monitoring. However, some are also designed with more specific functions, such as access to Facebook, e-mail, and messaging. Most of these devices can be wirelessly synchronized with a smartphone App (Android, iOS, or Windows) for additional personal logging and health information. Some of these fitness and lifestyle devices also incorporate physiological sensors, but the majority are developed for monitoring physical activity using accelerometers (movement sensors) that are in turn used to estimate other parameters.

The latest models of these fitness trackers and wristband wellness monitoring devices can monitor not only the activity of the user (step counts) but also heart rate, sleep patterns, diet tracking, and other parameters via their own apps. Some users of such devices (Digital Self, Quantified Self) hail these technologies as the next revolution in personal health tracking and monitoring. Some hospitals and healthcare

TABLE 2.2   Examples of Marketed Fitness and Health-Tracking Wearable Devices

Wearable Device	Company (Manufacturing Source)	Functions
Apple Watch	*Apple* (www.apple.com/uk/watch)	Track activity, sleep, and heart rate and synchronizes these with IPhone
FitBit Charge	*FitBit* (https://www.fitbit.com/uk)	Tracks all-day activity, measures steps, and automatically tracks sleep
Samsung Gear Fit	*Samsung* (http://www.samsung .com)	Wireless heart rate monitoring and other fitness tracking functions synchronized with Samsung smartphones
Nike+ FuelBand SE	*Nike* (www.nike.com)	Uses range of smart activity measurements and advises on appropriate actions with enhanced social features (e.g., links with user's Facebook friends list)
Garmin Vivofit	*Garmin* (sites.garmin.com/en-GB/ vivo/)	Waterproof device that measures heart rate with accurate tracking and physical activity functions
Jawbone UP3	*Jawbone* (https://jawbone.com/up)	Measures heart rate with improved sensor technology in addition to advanced sleep tracking autoactivity measurements
Microsoft Band	*Microsoft* (www.microsoft.com/ Microsoft-Band)	Tracks heart rate, steps, calorie burn, sleep quality, and other activity measurements. It also connects users with e-mails and messages

clinics support this view to justify the use of wearable devices for different services, including surgery care (Shantz and Veillette, 2014), elderly care in postsurgery hospital environments (Cook et al., 2013), and employees' wellness monitoring (Gutierrez and Rogen, 2014). However, recent reports cited potential health concerns from the long-term usage of wearable computers with their cellular phone connectivity (New York Times, 2015).

Powerful commercial appeal of these devices, combined with their polished design, sleek functionality, and ubiquitous health tracking and monitoring capabilities, could open new horizons for more personalized m-Health monitoring. However, there are a number of issues and challenges that need further consideration before these devices are adopted for broader medical settings:

- Commercial wearable systems and their associated technologies are mainly proprietary devices marketed by individual manufacturers. Their wider adoption for m-Health will be potentially restricted by the proprietary requirements of each manufacturer; this represents a challenge for those advocating open source and open m-Health systems and applications.
- These are relatively costly devices that target mainly consumer markets. The cost constraints may inhibit their wider use in healthcare services, for which cost-effectiveness and adherence to relevant business models are invariably necessary.
- These devices are capable of producing massive user-generated or reported health data. The ownership, privacy, and security of these "big health data" represent challenging issues.
- The complex cyber security threats and attacks from using these devices and their next generations are yet to be considered and understood from the m-Health perspective.

2.8   DESIGN CONSIDERATIONS FOR WIRELESS HEALTH SENSING AND MONITORING

As discussed earlier, the main function of wearable sensors and their wireless network connectivity is to enable the monitoring of health, fitness, and well-being during normal, everyday activity and working environments. We present this as the first of four main categories of m-Health monitoring; the others are diagnostic sensors, prognostic and treatment sensors, and assistive sensors. Also as mentioned earlier, commercial systems are invariably ad hoc designs for specific applications, such as wellness and sport and fitness tracking applications. Many sensors are already worn by ambulatory patients to monitor specific medical conditions, or by athletes to monitor their performance, notably their heart rate, but in the near future, when it is likely that more and more people (e.g., Quantified Self, Millennial Generation) will want to monitor their own wellness, activity, diet, and other activities in more detailed and extensive ways, it is important to think about design features that make this technology more acceptable for these and other healthcare-tailored monitoring purposes (Hori et al., 2000; Rasid and Woodward, 2005; Woodward et al., 2006; Darwish and Hassanien, 2011):

2.8.1   Unobtrusiveness and Wearability

No one wants to carry around a large box of electronics dripping with wires and electrodes that make it look like a mobile intensive care unit. The first design criterion is to make wearable sensors and networks as unobtrusive and comfortable as possible, and therefore wireless, small, and lightweight. For example, earlier body-worn fall detection systems, including LiveNet (Sung et al., 2005), FireLine (Baker et al., 2007), and triaxial accelerometers (Purwar et al., 2007), were heavy as well as intrusive, while watch-like activity recorders and bandage-type ECG sensors (Yoo et al., 2010) are easier and more discreet to wear.

2.8.2   Ease of Use

Any type of body-worn sensor or network should be straightforward to put in place, and easy to use. The wearer may, of course, have to place or replace these from time to time, for example, by sticking ECG sensors at particular sites on the chest. For elderly, disabled, or uneducated people, such procedures would normally be carried out by a caregiver or nurse. User-friendliness has been modeled to try to improve user interaction (Casas et al., 2008), while comfort and compliance especially among elderly people using at-home monitoring has been further studied (Jasemian, 2008). The ease of use or degree of acceptability varies for different groups of people. For example, a patient with a cognitive disability would be expected to have a different reaction to wearing attached sensors than a patient with a heart problem. For a monitoring system that is not automatic, any kind of interaction with it by handicapped or frail elderly people needs to be based on activation by voice, gesture, or visual cues rather than the exercise of skills requiring manual dexterity. In reality, such people would generally be looked after by caregivers who would be familiar with the system, and would be able to initiate the transmission of data, especially in an emergency.

2.8.3   Sensitivity

The sensitivity of sensors on the body can be impaired by poor placement, sweating, water from showering, harsh environments, temperature, and many other factors. As a consequence, body-worn sensors may need to be replaced or at least recalibrated, which for some devices may be done automatically (Gietzelt et al., 2008). It is obvious that malfunctioning, insensitive, or uncalibrated sensors can produce erroneous data that could lead to false diagnoses.

2.8.4   Power Harvesting and Battery Capacity

In general, a WBAN has limited power and battery capacity, which is typically measured in ampere-hours (Ah) and is invariably proportional to the physical size and weight of the battery or cell used (Anastasi et al., 2009; Li et al., 2010). A lead-acid car battery with a rating of 100 Ah would power a sensor network for decades, but it is useless for many of today's m-Health wearable applications. At the other extreme, tiny button cells of 10 mAh rating are much more suitable, and these have been used for many years in such applications as implanted heart defibrillators. Even these cells present a challenge for body-worn networks because several may be needed in different locations, and the cells need to be recharged or replaced. Even one discharged or dead sensor cell may cause a network to malfunction, so the design of robust, low-power, long-life cells remains a challenge. Approaches to tackle the improved design of body-worn wireless sensors, which would usually be located under clothing or be part of smart clothing, include recharging by motion and body heat (Renaud et al., 2008; Lauterbach et al., 2002).

The transmission bandwidth requirement of a WBAN is typically about 1.2 MHz. From the power consumption point of view, low-energy Bluetooth (Bluetooth 4.0) and UWB offer better performance than the earlier versions of Bluetooth and ZigBee.

Battery and energy-saving technology is an ongoing issue, but miniaturization of cells and integrated circuits relates to wearability and acceptability, which are critical for any future wearable and other m-Health applications. The recent introduction of low-power M2M communications devices might alleviate some of these challenges (Ullah et al., 2011).

2.8.5 Data Collection and Big Health Data Processing

The amount of biomedical data that can be collected from a single patient using the sensors discussed above can amount to many megabytes in an hour, so for multiple patients over days or weeks, the amount can be enormous. The problem is compounded by the addition of more sensors in a typical m-Health network. How to rationalize data collection—which parameters to monitor, whether to collect data continuously or to collect it intermittently, what data to suppress, what to send, and when to send it—is an important consideration. Relevant signal processing and data compression are also important considerations. For example, rather than transmitting raw ECG data, some degree of feature extraction can be carried out to identify a particular event, and only the data representing each such event are transmitted. If the reduction in data handling is not offset by an increase in computation before transmission, there is an added bonus of reducing battery drain, which therefore increases battery life. A further consideration is the integration of different types of sensors, such as radio frequency identification (RFID) tags, implantable sensors, and wireless body sensors (Darwish and Hassanien, 2011).

In recent years, the term *big health data* has become a buzzword in the field of health information technology. This reflects the challenges described above, from processing massive amounts of medical and behavioral data obtained from billions of smartphones and wellness sensors in a secure and efficient way. These challenges will be computationally and storage intensive in nature, and are further complicated by the increasing emphasis on security, privacy, and ownership of these data. These developments constitute a new and emerging area of research: "wearable big m-Health data" science. These issues will be discussed in further detail in Chapter 3.

2.8.6 Communication Links

Reliable communication between a network's sensors is vital, otherwise the data transmitted to the outside world will be meaningless. The transmission from different sensors requires various sampling rates, from less than 1 Hz for body temperature to

1 kHz for an ECG, as suggested earlier. For a body-worn network, signals are transmitted at low power between sensors, which need to be physically compact, with small batteries and antennas. The signal-to-noise ratios may be low enough to cause significant bit error rates and reduce the effective area of coverage. Coding schemes may improve the reliability of transmission. One scheme that has been tried for ECG and EEG data was based on a time-division multiple access protocol, which allowed each sensor to transmit data via two "Exclusive-OR" relays before sending it to the outside world (Marinkovic and Popovici, 2009). This showed a reduced error rate, but it also demonstrated the need for further research.

More recent versions of low-power Bluetooth and ZigBee dominate the current communications between sensor nodes and smartphones. However, M2M communications will potentially replace these links with more efficient Internet-based wireless protocols to control direct communication between sensor nodes. This will be discussed in further detail in Chapter 4.

2.8.7 Bandwidth and Sensor Data Rates

This topic has been the focus of extensive debate since the introduction of m-Health a decade ago. To illustrate this fact, Table 2.3 shows data rates, bandwidth, and latency of the main physiological and biomedical parameters involved in wireless monitoring applications. In the era of big data, it is most likely that a large amount of medical data will be transmitted over different communication network channels. In future scenarios, these will be transmitted simultaneously by possibly hundreds of millions of smartphone users, with their personal data acquired from medical sensory devices in one form or another. Some will use higher bandwidth than others, such as the case with diagnostic images, like X-rays or ultrasound, which demand data rates on the order of megabits per second for transmitting large volumes of data. These applications also require the use of more efficient and smart networking methods, and perhaps more effective compression algorithms tailored for these healthcare scenarios. Similarly, the quality of service and quality of user experience offered by these networks, resulting in better medical quality of service (m-QoS) and medical quality

TABLE 2.3   Data Rates and Bandwidth of Key Biomedical Wireless Monitoring Parameters

Physiological/Biomedical Parameter	Bandwidth/	Data Rate/Latency
ECG (12 leads)	0.1–1 kHz	~144 kbps/<200 ms
EEG (12 leads)	0.5–0.2 kHz	~40 kbps/<300 ms
EMG	0–10 kHz	~350 kbps/<200 ms
Body temperature	0–1 Hz	~0.1 kHz
Medical imaging and video streaming data	–	~>10 Mbps/<100 ms
Speech and voice	–	~>50–100 kbps/<10 ms
Accelerometer and motion sensing	0–0.5 kHz	~30 kbps
Blood glucose monitoring	0–40 Hz	~1.5 kbps
Blood pressure	0–1 Hz	~15 Hz

of experience (m-QoE), will require further research (Istepanian et al., 2013). Further communication details on these issues will be explained in Chapter 4.

2.8.8   Security and Privacy

The security and integrity of the data from sensors and networks is an important issue for any medical monitoring device. Security measures usually dictate that only authorized doctors or other qualified healthcare personnel have access to that medical data (Bao et al., 2005; Espina et al., 2009). This ensures the integrity, authentication, and identity of the data received over a secure connection, a topic that has been thoroughly researched; however, major challenges remain.

In an era of smart hacking and the near-collapse of Internet privacy, security and privacy in wireless health monitoring become more critical and important (Misic, 2008). For example, users will increasingly want to authorize control over the transmission of their biomedical data from sensors they are wearing (Ikonen and Kaasinen, 2008). An interesting security situation arises for an unconscious patient or elderly person with wearable devices who is unable to provide authorization or use a password, assuming one is necessary. In this case, a biometric method may be used for accountability, although finding uniquely identifying biometric features is still a big challenge.

A further interesting and potentially damaging situation in a legal sense is the protection of data representing diagnostic images, such as X-rays and ultrasonograms that can be transmitted over smartphones. Another example is that of a person who has a medical condition, say lung cancer, which he or she wants to keep private, perhaps from an employer. Interception of a chest X-ray revealing the disease could be disastrous. Although image processing and encryption is highly developed, it is not fool-proof. For example, a privacy leak called "fingerprint and timing-based snooping" has been reported that can occur even for encrypted signals (Srinivasan et al., 2008). This can be a potentially serious privacy and security breach for patients wearing medical devices in future smart home environments. Today, with billions of smartphone and social media users, security and privacy breaches, risks with third party tracking, hacking of personal medical data, and fingerprinting of vital information patterns constitute major challenges for future uses of wireless medical monitoring networks.

In general, the security threats in wireless health networks include the following (Kumar and Lee, 2012):

- Monitoring and eavesdropping of patients' vital signs.
- Threats to information during transmission.
- Routing threats in networks.
- Location threats.
- Activity tracking and denial of service (DoS) threats.

While privacy issues include, among others, patient privacy, misuse of medical information, leakage of prescriptions, eavesdropping on medical data, and social implications for patients, certain security measures may be adopted (Srinivasan et al., 2008; Kumar and Lee, 2012); these include cryptography, key management, secure routing, resilience to node capture, trust management and secure localizations, and robustness of DoS communications. Security and privacy issues are ongoing research topics and subject to further debate for the development of m-Health sensors and wearables.

2.8.9 Compatibility, Interoperability, and Standards

This is an important element, which is work in progress by the standards and regulatory bodies. The integration of networks of sensors that may be working at different frequencies increases the problem of compatibility. Transmission between sensors may employ a number of bands and use different protocols, which can lead to interference between different sensors. It is particularly the case for the unlicensed ISM band, which is suitable for m-Health. This is just one example where the interoperability of sensors and networks needs to be standardized, which is a challenge for an agreed-upon agenda by researchers, manufacturers, and medical experts. The formation of global industrial alliances has successfully addressed this challenge, including the Continua Alliance, whose aim is to recommend inter-operability standards for wireless sensors used for m-Health monitoring applications. The FDA and FCC issued specific regulations on wireless medical devices that define and regulate the frequency spectra of wireless medical devices used in short-range monitoring and diagnostics, together with long-term telemetry applications (Fish and Richardson Regulatory, 2013). Details of the different IEEE-1073 standards for medical device communications are cited in the relevant IEEE standards link (IEEE Standard Association, 2014). One of these standards that reflects the potential importance of wireless wearable devices is IEEE-1708-2014, which is designed specifically for wearable cuff-less blood pressure measuring devices.

Finally, future developments in the next generation of wireless health sensing and monitoring systems should be an exciting and active research area, with great potential for the future of m-Health. One such advance is the sensing concept of body area nano-networks (BANN) and molecular communications devices, where tiny messenger molecules on nanolevels are used as the communication carrier from a sender to a receiver (nanomachine) (Atakan and Akan, 2012) inside the bloodstream. Other developments also include nanorobotics and M2M communications sensors for m-Health and their connectivity with future 5G mobile systems.

2.9 DIAGNOSTIC SENSORS

The second main category in the new m-Health sensor taxonomy is diagnostic sensors. The adoption of these sensors for wireless, implantable applications for diagnostic purposes is evolving but challenging at the same time. These smart sensors, which target different diseases and illnesses, need to be capable of clinical accuracy to be effective. Earlier implantable diagnostic biosensors represented specialized sensors to provide the necessary medical information with accurate diagnostic capability to continuously measure metabolite levels without the need for patient intervention, and regardless of whether the patient was awake or sleeping (Shults et al., 1994). These earlier examples of diagnostic biosensors signaled the way forward for later generations of noninvasive sensors used for the early detection and diagnosis of different chronic diseases, such as pre-diabetes, heart failure, and cancer. In diabetes, collecting glucose data by obtaining blood samples from finger pricking cannot determine any trends or patterns associated with daily eating habits, exercise, and blood glucose levels (Santhisagar et al., 2010).

Recent developments in diabetic sensing technologies, particularly for type 1 diabetic patients, include a smart wireless artificial pancreas. This device can diagnose the levels of insulin injected and control a CGM wirelessly. Furthermore, it reacts to blood sugar trends while "remembering" how much insulin has been injected in the last few hours to avoid any hypoglycemic events. It also allows diabetic patients to view their glucose levels over the smartphone element of the device (McElwee, 2014).

Other examples in this category include the wireless diagnosis and prediction of the onset of any early cardiovascular illness patterns. These include the monitoring of implanted defibrillators and pacemakers that increase the life expectancy of heart patients by diagnosing and treating bradycardia events. Using m-Health technology to monitor these implantable devices remotely offers improved clinical care for huge numbers of heart failure patients, who mostly have impaired systolic left ventricle function, which, if untreated, leads to the onset of heart failure or life-threatening arrhythmias (erratic fast heart beat), atrial fibrillation, or stroke (Boriani et al., 2013).

Figure 2.7 shows what is considered to be the world's smallest pacemaker (Micra TPS, Medtronic Plc). This device is less than one-tenth the size of traditional pacemakers (~1cc) and is self-contained to deliver the most advanced pacing technology available to patients via a minimally invasive approach. It is still in the investigational phase and is not yet approved in the United States.

Further sensor examples in this category include point of care diagnostic sensors (Langer and Peppas, 2003). It is likely that the use of this category will increase significantly in the future, because of the rapid developments in microelectrome-chanical systems technology, combined with their wireless communications capabilities and digital electronics connectivity (Leonov et al., 2005).

For developing countries, there is interest in point of care sensors for low-cost diagnostics, a new class of portable easy-to-use products to increase access to healthcare worldwide (Bates, 2015). The aim is to carry out a simple test with a disposable sensor, a DxBox, which is a small mylar card containing dehydrated reagents that can withstand warm temperatures indefinitely, without the use of refrigeration or electricity. This sensor can analyze a patient's blood sample and differentiate between six pathogens to identify various fevers. The recent reported epidemics and spread of less-known viruses especially in remote areas such as the Zika virus in the Americas or the Camel Flu (MERS) virus in the Middle East warrant further research studies in this area.

Other diagnostic sensors include portable wireless ultrasound scanning devices that can provide remote diagnosis with connections to smartphones for transmitting the

FIG. 2.7 World's smallest pacemaker (Micra TPS). (Photo courtesy of Medtronic PLC.)

images. In the near future, we might witness a new generation of smart diagnostic m-Health sensors that can potentially diagnose and intelligently analyze diseases and predict their onset, perhaps by months if not years, and be able to aggregate the data to provide early warning alerts for any complications before disease onset. The development of ultrathin, flexible, and skin-mounted "Biostamp" wearable diagnostic devices has been reported recently (Chandler, 2016). These sensors can be used if clinically validated in wider diagnostic applications such as the prediction of pre-diabetes, cancer, and other chronic conditions. The use of these types of sensors will also probably be popular among the "quantified self" community, which advocates advanced ways of using evolving technology to collect data about an individual's life diagnosis.

2.10   PROGNOSTIC AND TREATMENT SENSORS

The third category of m-Health sensors comprises prognostic sensors. These represent the most exciting and challenging type, since they combine recent technological developments with bioscience, biotechnology, nanoscience, and nanotechnologies.

These sensors can provide intelligent and personalized self-treatment and provide wirelessly the necessary prognostic information on a particular illness or disease. Their functionalities can be closely associated with monitoring but are differentiated from them by their capability of providing dosages for treatment delivery and for prognostic data. An example of these includes *in vivo* smart controlled drug release sensors that can be ingested into the body, which act to deliver the appropriate quantity of drugs (LaVan et al., 2002, 2003). Others include sensors that can be swallowed and digested for smart medicine therapies, and devices that allow the wireless monitoring and release of drugs at the appropriate time and communicate with an outside hub. Another sensing concept in this category is the "smart dust" sensor (Kahn et al., 1999, 2000).

Further examples in this category are the miniaturized wireless capsules that can be swallowed during an endoscopy procedure; these are equipped with tiny video transmission systems, with processing capabilities to convey the diagnostic imaging information from inside the body and gastrointestinal tracts to an outside portable hub for further clinical decision making and analysis (Pan and Wang, 2012). Most of these developments are currently either early prototypes or precommercial versions, or in clinical evaluation and research phases. Their full wireless connectivity and real-time performances are yet to be clinically validated, which may take several years. Futuristic miniaturized micromolecular medical sensors can also be classified within this category.

2.11   ASSISTIVE SENSORS

The fourth category of sensors are those applied for ambient assisted living, cognitive tele-rehabilitation, and other smart home, disability, and mental health assistive m-Health applications (Basten et al., 2003). For care applications, further examples include event sensors in smart homes, special tracking and alarm aid sensors for dementia patients, and environmental sensors for domestic assistive robotics. Other applications include mobile RFID tracking devices for medication compliance and wearable haptics (tactile devices to enable interaction with a computer). Sensors that involve medical robotic applications can also be classified into this category. Further details of recent advances are cited in various publications (Sazonov and Neuman, 2014; Manuel and Pereira, 2014).

2.12   SUMMARY

In this chapter, we have described the general taxonomy of m-Health sensors. There is confusion in m-Health sensing, resulting in a misunderstanding based solely on wireless health monitoring, therefore without consideration of the wider spectrum that encompasses other m-Health domains.

The four main categories of m-Health sensors are health and wellness monitoring sensors, diagnostic sensors, prognostic and treatment sensors, and assistive sensors.

These have been described with emphasis on the first category, due to the current interest, commercial appeal, and their potential wider use in different health and well-being areas.

Low-power integrated circuits and wireless communications have allowed the design of inexpensive, miniature, lightweight, and intelligent physiological sensor nodes. These have the capability of sensing, processing, and communicating one or more biomedical signals, and they can be integrated into wireless body area networks that are usually stuck onto the body, but in the future may in some cases be implanted or ingested. Advances in sensor architectures and their connectivity in future wireless networks will revolutionize diagnosis and healthcare by allowing non-invasive, continuous, ambulatory monitoring with near-real-time updates of medical records via the Internet. These will be compounded by similar advances in newer generations of wearable medical sensors embedded in smart glass, watches, and smart clothing.

Smartphones are increasingly acting to sense, collect, process, and distribute medical data. The concept of "smartphone m-Health sensing" is becoming a reality, with recent advances of embedded medical sensors in phones. However, it is vital to validate these new sensing technologies in medical trials, especially their clinical efficacy and effectiveness. Technological advances are always faster than can be accommodated by the medical profession, mainly due to the extreme caution of doctors, who naturally lack training in technology (just as engineers lack training in anatomy, physiology, pathology, and surgery), but all kinds of sensing devices are now in use for many applications.

Today, we see numerous commercial wearable systems targeting mostly the consumer health and well-being market, and the latest figures of potential revenues for these systems are staggering. Commercially, according to industry estimates, the global market for wearable medical devices was valued at US$2 billion in 2012, and is expected to reach US$5.8 billion in 2019, growing at a 16.4% from 2013 to 2019 (TMR Report, 2013). The current average market prices of mobile, wearable wellness devices are typically in the range of $100–150; these can potentially usher a new era of wearable mobile health. As already emphasized, these devices are not yet of proven acceptability by both patients and clinicians. There are major opportunities for consumer-based m-Health applications in the foreseeable future, notably smart wearable trackers and wristwatches. These are equipped with embedded sensors that connect with smartphone Apps to monitor daily activities and "wellness patterns," such as sleeping habits, exercise regimes, and general health status. There is also increasing interest in using wearable technologies for lifestyle changes, for early detection of disorders, and for health changes that might require clinical interventions. It is evident that large-scale clinical studies must precede any widespread adoption of these technologies.

Technological advances in the IoT and M2M communications are being hailed as another major transformative development in future smart m-Health sensing. These emerging Internet-based technologies will enable more robust monitoring architectures with IoT wearable sensors. Potentially, patients may then feel not only better but perhaps also safer, as doctors and care providers could be constantly updated on their specific condition. In the next few years, we will witness more and more of IOT-based m-Health sensors within these categories. Further details on IoT connectivity are presented in Chapter 4.

The role of the big health data generated from these connected devices and their analytics will be another critical factor in these scenarios. The development of new intelligent data processing for extracting meaningful data for presentation and analysis represents a crucial milestone for future smart m-Health monitoring. This topic will be discussed in more detail in Chapter 3.

There is still uncertainty in the consumer-led market for these technologies. The most recent example is the announcement by Google to stop selling its Google Glass to individual customers (Luckerson, 2015). This news came after the much-hyped benefits of this state-of-the-art wearable product and its future applications.

From the global health economic perspective, most of the commercial success of smart m-Health sensing technology is focused on the wellness and sports sectors, which target well-off people who live and work in the developed world. There is little or no work to date on the economic drivers for developing easy-to-use, cost-effective versions of smart sensing technologies for the developing world, for which there is a major demand. For example, there is much demand in developing reliable, accurate, low-cost, secure m-Health sensors for early warning and detection of global disease outbreaks and healthcare crises. In resource-poor countries, with potential communicable diseases that could escalate into a global health crisis, these technologies are still absent. Many of these poor and remote regions, although equipped with minimum healthcare resources, are probably also hot spots of high-level smartphone penetration and usage that are usually not exploited for such scenarios. However, there are numerous technical and regulatory challenges that need to be carefully addressed and tackled:

- The clinical cost and efficacy of low-cost, smart sensor technologies and their ability to provide better and more accurate information, especially in remote regions.
- The capability of sensors to accurately capture medical data in adverse mobility conditions.
- Low-cost intelligent analysis of data in real-time scenarios to provide timely warning of, and response to, any potential global health crisis.
- Considerations of security and privacy in developing smart wireless sensor architecture linked to cloud architectures and services.
- Training healthcare professionals to translate the basic knowledge of using these technologies to patients.

Developments in sensing technologies are one of the fast evolving areas of m-Health. These will be closely linked in the near future with similar developments in the IoT and M2M technologies for healthcare, leading to a new generation of mobile devices designed and developed specifically for different m-Health applications.

REFERENCES

Akyildiz IF, Su W, Sankarasubramaniam Y, and Cayirci E (2002) Wireless sensor networks: a survey. *Computer Networks* 38: 393–422 (also *Sensors* (2011), 11: 5590).

Aleksandar M, Chris O, and Emil J (2006) Wireless sensor networks for personal health monitoring: issues and an implementation. *Computers and Communications* 29: 2521–2533.

Alemdar H and Ersoy C (2010) Wireless sensor networks for healthcare: a survey. *Computer Networks* 54: 2688–2710.

AliveCor (2016) Available at www.AliveCor.com. (accessed January 2016).

Anastasi G, Conti M, Di Francesco M, and Passarella A (2009) Energy conservation in wireless sensor networks: a survey. *Ad Hoc Networks* 7: 537–568.

Apple (2016) Available at www.apple.com. (accessed February 2016).

Atakan B and Akan OB (2012) Body area nano-networks with molecular communications in nanomedicine. *IEEE Communications Magazine* 50(1):28–34.

Aziz O, Lo B, King R, Darzi A, and Yang GZ (2006) Pervasive body sensor network: an approach to monitoring the postoperative surgical patient. In: *Proceedings of the International Workshop on Wearable and Implantable Body Sensor Networks*, Cambridge, MA, pp. 13–18.

Baker C, Armijo K, Belka S, Benhabib M, Bhargava V, Burkhart N, and Der Minassians A (2007) Wireless sensor networks for home health care. In: *Proceedings of the 21st International Conference on Advanced Information Networking and Applications Workshops (AINAW'07)*, Niagara Falls, Canada, vol. 2, pp. 832–837.

Bao S, Zhang Y, and Shen L (2005) Physiological signal-based identity authentication for body area sensor networks and mobile healthcare systems. In: *Proceedings of the 27th Annual International Conference of the IEEE Engineering in Medicine and Biology Society (EMBS)*, Shanghai, China, pp. 2455–2458.

Bashshur RL, Shannon G, Krupinski EA, and Grigsby A (2013) Sustaining and realizing the promise of telemedicine. *Telemedicine and e-Health* 19(5):339–345.

Basten T, Geilen M, and De Groot H (Eds.) (2003) *Ambient Intelligence: Impact on Embedded System Design*. Berlin, Germany: Springer, pp. 51–67.

Bates M (2015) IEEE Pulse. Available at pulse.embs.org/november-2015/the-present-and-future-of-low-cost-diagnostics (accessed January 2016).

Bazaka K and Jacob MV (2013) Implantable devices: issues and challenges. *Electronics* 2: 1–34.

Benhaddou D, Balakrishnan M, and Yuan X (2008) Remote healthcare monitoring system architecture using sensor networks. In: *Proceedings of the IEEE Region 5 Conference*, Kansas City, MO, pp. 1–6.

Bluetooth (2014) Specification guide. Available at www.bluetooth.com. (accessed September 2014).

Boriani G, Da Costa A, Ricci RP, Quesada A, Favale S, Iacopino S, Romeo F, Risi A, Mangoni di S Stefano L, Navarro X, Biffi M, Santini M, and Burri H (2013) The MOnitoring Resynchronization dEvices and CARdiac patiEnts (MORE-CARE) randomized controlled trial: phase 1 results on dynamics of early intervention with remote monitoring. *Journal of Medical Internet Research* 15(8):e167.

Braem B, Latre B, Moerman I, Blondia C, and Demeester P (2006) The wireless autonomous spanning tree protocol for multi-hop wireless body area networks. In: *Proceedings of the 3rd Annual International Conference on Mobile and Ubiquitous Systems—Workshops*, San Jose, CA, pp. 1–8.

Cao H and Leung V (2009) Enabling technologies for wireless body area networks: a survey and outlook. *IEEE Communications Magazine* 47:84–93.

Casas R, Blasco MR, Robinet A, Delgado AR, Yarza AR, Mcginn J, Picking R, and Grout V (2008) User modelling in ambient intelligence for elderly and disabled people. In: *Proceedings of the 11th International Conference on Computers Helping People with Special Needs*, Linz, Austria, pp. 114–122.

Casson AJ and Rodriguez-Villegas E (2013) Advances in circuits and algorithms for high-quality EEG recording in non-controlled environments. Available at http://embc.embs.org/files/2013/1150_FI.pdf. (accessed March 2016).

Chandler DL (2016) John Rogers and the ultrathin limits of technology. *IEEE Pulse* 7(1):9–12.

Chen K and Bassett DJ (2005) The technology of accelerometry-based activity monitors: current and future. *Medicine and Science Sports Exercise* 37: 490–500.

Cook DJ, Thompson JE, Prinsen SK, Dearani JA, and Deschamps C (2013) Functional recovery in the elderly after major surgery: assessment of mobility recovery using wireless technology. *The Annals of Thoracic Surgery* 96: 1057–1061.

Coosemans J and Puers R (2005) An autonomous bladder pressure monitoring system. *Sensors and Actuators A* 123–124: 155–161.

Darwish A and Hassanien AE (2011) Wearable and implantable wireless sensor network solutions for healthcare monitoring. *Sensors* 11: 5561–5595.

Dervisoglu G, Gutnik L, Haick M, Ho C, Koplow M, Mangold J, Robinson S, Rosa M, Schwartz M, Sims C, Stoffregen H, Waterbury A, Leland E, Pering T, and Wright P (2007) Wireless sensor networks for home health care. In: *Proceedings of the 21st International Conference on Advanced Information Networking and Applications Workshops*, Niagara Falls, Canada, May 21–23, 2007, pp. 832–837.

Diabetes UK (2013) Available at www.diabetes.co.uk. (accessed March 2013).

Diamond D, Coyle S, Scarmagnani S, and Hayes J (2008) Wireless sensor networks and chemo-biosensing. *Chemical Reviews* 108: 652–679.

D-MAS HyperSat (2016) Available at www.DanMedical.com. (accessed January 11, 2016).

Eklund JM, Hansen TR, Sprinkle J, and Sastry S (2005) Information technology for assisted living at home: building a wireless infrastructure for assisted living. In: *Proceedings of the 27th Annual International Conference of IEEE Engineering in Medicine and Biology Society*, Shanghai, China.

Espina J, Baldus H, Falck T, Garcia O, and Klabunde K (2009) Towards easy-to-use, safe, and secure wireless medical body sensor networks. In: Olla P and Tan J (Eds.), *Mobile Health Solutions for Biomedical Applications*. Medical Information Science Reference, pp. 159–179.

FCC (Federal Communications Commission) (2011) Medical Devices Radiocommunications Service (MedRadio). Available at https://www.fcc.gov/general/medical-device-radio communications-service-medradio. (accessed January 18, 2016).

Filho RV, Neto R, Silvestre B, and de Oliveira, GW Jr. (2013) An evaluation method of research on wearable wireless body area networks in healthcare. *International Journal of Computer Science and Information Technology* 5(1):65–78.

Firstbeat Technologies (2012) *An energy expenditure estimation method based on heart rate measurement*, White Paper, 2–5. Available at www.firstbeat.com/userData/firstbeat/Energy_Expenditure_Estimation.pdf. (accessed October 2014).

Fish and Richardson Regulatory (2013) *Wireless medical technologies: navigating government regulation in the new medical age*. Fish's Regulatory & Government Affairs Group. Available at http://www.fr.com/files/uploads/attachments/finalregulatorywhitepaperwirelessmedicaltechnologies.pdf. (accessed September 2014).

Fitbit (2016) Available at www.fitbit.com. (accessed February 2016).

Free C, Phillips G, Watson L, Galli L, Felix L, Edwards P, Patel V, and Haines A (2013) The effectiveness of mobile health technologies to improve health care service delivery processes: a systematic review and meta-analysis. *PLoS Medicine* 10(1):1–26.

Gietzelt M, Wolf KH, Marschollek M, and Haux R (2008) Automatic self-calibration of body worn triaxial-accelerometers for application in healthcare. In: *Proceedings of the 2nd International Conference on Pervasive Computing Technologies for Healthcare*, Tampere, Finland, pp. 177–180.

González-Valenzuela S, Liang X, Cao H, Chen M, and Leung M (2012) Body area networks. In: Filippini D (Ed.), *Autonomous Sensor Networks: Collective Sensing Strategies for Analytical Purposes*. Springer, pp. 17–37.

Gopalsamy C, Park S, Rajamanickam R, and Jayaraman S (2005) The wearable motherboard TM: the first generation of adaptive and responsive textile structures (ARTS) for medical applications. *Virtual Reality* 4: 152–168.

Gutierrez J and Rogen B (2014) *Managing risk: Cleveland Clinic's population management of employees and their families*. Cleveland Clinic Presentation, pp. 1–63. Available at www.amga.org/docs/Meetings/IQL/2014/Breakouts/ClevelandClinic-GutierrezRogenUpdated.pdf. (accessed December 2014).

Habetha J and Reiter H (2008) MyHeart Project. Available at http://www.hitech-projects.com/uprojects/myheart/. (accessed December 2013).

Halin N, Junnila M, Loula P, and Aarnio P (2005) The Life Shirt system for wireless patient monitoring in the operating room. *Journal of Telemedicine and Telecare* 11: 41–43.

Heikenfeld J (2014) *Sweat sensors will change how wearables track your health*. IEEE Spectrum. Available at http://spectrum.ieee.org/biomedical/diagnostics/sweat-sensors-will-change-how-wearables-track-your-health. (accessed October 2015).

Herzog R (2004) MobiHealth Project. Available at http://www.mobihealth.org/. (accessed December 2013).

Hirt E and Scheffler M (2008) Personal supervision and alarming systems. In: Xiao Y and Chen H (Eds.), *Mobile Telemedicine: A Computing and Networking Perspective*. Auerbach Publications, pp. 3–28.

Hoffmann KP and Solzbacher F (2011) Recording and processing of biosignals. In: Kramme R, Hoffmann K, and Pozos RS (Eds.), *Springer Handbook of Medical Technology*. Berlin: Springer, pp. 923–946.

Hori T, Nishida Y, Suehiro T, and Hirai S (2000) SELF-Network: design and implementation of network for distributed embedded sensors. In: *Proceedings of the IEEE/RSJ International Conference on Intelligent Robots and Systems*, Takamatsu, Japan, pp. 1373–1378.

IEEE Standard Association (2014) *IEEE-11073 standard for medical device communications*. Available at http://standards.ieee.org/findstds/standard/healthcare_it_all.html. (accessed September 2013).

IHS Electronics and Media (2013) *Wearable technology: market assessment.* IHS White Paper. Available at http://www.ihs.com/pdfs/Wearable-Technology-sep-2013.pdf. (accessed October 2014).

Ikonen V and Kaasinen E (2008) Ethical assessment in the design of ambient assisted living. In: *Proceedings of the of Assisted Living Systems: Models, Architectures and Engineering Approaches,* Schloss Dagstuhl, Germany, pp. 14–17.

Intanagonwiwat C, Govindan R, and Estrin D. (2000) Directed diffusion: a scalable and robust communication paradigm for sensor networks. In: *Proceedings of the ACM MobiCom'00,* Boston, MA, pp. 56–67.

Istepanian RSH, AliNejad A, and Philip N (2013) Medical quality of service (m-QoS) and medical quality of experience (m-QoE) for 4G health systems. In: Farrugia R and Debono C (Eds.), *Multimedia Networking and Coding.* IGI Global, pp. 359–376.

Jasemian Y (2008) Elderly comfort and compliance to modern telemedicine system at home. In: *Proceedings of the 2nd International Conference on Pervasive Computing Technologies for Healthcare,* Tampere, Finland, pp. 60–63.

Jawbone (2016) Available at www.jawbone.com. (accessed February 2016).

Joel JPCR, Pereira OE, and Neves PCS (2011) Biofeedback data visualization for body sensor networks. *Journal of Network and Computer Applications* 34: 151–158.

Jones DP (2010) *Biomedical Sensors.* New York: Momentum Press,

Jovanov E, Price J, Raskovic D, Kavi K, Martin T, and Adhami R (2000) Wireless personal area networks in telemedical environment. In: *Proceedings of the 2000 IEEE EMBS International Conference on Information Technology Applications in Biomedicine,* Arlington, VA, pp. 22–27.

Jovanov E, Lords A, Raskovic D, Cox P, Adhami R, and Andrasik F (2003) Stress monitoring using a distributed wireless intelligent sensor system. *IEEE Engineering in Medicine and Biology Magazine* 22: 49–55.

Jovanov E, Milenkovic A, Otto C, and de Groen P (2005) A wireless body area network of intelligent motion sensors for computer assisted physical rehabilitation. *Journal of Neuro Engineering and Rehabilitation* 2: 6.

Jovanov E, Poon CCY, Yang G-Z, and Zhang YT (2009) Guest Editorial. Body sensor networks: from theory to emerging applications. *IEEE Transactions on Information Technology in Biomedicine* 13(6):859–863.

Kim D-Y and Cho J (2009) WBAN meets WBAN: smart mobile space over wireless body area networks. In: *Proceedings of the 7th IEEE Vehicular Technology Conference,* Fall (VTC 2009-Fall), September 20–23, Anchorage, Alaska.

Kahn JM, Katz RH, and Pister KSJ (1999) Next century challenges: mobile networking for smart dust. In: *Proceedings of the ACM MobiCom'99,* Washington, DC, pp. 271–278.

Kahn JM, Katz RH, and Pister KSJ (2000) Emerging challenges: mobile networking for smart dust. *Journal of Communication Networks* 2: 188–196.

Khan P, Hussain MA, and Kwak KS (2009) Medical applications of wireless body area networks. *International Journal of Digital Content Technology and Its Applications* 3: 185–193.

Kraemer R and Katz MD (Eds.) (2009) *Range Wireless Communications: Emerging Technologies and Applications.* John Wiley & Sons, Inc.

Kumar P and Lee HJ (2012) Security issues in healthcare applications using wireless medical sensor networks: a survey. *Sensors* 12(1):55–91.

Kumar S, Nilsen W, Pavel M, and Srivastava M (2013) Mobile health: revolutionizing healthcare through trans-disciplinary research. *Computer* 28–35.

Langenhove LV (2007) *Smart Textiles for Medicine and Healthcare: Materials, Systems and Applications.* Elsevier.

Langer R and Peppas NA (2003) Advances in biomaterials, drug delivery, and bionanotechnology. *Bioengineering, Food and Natural Products* 49: 2990–3006.

Latre B, Braem B, Moerman I, Blondia C, Reusens E, Joseph W, and Demeester P (2007) A low-delay protocol for multi-hop wireless body area networks. In: *Proceedings of the 4th Annual International Conference on Mobile and Ubiquitous Systems: Networking and Services,* Philadelphia, PA, pp. 1–8.

Lauterbach C, Strasser M, Jung S, and Weber W (2002) Smart clothes self-powered by body heat. In: *Proceedings of the Avantex Symposium,* Frankfurt, Germany, pp. 5259–5263.

LaVan DA, Lynn DM, and Langer R (2002) Moving smaller in drug discovery and delivery. *Nature Reviews* 1: 77–84.

LaVan DA, McGuire T, and Langer R (2003) Small-scale systems for *in vivo* drug delivery. *Nature Biotechnology* 21: 1184–1191.

Leonov V, Fiorini P, Sedky S, Torfs T, and van Hoof C (2005) Thermoelectric MEMS generators as a power supply for a body area network. In: *Proceedings of the 13th International Conference on Solid-State Sensors, Actuators and Microsystems,* Seoul, Korea, pp. 291–294.

Li P, Wen Y, Liu P, Li X, and Jia C (2010) A magnetoelectric energy harvester and management circuit for wireless sensor network. *Sensors and Actuators A* 157: 100–106.

Li G, Lee BL, and Chung WY (2015) Smartwatch-based wearable EEG system for driver drowsiness detection. *IEEE Sensors Journal* 15(12):7169–7180.

Lin G and Tang W (2000) *Wearable Sensor Patches for Physiological Monitoring.* New York, NY: NASA Tech Briefs, pp. 354–2240.

Lo BPL and Yang GZ (2005) Technical challenges and current implementations of body sensor networks. In: *Proceedings of the 2nd International Workshop on Wearable and Implantable Body Sensor Networks,* London, UK, vol. 1.

Loew N, Winzer K-J, Becher G, Schönfuß D, Falck Th, Uhlrich G, Katterle M, and Scheller FW (2007) Medical sensors of the BASUMA body sensor network. In: *Proceedings of the 4th International Workshop on Wearable and Implantable Body Sensor Networks (BSN'07),* IFMBE Proceedings, vol.13, pp. 171–176.

Luckerson V (2015) *Google will stop selling glass next week.* Time. Available at Time.com/3669927/google-glass-explorer-program-ends (accessed January 2015).

Luke D. (2014) *Using brainwaves to control technology through wearables.* Available at http://ieeexplore.ieee.org/xpl/login.jsp?tp=&arnumber=6824740&url=http%3A%2F%2Fieeexplore.ieee.org%2Fxpls%2Fabs_all.jsp%3Farnumber%3D6824740. (accessed March 2016).

Lymberis A and de Rossi D (Eds.) (2004) *Wearable, e-Health Systems for Personalised Health Management.* Amsterdam, The Netherlands: IOS Press.

Lymberis A and Gatzoulis L (2011) Wearable health systems: from smart technologies to real applications. In: *Proceedings of the IEEE Engineering in Medicine and Biology Society,* New York, NY 2006 (*Sensors,* 11: 6789–6792, 5591–5595).

Lymberis A and Paradiso R (2008) Smart fabrics and interactive textile enabling wearable personal applications: R&D state of the art and future challenges. In: *Proceedings of the 30th*

Annual International Conference of IEEE Engineering in Medicine and Biology Society (EMBS), Vancouver, BC, Canada, August 20–24, pp. 5270–5273.

Malan D, Fulford-Jones T, Welsh M, and Moulton S (2004) Codeblue, an ad hoc sensor network infrastructure for emergency medical care. In: *Proceedings of the 1st International Workshop on Wearable and Implantable Body Sensor Networks*, London, UK, pp. 55–58.

Manuel A and Pereira J (2014) Ambient assisted living (AAL): sensors, architectures and applications. In: *Sensors*, Special Issue, March 2014. Available at http://www.mdpi.com/journal/sensors/special_issues/aal. (accessed May 2015).

Marinkovic S and Popovici E (2009) Network coding for efficient error recovery in wireless sensor networks for medical applications. In: *Proceedings of the International. Conference on Emerging Network Intelligence*, Sliema, Malta, pp. 15–20.

Mark AH, Harry C.P. Jr., Adam TB, Kyle R, Benton HC, James HA, and John L (2009) *Body Area Sensor Networks: Challenges and Opportunities*. Atlantic City, NJ: IEEE Computer Society, pp. 58–65.

McCann J and Bryson D (Eds.) (2009) *Smart Clothes and Wearable Technology*. Woodhealth Publishing/CRC Press.

McElwee M (2014) *Type 1 and working on an artificial pancreas*. Insulin Nation. Available at //insulinnation.com/treatment2/artificial-pancreas/type-1-and-working-on-an-artificial-pancreas/ (accessed September 2014).

McFadden T and Indulska J (2004) Context-aware environments for independent living. In: *Proceedings of the 3rd National Conference of Emerging Researchers in Ageing*, Brisbane, Australia, pp. 1–6.

Mihajlović V, Grundlehner B, Vullers R, and Penders J (2014) Wearable, wireless EEG solutions in daily life applications: what are we missing? *IEEE Journal of Biomedical and Health Informatics* 19(1):6–21.

Misic J (2008) Enforcing patient privacy in healthcare WSNs using ECC implemented on 802.15.4 beacon enabled clusters. In: *Proceedings of the 6th Annual IEEE International Conference on Pervasive Computing and Communications*, Hong Kong, pp. 686–691.

Mohseni P and Najafi K (2005) A 1.48-mw low-phase-noise analog frequency modulator for wireless biotelemetry. *IEEE Transactions on Bio-Medical Engineering* 52: 938–943.

Mohseni P, Najafi K, Eliades S, and Wang X (2005) Wireless multi-channel biopotential recording using an integrated FM telemetry circuit. *IEEE Transactions on Neural Systems and Rehabilitation Engineering* 13: 263–271.

Mundt W, Montgomery KN, Udoh UE, Barker VN, Thonier GC, Tellier AM, Ricks RD, Darling RB, Cagle YD, Cabrol NA, Ruoss SJ, Swain JL, Hines JW, and Kovacs GTA (2005) A multi-parameter wearable physiologic monitoring system for space and terrestrial applications. *IEEE Transactions on Information Technology in Biomedicine* 9: 382–391.

Muro-de-la-Herran A, Garcia-Zapirain B, and Mendez-Zorrilla A (2014) Gait analysis methods: an overview of wearable and non-wearable systems, highlighting clinical applications. *Sensors* 14: 3362–3394.

Neihart N and Harrison R (2005) Micropower circuits for bidirectional wireless telemetry in neural recording applications. *IEEE Transactions on Bio-Medical Engineering* 52: 1950–1959.

New York Times (2015) *The health concerns in wearable Tech*. Available at http://www.nytimes.com/ (accessed December 2015).

Noury N, Herve T, Rialle V, Virone G, Mercier E, Morey G, Moro A, and Porcheron T (2000) Monitoring behavior in home using a smart fall sensor. In: *Proceedings of the 1st Annual*

International Conference on Microtechnologies in Medicine and Biology, IEEE, Lyon, France, October 12–14, 2000, pp. 607–610.

Otto CA, Jovanov E, and Milenkovic EA (2006) WBAN-based system for health monitoring at home. In: *Proceedings of the IEEE/EMBS International Summer School on Medical Devices and Biosensors*, Boston, MA, pp. 20–23.

Pai A (2014) *ABI: 90M wearable devices to ship in 2014*. Mobile Health News. Available at http://mobihealthnews.com/29532/abi-90m-wearable-devices-to-ship-in-2014/ (accessed December 2014).

Pan G and Wang L (2012) Swallowable wireless capsule endoscopy: progress and technical challenges. *Gastroenterology Research and Practice* 2012. doi: 10.1155/2012/841691.

Pandian PS, Mohanavelu K, Safeer KP, Kotresh TM, Shakunthala DT, Gopal P, and Padaki VC (2007) Smart vest: wearable multi-parameter remote physiological monitoring system. *Medical Engineering & Physics* 30: 466–477.

Pandian PS, Safeer KP, Gupta P, Shakunthala DT, Sundersheshu BS, and Padaki VC (2008) Wireless sensor network for wearable physiological monitoring. *Journal of Networks* 3: 21–28.

Paradiso R, Loriga G, and Taccini N (2004) Wearable system for vital signs monitoring. *Studies in Health Technology and Informatics* 108: 253–259.

Paradiso R, Loriga G, and Taccini N (2005) A wearable health care system based on knitted integrated sensors. *IEEE Transactions on Information Technology* B9: 337–344.

Park A (2015) *The 25 best inventions of 2015: the superior stethoscope*. Available at http://time.com/4115398/best-inventions-2015/ (accessed January 10, 2016).

Park S and Jayaraman S (2004) e-Health and quality of life: the role of the wearable motherboard. *Studies in Health Technology and Informatics* 108: 239–252.

Patel S, Park H, Bonato, CL, and Rodgers M (2012) A review of wearable sensors and systems with application in rehabilitation. *Journal of NeuroEngineering and Rehabilitation* 9 (21): 1–17.

Pennic F (2015) *Survey: Only 15% of Physicians Discuss Wearables with Patients*. Available at http://hitconsultant.net/, http://hitconsultant.net/2015/06/22/survey-15-physicians-discuss-wearables-patients/ (accessed July 2015).

Pentland A (1996) Smart rooms, smart clothes. *Scientific American* 274(4):68–76.

Poon CCY and Zhang YT (2008) Perspectives on high technologies for low-cost healthcare. *IEEE Engineering in Medicine and Biology Magazine* 27: 42–47.

Purwar A, Jeong DU, and Chung WY (2007) Activity monitoring from real-time tri-axial accelerometer data using sensor network. In: *Proceedings of the. International Conference on Control, Automation and Systems*, Hong Kong, March 21–23, pp. 2402–2406.

Quellrelief (2016) Available at www.quellrelief.com. (accessed January2016).

Quwaider M and Biswas S (2008) Physical context detection using multi-modal sensing with wearable wireless networks. *Journal of Network and Systems Management* 4: 191–202.

Rajan RD (2013) *Wireless-Enabled Remote Patient Monitoring Solutions, mHIMSS*. Available at http://www.mdtmag.com/articles/2013/05/wireless-enabled-remote-patient-monitoring-solutions (accessed September 2014).

Rasid MFA and Woodward B (2005) Bluetooth telemedicine processor for multi-channel biomedical signal transmission via mobile cellular networks. *IEEE Transactions on Information Technology in Biomedicine* 9(1):35–43.

Raskovic D, Milenkovic P, deGroen A, and Jovanov E (2007) From telemedicine to ubiquitous m-Health: the evolution of e-Health systems. In: Feng DD (Ed.), *Biomedical Information Theory*. Elsevier/Academic Press, pp. 479–497.

Renaud M, Karakaya K, Sterken T, Fiorini P, Hoof CV, and Puers R (2008) Fabrication, modelling and characterization of MEMS piezoelectric vibration harvesters. *Sensors and Actuators A* 145–146: 380–386.

Santhisagar V, Ioannis T, Diane JB, Faquir CJ, and Fotios P (2010) Emerging synergy between nanotechnology and implantable biosensors: a review. *Biosensors and Bioelectronics* 25: 1553–1565.

Sazonov E and Neuman, MR (2014) *Wearable Sensors: Fundamentals, Implementation and Applications*. Academic Press.

Schenker EB, Catarsi F, Coluccini L, Belardinelli A, Shklarski D, Alon M, Hirt E, Schmid R, and Vuskovic M (2004) AMON: a wearable multi-parameter medical monitoring and alert system. *IEEE Transactions on Information Technology in Biomedicine* 8: 415–427.

Schumacher F (2014) *Apple Watch: a new era of wearable computing*. Wearable Technologies. com, September10, 2014. Available at http://www.wearable-technologies.com/2014/09/apple-watch-a-new-era-of-wearable-computing/ (accessed October 1, 2014).

Schwiebert L, Gupta SKS, and Weinmann J (2001) Research challenges in wireless networks of biomedical sensors. In: *Proceedings of the 7th Annual International Conference on Mobile Computing and Networking*, Rome, Italy, 151–165.

Sensatex (2016) Available at http://www.sensatex.com. (accessed January 2016).

Shantz JA and Veillette CJ (2014) The application of wearable technology in surgery: ensuring the positive impact of the wearable revolution on surgical patients. *Frontier in Surgery* 1(39):1–4.

Shults MC, Rhodes RK, Updike SJ, Gilligan BJ, and Reining WN (1994) A telemetry instrumentation system for monitoring multiple subcutaneously implanted glucose sensors. *IEEE Transactions on Biomedical Engineering* 41: 937–942.

Sneed A and Scheer R (2014) Safety in a sock: a teenager wins big for an invention that monitors Alzheimer's patients. *Scientific American* 311(4):10.

Spree (2016) Available at spreewearables.com. (accessed February 2016).

Srinivasan V, Stankovic J, and Whitehouse K (2008) Protecting your daily in-home activity information from a wireless snooping attack. In: *Proceedings of the UbiComp'08 Conference*, September 21–24, Seoul, Korea, pp. 21–24.

Sund-Levander M, Forsberg C, and Wahren LK (2002) Normal oral, rectal, tympanic and axillary body temperature in adult men and women: a systematic literature review. *Scandinavian Journal of Caring Sciences* 16(2):122–128.

Sung M, Marci C, Pentland A (2005) Wearable feedback systems for rehabilitation. *Journal of NeuroEngineering and Rehabilitation* 1: 2–17.

Tamura T, Maeda Y, Sekine M, and Yoshida M (2014) Wearable photoplethysmographic sensors: past and present. *Electronics* 3: 282–302.

Tang Q, Tummala N, Gupta S, and Schwiebert L (2005) Communication scheduling to minimize thermal effects of implanted biosensor networks in homogeneous tissue. *IEEE Transactions on Biomedical Engineering* 52: 1285–1294.

Tao W, Liu T, Zheng R, and Feng H (2012) Gait analysis using wearable sensors. *Sensors* 12(2):2255–2283.

Tarassenko L (2014) *Proceedings of the World Economic Forum Annual Meeting*, Davos, Switzerland. Available at http://www.ibme.ox.ac.uk/news-events/news/professor-tarassenko-speaks-at-the-world-economic-forum-in-davos (accessed January 11, 2016).

Thotahewa KMS, Redoute JM, Yuce MR (2014) A low power wearable dual-band wireless body area network system: development and experimental evaluation. *IEEE Transactions on Microwave Theory and Techniques* 62(11):2802–2811.

Time (2014) The smarter home. *Time Magazine*, July 14, P 36-L 47.

TMR (Transparency Research Market) Report (2013) *Wearable medical devices: global market analysis, size, share, growth and trends and forecast, 2013–2019*. Available at http://www.academia.edu/5437132/Global_Wearable_Medical_Devices_Market_Industry_Analysis_Size_Share_Growth_Trends_and_Forecast_2013_2019) (accessed August 2014).

Topol E (2012) *The Creative Destruction of Medicine*. New York: Basic Books.

Ullah N, Khan P, and Kwak KS (2011) A very low power MAC (VLPM) protocol for wireless body area networks. *Sensors* 11(4):3717–3737.

US FDA (Food and Drug Administration) (2013) *Medical Devices: GlucoWatch G2 Biographer*, August 26, 2002. Available at www.fda.gov/medicaldevices/productsandmedicalprocedures. (accessed August 19, 2013).

US FDA (Food and Drug Administration) (2014) *Medical devices: subcutaneous implantable defibrillator (S-ICD) system—P110042, 2012*. Available at www.fda.gov/medicaldevices/productsandmedicalprocedures. (accessed August 2014).

Weber W, Rabaey JM, and Aarts E (Eds.) (2005) *Ambient Intelligence*. Berlin, Germany: Springer, pp. 499–508.

Woodward B, Rasid MFA, Gore L, and Atkins P (2006) GPRS-based mobile telemedicine system. *Journal of Mobile Multimedia* 2(1):2–22.

Yanowitz FG (2012) *Introduction to ECG interpretation*. Available at http://ecg.utah.edu/pdf. (accessed August 5, 2013).

Yoo J, Yan L, Lee S, Kim Y, and Yoo HJ (2010) A 5.2 mW self-configured wearable body sensor network controller and a 12 W wirelessly powered sensor for a continuous health monitoring system. *IEEE Journal of Solid-State Circuits* 45: 178–188.

Yoon Y, Gil SL, Yoo K, and Lee JB (2013) Fabrication of a microneedle/CNT hierarchical micro/nano surface electrochemical sensor and its *in-vitro* glucose sensing characterization. *Sensors* 13: 16672–16681.

Zimetbaum P and Goldman A (2010) Ambulatory arrhythmia monitoring: choosing the right device. *Circulation* 22: 1629–1636.

Zimmerman TG (1996) Personal Area Networks: near-field intra-body communication. *IBM Systems Journal* 35: 609–617.

3

m-HEALTH COMPUTING: m-HEALTH 2.0, SOCIAL NETWORKS, HEALTH APPS, CLOUD, AND BIG HEALTH DATA

> m-Health is the biggest technology breakthrough of our time.
> Kathleen Sebelius, U.S. Secretary of Health and Human Services, 2011

3.1 INTRODUCTION

As described in Chapter 1, advances in computing and Internet technology, compounded with the exponential growth in Internet usage, social media, and smart applications (Apps) are revolutionizing m-Health systems, their future developments, and applications. The main drivers for this change are the following transformational technologies:

- Smart m-Health applications (m-Health Apps)
- Web 2.0 (Health 2.0, Medicine 2.0)
- Social networking for healthcare
- Cloud computing for healthcare
- "Big health data" and analytics

These developments are increasingly changing the way people communicate, and how they connect, interact, and socialize. In the pre-smartphone era, m-Health and the other main IT for healthcare domains (telemedicine, telehealth, and e-Health) were all

m-Health: Fundamentals and Applications, First Edition. Robert S. H. Istepanian and Bryan Woodward.
© 2017 by The Institute of Electrical and Electronics Engineers, Inc. Published 2017 by John Wiley & Sons, Inc.

based on the Web 1.0 Internet connectivity model. In this model, a mobile phone (m-Health), a desktop, or a laptop computer (telemedicine, telehealth, and e-Health) were equipped with a tailored software application program and storage capabilities that connected medical and personal data from a patient to a remote healthcare center via an Internet hub.

The first transformative step was the development of smartphone technology and its associated software platforms. The smartphone has basically redefined mobile health, becoming the core element for integrating the combined powers of computing, sensors, and high-speed Internet communications into one integrated mobile platform. This convergence into one smart, compact, easy-to-use mobile device had a profound impact on the thinking behind m-Health, bringing with it both a blessing and a curse to the concept. The benefits were obvious with the introduction of the m-Health Apps era, where currently more than 100,000 of these are available to download and are used by millions of people. The curse is the misinterpretation and ambiguity that this process has caused. The concept has diverged, perhaps to an irrevocable degree, with its link to smartphone applications, driven by massive interest by mobile business and medical industry stakeholders. "What is m-Health all about?" is therefore now a topic dominated by commercial rather than scientific advances.

Another major computing driver influencing the evolution of m-Health is the introduction of Web 2.0, and the subsequent developments in social networking and social media. These technologies have made a profound impact on how people interact and exchange information about their health status, in a way that was not envisaged before. Thus, the introduction of smartphones and Web 2.0 technologies has changed the landscape of how remote healthcare delivery models and services are configured and implemented.

Cloud computing has been a further major development that has triggered new service delivery models and more flexible access approaches to m-Health data storage. Finally, the development in big health data, generated from the massive quantity of data being acquired from these developments, has paved the way for an emerging digital health data era. All these technological developments are creating new horizons and possibilities for m-Health, but at the same time establishing new challenges with them.

In this chapter, we highlight some of these important computing developments, and their role in transforming current m-Health systems into the next generation, the "m-Health 2.0" era.

3.2 THE EVOLUTION OF m-HEALTH WITH WEB 2.0 AND MEDICINE 2.0: m-HEALTH 2.0

In the first years of the introduction of m-Health, and particularly before 2007, the pre-smartphone era, the computing science of m-Health included specialist computing programs tailored for specific applications. They included, for example, mobile data and image transmission (Busch et al., 2004; Alvaoro et al., 2006), mobile access to electronic health records (EHR) and public health records (PHR) (Eddabbeh and Drion, 2006), mobile medication compliance applications (Wang and Istepanian, 2005), and numerous other applications (Istepanian et al., 2006). These examples were essentially centered on the transition from earlier e-Health desktop-based computing platforms to corresponding mobile access ones.

These m-Health architectures changed with the introduction of Web 2.0 and smartphone technologies and associated platforms. Web 2.0 is essentially based on the development of new Internet models that have transcended from the traditional unidirectional approach to a more interactive one in which users can add information and content, and interact with each other. To the best of our knowledge, Web 2.0 was first defined as "a set of economic, social, and technology trends that collectively form the basis for the next generation of the Internet, a more mature, distinctive medium characterized by user participation, openness, and network effects" (O'Reilly, 2005). With a vague definition like that it is not surprising that there is an ongoing debate as to what exactly Web 2.0 is!

The basic aspects and functionalities brought by Web 2.0 may be summarized by the following developments:

- Interpersonal computing, that is, linking people to people, for example, Facebook, Twitter, Instagram, and LinkedIn.
- Linking Web Services, which means interconnecting Web sites, for example, PayPal and Apple Pay.
- Mass translation of the software as a service, such as linking people to applications, for example, cloud computing.

In the Web 2.0 sphere, users can not only generate their own content and information, such as videos, photos, or blogs, but may also develop, customize, and edit their own web pages. Consequently, the terms *Health 2.0* and *Medicine 2.0* were originated by combining health and medicine services with Web 2.0 technologies (Van De Belt et al., 2010). In general, these reflect the broad terms that classify how different Web 2.0 technologies are associated with healthcare or medical domains.

The debate on the differences between the two, if any, is ongoing, so there is still no agreement as to their definition. However, the general consensus is that they both represent the same concept, that is, they use different Web 2.0 and social media technologies for different healthcare delivery processes and medical services.

The introduction of the first smartphone complemented this Web 2.0 transformative process by incorporating the mobility, personalization, and healthcare Apps into a single mobile communication device. We envisage in the near future, with further developments such as Web 3.0, predictive analytics, cognitive mobiles, and the emergence of smarter generations of m-Health models with architectures that will perhaps redefine "smart"! In particular, these developments will transform m-Health into a new era of *m-Health 2.0*. We can define m-Health 2.0 as "the convergence of m-Health with emerging developments in smart sensors, 5G communications systems

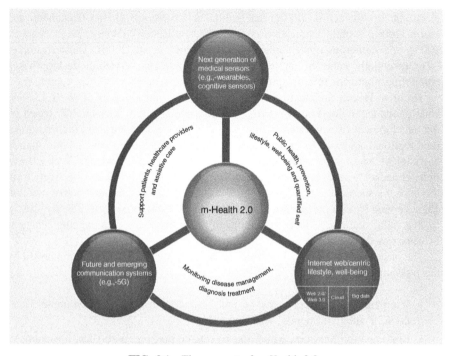

FIG. 3.1 The concept of m-Health 2.0.

with the functional capabilities of Web 2.0, cloud computing, and social networking technologies, toward personalized patient-centered healthcare delivery services." Figure 3.1 illustrates the m-Health 2.0 concept.

As shown in Fig. 3.1, m-Health 2.0 embraces an evolution of the original defining elements of m-Health that constituted their basic scientific and technological principles:

- Smarter m-Health sensing, driven by advances in wearable and other emerging medical sensing technologies, as described in Chapter 2. These developments represent the evolution of the original sensors' building block.
- Future mobile communications and wireless connectivity, notably 5G mobile communication systems, machine-to-machine (M2M), and Internet of things (IoT). These developments represent the evolution of mobile communications and wireless connectivity.
- Internet-centered computing, which represents the evolution of the original mobile computing principle by adopting the functional capacities of Web 2.0 (and beyond), cloud computing, and social networking.

These evolving building blocks can formulate the basic principles for developing the next generation of systems and their associated architectures and services, ushering a new era of m-Health. It is also imperative, in developing future strategies and implementation models around m-Health 2.0, to consider the best trade-off to exploit these innovations and minimize or optimize the current asymmetry between the m-Health markets and consumer-driven demands on the one hand and the clinical requirements and global healthcare challenges on the other. Some of these issues will be discussed in Chapters 5–7.

3.2.1 m-Health and Social Networking

One of the most remarkable outcomes of Web 2.0 developments was the introduction of social networking and social media applications for healthcare. Within less than a decade since the advent of social networking, its appeal and global usage is rivaled only by the global success of smartphones. Today, social networks include a plethora of massively popular systems, notably *Facebook, Twitter, LinkedIn, Instagram, Flickr, Tumblr, Pinterest*, and others, with hundreds of millions of users globally.

For example, Facebook remains the most commercially successful and popular social media example, with an estimated 1.44 billion active users globally. Nearly 71% of online American adult Internet users are on Facebook, followed by 28% using LinkedIn and Pinterest sites, 26% using Instagram, and 23% using Twitter (Duggan et al., 2014).

A social networking site, in general, can be defined as "the digital representation of its users and their social connections or relationships in the physical or virtual world, plus providing networking services for messaging and socializing among its users" (Boyd and Ellison, 2007; Zhang et al., 2010). The following are its purposes:

- Allow users to construct digital representations of themselves (i.e., profiles) and articulate their social connections with other users (i.e., lists of contacts).
- Support the maintenance and enhancement of preexisting social connections among users in the physical or virtual world.
- Assist in forging new connections based on common interests, locations, activities, and other functions.

Others define social media as "a group of Internet-based applications that build on the ideological and technological foundations of Web 2.0, and that allow the creation and exchange of user-generated content" (Kaplan and Haenlein, 2010).

Similarly, for m-Health, the same functionalities and social media contexts defined earlier are built around healthcare, and the user's interests and perceptions of their health, wellness, or illness. However, there are concerns that social media can also pose significant problems for psychological well-being (Morris and Aguilera, 2012). Furthermore, the large number of health domains affected by recent research on the spread of behaviors has made social diffusion a topic of growing interest for an increasing number of researchers and practitioners who are concerned with understanding the social dimensions of health (Centola, 2013).

In a survey of 257 U.S. physicians on their usage and perceptions of social media in their workplace, 44% did not use social media for work, and 29% said they did not use it at all. Among the users, 40% used it for keeping up with health news, followed by 33% for discussions with peers (MedData Group, 2014). In another survey of 485 U.S. physicians, the aim was to understand the factors that can influence the adoption and meaningful use of social media by physicians. The conclusion was that 85% of the oncologists and primary care physicians surveyed use social media at least daily to scan information or to explore health information, and 65% said it improved their care delivery (McGowan et al., 2012). It is also estimated that 40% of the millennials use social media for any health-related purpose and 16% use social media to learn more about a specific health problem (World of Marketing Online)

These studies indicate that overall social media can be seen as an efficient and effective method for physicians and consumers to keep up-to-date with medical information and to share newly acquired medical knowledge with their peers, with potential improvements in the quality of patient care. Although these studies were restricted mainly to the opinions of U.S. physicians, further work is required to better understand the impact of social media on physicians' work and practice in a global context.

In general, social networks can be seen to provide potential benefits by enabling popular access and communication to provide patients, specialists, and others with healthcare information. These networks can therefore provide the most relevant information, and allow the exchange of healthcare information between patients (Pearson et al., 2011). Consequently, there is increasing trend in the use of social networking sites for different healthcare conditions and wellness applications.

Some of the existing healthcare applications to benefit from social networking participation include (Keckley and Hoffmann, 2010):

- *Maintaining Health and Wellness:* Examples of these include social networks that offer "wellness moderation platforms" for patients and physicians (e.g., WebMD, 2016) or using social networks for health activists to advise patients about their health on one site (WEGO Health, 2016).

- *Disease Management:* Examples include the popular network *PatientsLikeMe* (2014) that offers a 24/7 secure login for health plans to enable disease management, and Inspire (2016) that offers patients 24/7 access to peer communities.

- *Clinical Trial Recruitments:* These include using social networks for recruiting patients for clinical trials for different diseases, for example Novartis' use of networks (*PatientsLikeMe*) for their FTY720 multiple sclerosis trials.

- *Public Health Records:* These include, for example, participants in sites such as *PatientsLikeMe* and *MedHelp* to upload detailed information about their conditions, and also to receive information from other patients.

- *Health Professional Training:* Examples include networks such as *RadRounds*, which offer radiology professionals a community to collaborate, share cases, and receive opinions. Other networks such as *Sermo* and *Ozmosis* offer physicians the opportunity to submit cases for community discussions.

- *Public Health Announcements and Campaigns:* Examples include *DailyStrength* and *Sermo* networks to inform the public and physicians about any flu outbreaks.
- *Treatment, Physician, or Hospital Selection:* Examples of these include *DailyStrength* and *FacetoFace Health* that offer a search engine so that patients can find matches and ask for advice. Also, the Mayo Clinic's *Sharing Mayo Clinic* blog site allows patients and physicians to share their stories, together with their *Medical Edge* site, a patient information podcast and news site.

Another survey for comparing healthcare social networking sites illustrated the impact of social networking on different healthcare outcomes and service models. This study classified the social networking services into three categories (Swan, 2009): healthcare social networking; consumer personalized medicine; and quantified self-tracking. The study also listed four major healthcare services offered through these social networks: clinical trial access; emotional support and information sharing; quantified self-tracking; and questions and answers with a physician.

As already mentioned, one of the most popular and widely used social health networks is *PatientsLikeMe*. This network offers an online quantitative personal research platform for patients with life-changing illnesses to share their experiences using patient-reported outcomes, to find other patients like them matched by demographic and clinical characteristics, and to learn from the aggregated data reports of others how to improve their own outcomes (Wicks et al., 2010). This network also has 16 disease communities, which in turn represent information from over 40,000 patients. Other popular social health networking sites include *CureTogether*, *MedHelp*, and *mCare*, which also provide different healthcare services supported by their own tools and models (Al Anzi et al., 2014a).

A recent study conducted by the IMS Institute for Health Informatics on engaging patients through social media indicated the following outcomes (Aitken et al., 2014):

- The usage and presence of social media channels are rising, although they still lag among the population segment that uses healthcare services the most, such as patients over 65 years of age, and those with multiple chronic conditions.
- Healthcare providers are currently ambivalent about the importance of social media on their patients and practice. However, a well-established social media presence can provide an important forum for patient engagement.
- Pharmaceutical companies have greater hurdles to overcome in using social media, in part because of the regulatory requirements and constraints outside of the United States in reaching patients directly and efficiently. Among the top 50 pharmaceutical companies included in the study, half did not engage with consumers or patients through social media on healthcare-related topics.
- Heterogeneous market regulations and the nonexistence of Internet borders require a call for regulations that ensures consistent information, and a stable environment for healthcare information contributors. Regulators have been slow and tentative so far in providing regulatory guidance.

- Wikipedia is the leading single source of healthcare information for patients and healthcare professionals, with visits to Wikipedia pages being higher for rare diseases than for common diseases. The most visited diseases from a total of 25 identified in this survey were tuberculosis, the most common disease with over 4.2 million visits in a 12-month period, followed by Crohn's disease (4.1 million), pneumonia (3.9 million), multiple sclerosis (3.8 million), and diabetes mellitus (3.4 million).

The study concluded with the following several recommendations:

- The need for expediting the release and implementation of a framework to regulate the increased improvement in the quality of information available online, and increased social media monitoring. The major hurdle in this process will be to define a regulatory environment for an increasingly borderless digital world.
- Effective engagement by healthcare providers with their patients on social media issues with which they feel most comfortable, with a better understanding of patient empowerment that these can bring. In addition to the learning benefits that healthcare providers can benefit from, their patients' engagement in social media help them to learn about their conditions and the realities of living with them.
- The need for pharmaceutical and other companies to find appropriate mechanisms to support social media groups, and ensure they remain aware of current trends and new technologies being developed.

Access to popular social networking sites such as *Pinterest, Instagram,* and *Tumblr* from smart mobile phones and other mobile platforms is gaining importance and influence, apart from *Facebook.* For example, while *Pinterest* and *Tumblr* networks have more or less similar functionalities, there are a few important differences that distinguish them:

- *Pinterest* is exclusively used for visual content, such as pictures and printed posters that can be either uploaded or re-pinned, which is favored by the majority of users. Every page in this site is identical; however, users can create their own unique "boards" to pin on the page, but the general layout is the same for each user.
- In *Tumblr* the content comprises photographs, text, links, and audio clips. This site basically offers a wider variety of options for content publishing, whereas the appeal of Pinterest is in the organization of the page. Tumblr users generally have the option to create original themes, and each page can be completely individualized.

Facebook remains the most popular social networking site for diabetes, with over 500 existing diabetes-related groups (McDarby et al., 2015). These social networking mechanisms will undoubtedly enhance the healthy engagement and awareness issues among patients. These can also provide popular and perhaps clinically effective platforms for the much discussed behavioral change challenges.

Although these sites offer health information and educational and social interaction benefits as mentioned earlier, it is interesting to note that there are so far no empirical studies on their effectiveness, efficacy, and healthcare impact. There is also an increasing global trend to use free cross-platform mobile text messaging and communication Apps such as *WhatsApp*. The eventual linkage of these platforms with popular social networking sites can offer exciting opportunities for new m-Health applications such as developing new messaging platforms with persuading context associated with subliminal effect.

The increasing use of social networking for healthcare is also fuelling the ongoing debate on patient data ownership and privacy, especially with an increasing number of young adults spending more time with different types of social and consumer media information sites.

The projected increase of social media for health users is generating massive amounts of "patient-reported data." These types of data are part of the big health data being generated on various mobile platforms, with Apps for many personal health and wellness applications. The privacy, security, and ownership of these data will constitute the next challenge for m-Health, with as yet many unresolved problems. This challenge, in addition to the regulatory governance, ethics, and confidentiality, will undoubtedly impact on the development of future m-Health social networking.

The benefits of social networking for healthcare do not preclude us from mentioning some of the "dark sides" of social media and mobile Apps on m-Health. One such concern, especially among healthcare professionals, is the increasing debate on the potential "Uber-ization" of healthcare. This is a reference to the controversial smartphone App (Uber) that connects drivers with passengers directly instead of through a centralized booking service or by just hailing a taxi in the street. The parallel notion is of a new digital health economy based on the global proliferation of social networks and mobile health Apps; this patient-led health economy warrants further reflection, with caution concerning potential adverse factors.

There is also increasing concern over the blurring of boundaries between the unauthorized surveillance and monitoring of social networking users on sites for healthcare applications; thus, there is a risk of ethical and privacy implications. Another concern is the mental and psychological impact of social media from the m-Health perspective that has not been fully understood yet. These and other issues constitute some of the challenges for further research.

The popularity of social networking among different patient groups is not reflected by parallel studies to evaluate their effectiveness and efficacy in comparison with conventional interventions. The role of social networking in healthcare education, and particularly in the public health services arena, is likely to increase globally. It is also most likely that such trends will increase in the developing world, particularly with recent studies indicating an exponential increase of mobile social networking for healthcare and well-being. This is especially the case among chronic patients, such as diabetics (Al Anzi et al., 2014b).

3.3 MOBILE HEALTH APPLICATIONS (m-HEALTH APPS)

What is the first thought that occurs to someone being asked the question: what is m-Health and what it is all about? Today, the answer from a lay person is most likely to be a smartphone or tablet with a healthcare App. This widespread perception of m-Health is due to two factors:

- The misconception and misunderstanding of m-Health, framing the concept only within the health Apps sphere.
- The unprecedented commercial success and proliferation of smartphone-based health or medical applications, widely known as "Health Apps."

The enthusiasm for heath Apps is driven by three powerful contributory factors: the unsustainability of current healthcare spending and costs; the accelerating growth in wireless connectivity; and the need for individualized medicine (Steinhubl et al., 2013).

The inception and rapid development of health Apps, sometimes referred to as "mobile medical Apps" or "m-Health Apps," is one of the most controversial aspects of m-Health. This conundrum originated shortly after the introduction of the first generation of smartphones in 2007, when m-Health started to become increasingly synonymous with health Apps. Today we witness an unprecedented proliferation of tens of thousands of mostly unregulated and clinically nonvalidated health Apps, downloadable globally from the Apple Store, Google Play, and other smartphone shopping sites for numerous health, wellness, and medical conditions, including the management of chronic diseases. For example, it was reported that in 2013 Apple celebrated its 50 billionth App download with Google trailing only slightly behind with 48 billion, and that its ecosystem enabled by its App Store has led to an overall payout to developers of more than $9 billion (Skillings, 2013).

Figure 3.2 illustrates the proliferation of these Apps, developed for numerous smart mobile platforms and mobile operating systems (iOS, Android, Windows Mobile OS, etc.).

As mentioned in Chapter 1, this important, albeit mainly market- and consumer-driven development has brought both a blessing and a curse to the m-Health concept. The blessing is the transformation of m-Health to a truly global phenomenon, mirroring smartphone success from the healthcare perspective. The curse is the false pretext that m-Health is just another App domain. This has eventually narrowed the broader view of the scientific core of the concept itself (Istepanian, 2014). This view has been strengthened by numerous misrepresentations of m-Health; as a result, m-Health has become depicted as an important vertical market of the Apps business ecosystem, complemented by ambiguous redefinitions and reinterpretations of the original concept.

To highlight these issues, a report by the American Health Information Management Association (AHIMA) defined m-Health as the use of devices such as smartphones or tablets in the practice of medicine, and the downloading of health-

FIG. 3.2   The proliferation of smart health Apps for chronic diseases.

related applications or Apps (AHIMA, 2013). Other studies advocating the interpretation of m-Health as smartphone-based wireless monitoring and intervention also highlighted similar views (Kumar et al., 2013; Rajan, 2013). This can be translated into the following formula:

$$\text{m-Health} = \text{smartphone} + \text{App software} + \text{remote healthcare service,}$$
$$\text{cloud access, and feedback processing}$$

We will argue throughout this book that this simplistic formulaic view needs to change in order to enable m-Health to evolve outside its current boundaries so as to transform healthcare to have better global impact. This visionary aim is a controversial topic that will remain the subject to ongoing debate and research in the foreseeable future.

We will next describe the most relevant issues of health Apps, namely, their current status, clinical efficacy, effectiveness, and regulatory issues. These issues are still considered open topics for further debate, especially by clinicians. Similar to the social networking trends, the current proliferation of health Apps is truly incredible (Bernhardt, 2015). This trend is reflected by numerous studies predicting that nearly 50% of mobile users will download a mobile health App by 2017 (Research2guidance, 2013). Others examined similar issues in more detail, and estimated that in 2013 there were 43,689 Apps available (from Apple's iTunes market only). Of these,

23,682 could be categorized as "healthcare and fitness" or "medical" Apps (IMS, 2013). In this report, the main functionalities of these Apps were selected from the following seven categories:

- *Inform:* Apps that provide information in a variety of formats, that is, text, photo, video.
- *Instruct:* Apps that provide instructions to the user.
- *Record:* Apps that capture user-entered data.
- *Display:* Apps that graphically display user-entered data or output user-entered data.
- *Guide:* Apps that provide guidance based on user-entered information, and may further offer a diagnosis, or recommend a consultation with a physician or course of treatment.
- *Remind/Alert:* Apps that provide reminders to the user.
- *Communicate:* Apps that provide communication with patients and/or provide links to social networks.

Furthermore, the same study highlighted that the most popular of these categories were Apps that provided information (two-thirds of the total). It must be noted that this market study was restricted to the Apple mobile phone market, and did not consider other smart Apps, platforms, and markets such as Google Play (Android) and Microsoft platforms. In addition, the study was concluded with a cutoff of October 2013; it is estimated that the number of these Apps has doubled since then.

An earlier study reviewed a total of 1056 iOS (Apple) mobile phone Apps in the medical category, with another 1004 in the healthcare and fitness category (Liu et al., 2011). The medical category in this study was classified in terms of drug or medical information databases; medical information references; decision support; educational tools; tracking tools; medical calculators; and others.

A further interesting study on a sample of U.S. adult App users concluded that half of smartphone owners use their devices to get health information, and nearly one-fifth of them have health Apps downloaded on their phones; the most popular are those that monitor exercise, diet, and weight (Fox and Duggan, 2012).

Another study reviewed mobile health Apps for the most prevalent conditions cited by the World Health Organization (Martínez-Pérez et al., 2013), listed as anemia, hearing loss, migraine, low vision, asthma, diabetes mellitus, osteoarthritis, and depressive disorders. This study identified that by April 2013 more than 247 published papers and 3673 Apps were related to these conditions. The study concluded that there was an uneven distribution of the development of these Apps for different conditions. Some conditions, such as diabetes and depression, had an overwhelming number of Apps developed, whereas there were few relating to the other conditions, such as anemia, hearing loss, or low vision. Other market statistics and predictions on health Apps are cited in many m-Health online fact sheet sites.

In a recent U.S. survey, physicians were asked how comfortable they were using patient data streamed from mobile health Apps and devices: 74% of the respondents said that they were comfortable with Apps designed for checking ear infection data, compared to 53% with those for analyzing urine data, and only 48% with those for checking vital signs (PwC, 2014). However, the same survey concluded that 86% of U.S. clinicians believe that mobile health Apps will become important to physicians for patients' health management over the next 5 years.

In addition to these market studies, there are also numerous commercial and specialized Web sites that annually tabulate their m-Health Apps "favorite" lists, or league tables (Conn, 2012; Internetmedicine, 2014; Butler, 2012). It is important to view these market-oriented Apps league tables for information purposes only, and caution is needed when looking at the categorizations. This is because the current market of these Apps is essentially business driven, with the majority of the most popular Apps being developed and introduced by small-to-medium-sized enterprises (SMEs), with some exceptions.

The basic m-Health Apps architecture is shown in Fig 3.3. This model corresponds closely to the "m-Health-monitoring" system described in Chapter 2, except that smartphones, with their intelligent data processing capabilities, and cloud computing, are added to this architecture.

The basic architecture of "mobile health" can be represented by the following formula:

$$\text{Mobile health} = \text{smartphone} + \text{App software} + \text{user (patient)}$$
$$+ \text{remote healthcare service provider (medical feedback)}$$

This formula justifies the ongoing concerns, particularly from the clinical perspective, about the promises and perils of so-called "m-Health Apps" because clinicians are increasingly worried about their clinical efficacy and effectiveness. Their arguments are supported by the lack of rigorous, large-scale, long-term clinical studies in the face of intense business-driven interests of commercial stakeholders.

However, in spite of the expanding market of consumer-driven m-Health Apps, which are downloaded by hundreds of millions of users worldwide, this market has not yet reached its expected clinical maturity. Typically, patients search for appropriate Apps that meet their perceived health status or medical needs, and then find a vast number of Apps on the current market, supplied by small businesses targeting different patient populations, healthcare providers, insurers, and so on.

In the midst of this controversy, there is also an increasing trend of clinicians encountering patients with questions about these Apps and their clinical effectiveness, rather than directly about their own health. This scenario is most likely to happen with younger patients who are more accustomed to health information access via their smartphones.

An example of this process is the ongoing debate on the increasing trend of "health App prescriptions" culture, and whether specific Apps can be prescribed to some patients with chronic diseases, which might well happen in the near future, especially in the developed world. There is an emphasis and perhaps near consensus from the medical community that unless larger scale and more robust clinical studies are

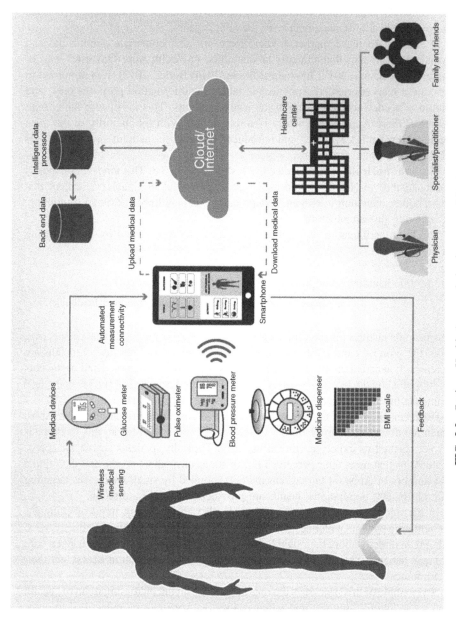

FIG. 3.3 Basic m-Health Apps and cloud access architecture.

conducted with ample medical proof of the effectiveness and efficacy of these Apps, supported with a clear and rigorous validation process, these scenarios could be some years away.

With the increase of embedded health sensors in the newer generation of smartphones, supplemented by their own personalized health and wellness Apps, we might witness an accelerated path toward such adoption, especially for chronic diseases. There is also increasing evidence that some m-Health Apps are playing an increasing role in improving and transforming healthcare services in the developing world; with poor resources, especially in remote areas, access to medical care is either nonexistent or scarce at best.

We will not aim here to provide a comprehensive classification of m-Health Apps, simply because new ones are being introduced on a daily basis. Instead, we aim to categorize these generally in terms of some of the most popular of existing mobile health applications and services, such as health and wellness, disease monitoring and management, diagnosis, prevention, education, hospital workflow, process delivery improvement, and other areas (IMS, 2013; Research2guidance, 2013; Free et al., 2013; Liu et al., 2011).

We can conclude from available studies and literature that existing m-Health Apps can be categorized as follows:

- *Health Prevention, Wellness, and "Worried Well"*: These include applications in different areas of public health and prevention, fitness, diet, appointment scheduling, weight loss, personal health record access, interactive and social networking, and so on.

- *Disease Management, Assessment, and Monitoring:* These can include Apps covering areas from the wireless monitoring of vital signs, global chronic diseases and their management tools (e.g., diabetes, COPD, cardiac, and heart failure), tools for medication reminders, health and trend analysis, and mobile interactive connections with carriers.

- *Ageing Well and Living Independently:* These include applications in areas such as smart home living, assisting and supporting patients with mental disease help, and monitoring and assisting elderly people.

- *Remote Diagnostics and Imaging:* These include applications such as remote point of care diagnostics of different diseases such as infectious diseases, cancer, and other medical and radiological imaging, for example, ultrasound diagnostic applications.

- *Clinical Data Access for Improving Global Healthcare and Service Delivery:* These include Apps targeting global health services, access to electronic patient records, disease prediction, and prevention.

- *Self-Support and Therapeutic Assistance:* These include Apps that provide self-assistive and behavioral change tools, psychological support, and interactive health games.

- *Medical Support, Education, and Clinical Assessments:* These include Apps that provide support for medical staff, education, and clinical assessments.

3.3.1 Clinical Efficacy and Effectiveness of m-Health Apps

The clinical efficacy and effectiveness of health-related Apps available on the market is an important, ongoing topic. Examining it is a difficult and complex task, and requires extensive further research that is beyond the scope of this book. In Chapter 5 we will present in more detail the wider clinical aspects and existing clinical evidence of the smartphone-centric m-Health interventions and "App Therapeutics." In this section we focus on an example of the clinical efficacy and effectiveness of diabetes Apps for completeness. Considering that the current health Apps market is commercially driven, and not backed by larger clinical studies or expert assessments, most users of these Apps are probably unaware of the relevant guidelines produced by medical and healthcare organizations and regulatory bodies, notably the U.S. Food and Drug Administration (FDA) and the AHIMA. There is also ongoing debate on whether mobile health Apps can provide the expected benefits to healthcare providers in areas such as improved access to care, better patient engagement, reduced errors, improved patient safety, and developing new healthcare business models.

The crystallization of these benefits in different healthcare applications depends on several factors. The most critical of these is a better understanding of their clinical efficacy and effectiveness in different healthcare scenarios. As mentioned earlier, we aim here to illustrate some of the important challenges of this topic by examining an exemplar case of one of the most prevalent chronic diseases anywhere, namely, diabetes. We have selected this disease as an example because it is both a major global concern and also a focus of one of the highest medical Apps markets. As verification, a study estimated that a search for the word "diabetes" on both Apple and Android mobile devices yielded more than 1000 hits. Of these, only one application for diabetes management has received clearance from the FDA (Eng and Lee, 2013). A similar market analysis also estimated that the number of diabetes Apps available in the current market is more than 1100 in 27 countries (Research2guidance, 2014a).

However, most commercial diabetes Apps seem to be designed on a generic basis, and are unable to distinguish between the key functional requirements and management protocols between type 1 and type 2 diabetes. Furthermore, in most of these Apps, the role of the healthcare provider is missing, and most are developed as tools acting as electronic logbooks for the patient's blood glucose levels. Other studies have illustrated a lack of user-centered and socio-technical design principles in their development process (El-Gayar et al., 2013), with yet larger scale clinical trials and evidence still missing.

It is interesting to note that most recent reviews and meta-analysis studies indicate potential clinical effectiveness and cost-benefits of self-management in using mobile diabetes management systems (Baron et al., 2012; Holtz and Lauckner, 2012). In particular, one of these studies reported recommended outcomes on the clinical usability of diabetes Apps that can be summarized as follows (Baron et al., 2012):

- The use of the Apps has showed more benefits in patients with poorly controlled diabetes.

TABLE 3.1   Main Diabetes App Functions for Self-Monitoring and Management

Diabetes Apps Function	Monitoring Functions	Management and Support Functions
1	Wireless blood glucose measurement and trends[a]	–
2	Medication reminders and alerts[a]	–
3	Physical exercise tracking and monitoring[a]	–
4	Diet and carbohydrate management and tracking[a]	–
5	Wireless blood pressure measurements	–
6	Weight tracking and measurements	–
7	Insulin measurements and adjustment tracking	–
8	–	Education
9	–	Patient Health Records (PHR)
10	–	Decision and therapeutic support
11	–	Social networking
12	–	Automated data forwarding with remote server/cloud connectivity
13	–	Security and privacy
14	–	Data analysis functions
15	–	Other documentation

[a]Key function in the diabetes management process.

- The use of the smart Apps by patients recently diagnosed (less than 5 years) had a significant reduction of glycated hemoglobin, referred to as HbA1c.
- The feedback from health professionals may not necessarily be considered as an Apps success factor.

Although these recommendations are not conclusive, they provide some insight into the potential impact of smart Apps on diabetes self-management. However, fewer if any of these studies list the most appropriate application functionalities required for proper clinical diabetes monitoring and management tasks. These functionalities are shown in Table 3.1 (El-Gayar et al., 2013; Chomutare et al., 2011).

It is still a difficult task to develop a successful App that can implement all the necessary functions, with both effective usability and clinical outcomes. Furthermore, there have been no studies so far on issues such as "Apps compliance," "mobile usability" and "App design," and their impact on long-term patient use (Istepanian, 2015).

In conclusion, medicine needs clinical evidence, and with the absence of such pivotal evidence, these Apps will remain largely market- and consumer-driven, and therefore largely unacceptable clinically. There are many open questions that need answering on the efficacy and clinical effectiveness of these Apps. It is therefore important to conduct further large-scale clinical and usability studies to address these questions, along with the asymmetry conundrum of the unprecedented consumer Apps growth, on the one hand, and the relative lack of clinical acceptance and lukewarm intake by clinicians on the other.

3.3.2   The Development and Regulation of m-Health Apps

The majority of the health Apps on the market today are not regulated. Developing a mobile health application, especially for successful deployment in specific health services, is a complex and difficult task with a web of interacting requirements and functions. The development process of an m-Health App involves two main interchangeable drivers, and these in turn consist of several sublayers and requirements as follows.

Healthcare Market and Business Requirements   In this category, the application process involves different market factors (Research2guidance, 2014b):

- Clear preference for "service" as the primary revenue model.
- Previous experience in the specific application market.
- Greater use of application development tools.
- Clearer preference for the current Android and iOS platforms as the most popular smartphone types used globally.

Moreover, further consideration should be given to the current e-Health market drivers (Foh, 2012), namely, consumer adoption, clinical adoption, evidence and effectiveness, cost of deployment, and regulatory environment.

Some of these factors can also be applicable for the m-Health Apps market. If we consider, for example, the earlier diabetes App case, it is clear that the existing evidence of the effectiveness and clinical adoption factors can affect the successful consumer adoption of one specific market App over another. The regulatory approval of an App is another factor; as few m-Health Apps are currently regulated for therapeutic purposes, this can potentially impact on the long-term acceptability and usability of these Apps over others in the market.

Clinical Areas and Their Targeted Considerations   For this category, the targeted medical and healthcare requirements might include the following (Marvel et al., 2014):

- *App Vision*, which includes conceptualization, defining objectives, and establishing the innovativeness of the application for specific medical scenarios.
- *Creation*, which relates to medical applications, from the main features to the beta testing process of the App before deployment.

- *Dissemination*, which includes the prerelease clearance of patient safety, followed by distribution to virtual marketplaces and social media outlets.
- *Determining Utility*, which includes regular updates of the specific application, and further automated functionality additions, as required medically.
- *Usability*, which relates how patients react to using the specific App.
- *Security and Privacy* measures and infringements issues.

Other likely considerations that need to be considered include the following (Scher, 2013):

- *Health App Development Rationale*: Most people who look for and download a health-related App are likely to have a health problem that the App purports to "solve." It is now fairly straightforward with current developments in commercial wireless sensors and their smartphone connectivity to monitor a medical parameter, such as heart rate or blood pressure, but it is not so easy for lay people to diagnose their own condition from the "evidence" generated by the App. Indeed, it may be entirely misleading and potentially dangerous to believe the evidence, which may not be consistent with the actual malady. The App may generate and store data, but it can still be, and often is, a "recipe for failure." Unless the data are analyzed carefully, they may not convey the correct message, which is often beyond the experience of lay people, and requires medical input and advice from a doctor. Even doctors consult other doctors when they are in need of a diagnosis!

In addition to these, the Apps development process also needs to embrace some of the following important functionalities offered by most smart mobile phone technologies:

- Simplicity and effective use of embedded sensors within these smartphones.
- The seamless wireless connectivity between external medical devices and sensors linked to the phone. Also, the proper design of real-time data uploading and downloading functions to and from a remote cloud, or server.
- Potential use of community and social networking interactions in the App design process.
- The ubiquitous and pervasive nature of medical data collection and user independence.
- Development of appropriate and easy-to-use smart alerting and reminder functions embedded in the smartphone.

Medical Input Omission   In a largely unregulated market, most of the current healthcare Apps are developed by nonmedical specialists who would benefit from medical experts' professional advice. Yet, it is patently obvious that a doctor–technologist collaboration is imperative in this development process. In rare cases, doctors may be "technology savvy," and can design Apps themselves, but usually someone in the commercial world comes up with what seems a brilliant idea and goes ahead with its development based on its presumed commercial potential, completely independent of any medical input.

Poor Usability Criteria   Usability can be defined as "the effectiveness, efficiency and satisfaction with which specific users can achieve a specific set of tasks in a particular environment" (HIMSS, 2009). This definition implies that a health-related App should be designed in such a way that it requires a certain minimum level of knowledge or experience in order to use it. As discussed earlier, recent studies indicate the lack of continued use of Apps after a while rather than over a sustained period of time, an issue that can become critical, especially in the case of the benefits envisaged by these Apps.

Ignorance of the Healthcare "Landscape"   Producing an App in complete isolation is to be unaware of the "healthcare landscape," an ingredient that is vital for the success in applying and adopting the App successfully in the medical environment. There is a need to know what other Apps or device technologies are available on the market, and whether it can be used by unskilled and non-specialist people, while at the same time being sold at prices that are affordable.

Regulatory and Standardization Requirements   Healthcare is a highly regulated industry, and the issues relating to m-Health and particularly for Apps, have seen important regulations and guidance procedures developed by regulatory agencies in recent years, particularly in the developed world (Cortez et al., 2014).

As a step forward in this process, the FDA issued their final guidance on Medical Device Data Systems (MDDS) and Mobile Medical Apps (MMA) in early 2015 (FDA, 2015). According to this guidance, a MMA is a mobile App that meets the definition of a device in Section 201(h) of the Federal Food, Drug, and Cosmetic Act (FD&C Act) and is either intended to be used as an accessory to a regulated medical device or to transform a mobile platform into a regulated medical device. For example, the first mobile Apps cleared for continuous glucose monitoring (CGM) in real-time using Apple mobile devices were also announced in early 2015 (FDA News Release, 2015). The FDA does not regulate wellness Apps, but some of these "walk a fine line" between wellness and what the FDA may consider a medical device.

On privacy issues in the United States, there are two major laws that relate to health information privacy in general. These are also relevant to mobile health and m-Health Apps in particular. The first is the "Health Insurance Portability Accountability Act" (HIPAA), which has been well understood by the medical community since 1996. The second is the "Heath Information Technology for Economic and Clinical Health" (HITECH), which became a law in 2009, and HITECH adds to HIPAA regulations.

The rules governing HIPAA compliance of m-Health Apps are based on two main issues (Green, 2011):

- HIPAA applies to "covered entities" only (healthcare providers, health plans, and healthcare clearing houses); therefore, avoiding data sharing with covered entities entails far fewer rules.
- HIPAA only applies to health-related data.

Hence, for any m-Health App in the United States to comply with HIPAA, developers need to analyze whether their software will be used by a covered entity, such as a physician, hospital, or health plan, and whether it will include any protected health information. This includes individually identifiable information about health, healthcare services, and payment for healthcare services (Green, 2011). "The Patient Protection and Affordable Care Act (ACA)," known as the "ObamaCare Act," represents a sweeping reform of the U.S. healthcare system; it will inevitably initiate further debate on the role of mobile health Apps, especially insurance, regulatory, and privacy issues. This debate is supported by recent statistics that indicate the growing U.S. consumer Apps market. Further details on these and other relevant legal, privacy, and reimbursement issues of m-Health Apps, particularly in the United States, are discussed extensively elsewhere (Malvey and Slovensky, 2014; Green, 2011).

From the European perspective, the European Union is proposing a framework for medical devices developed to define the regulatory boundaries of m-Health Apps (GSMA, 2012); this is controlled by the EU Medical Devices Directive (EU Directive 93/42/EEC) (European Commission, 2015a). The framework is undergoing revision, because most current "software-based devices" (where Apps can be defined in this category) fall under Class I of the directive.

There is increasing argument for revision and further clarification of the current EU Medical Devices Directive concerning diabetes Apps and their associated devices. Furthermore, the European Commission recently published the outcomes of their consultative Green Paper on mobile health; these include mobile health issues, App development, other standards, and personal privacy (European Commission, 2015b).

From the UK perspective, there are similar ambiguities and challenges with the current regulatory procedures of health-related Apps. In general, the software intended for use in a medical context can be classified as a "medical device," and health Apps therefore potentially fall within the regulatory remit of the UK's Medicines and Health Care Products Regulatory Agency (MHRA). In 2014, the MHRA published guidance on medical standalone software (including Apps), which aimed to clarify when it considered an App to be a medical device and how it would be classified (RAE, 2014). There are also ongoing efforts to develop a National Health Service (NHS) App Library, with discussions on the best approach to implement these through the National Institute for Health and Care Excellence (NICE).

These different regulatory positions of the United States, European Union, and the United Kingdom reflect the global uncertainty in the regulatory requirements for health Apps. This could limit the growth of the m-Health market (PwC, 2013), although economic indicators suggest otherwise.

In addition to these, there is also the computing and IT management of the services required to operate these Apps. These are typically administered by the healthcare service providers in hospitals or care centers as a multilayer computing infrastructure. The layers are usually presented, from top to bottom, as Application, Data, Runtime, Middleware, Operating System (O/S), Virtualization, Servers, Storage, and Networking.

With ongoing debate about security and privacy issues of Apps, the final outcome requires extensive studies in the future. Furthermore, these Apps produce massive volumes of data that are mainly stored in the cloud. The Internet connectivity and their storage formats are increasingly becoming part of the smart m-Health monitoring ecosystem, as shown in Fig. 3.2. These developments formulate the core of the current debate on the role "big medical data," or "smart medical data," in this ecosystem. However, more detailed issues on these computing and software development processes for Apps are beyond the scope of this book and are cited as more specialist references (Iverson and Eierman, 2014).

Security and Privacy Measures and Infringement Issues In a recent study analyzing 24,405 health-related Apps (iOS 21,953; Android, 2452), it was reported that 95.63% of them pose at least some potential damage through information security and privacy infringements (Dehling et al., 2015). Other similar studies also reveal an alarming and urgent need to address these challenges by developing robust, but acceptable, security and privacy procedures. These measures, if utilized and applied efficiently, can avoid any unnecessary security and privacy infringements of medical data and personal information. These issues, although very important, remain overlooked by clinicians and healthcare providers, with many obstacles to wider clinical acceptance of Apps.

Effective Use and Engagement by Clinicians and Patients There are generally many challenges associated with the effective use of health Apps by both patients and clinicians. For patients these include the "download and ignore" phenomenon and their long-term "App compliance." Also, if a specialist recommends the use of an App, there is no evidence that patients comply with the advice. For physicians, Apps can cause potential interruptions to clinical procedures in hospitals, and problems associated with management of the big data generated. In addition, they are faced with a confusing variety of available Apps on the market from which to select for specific diseases to achieve the best medical outcome.

There is also the engagement element in this process, which means incorporation of appropriate behavioral changes and reward mechanisms, which are not considered in the App design process; nor are personalization and relevance to daily activity. In summary, the App development process needs to strike a balance between the needs of both the patient and the clinician.

Finally, to provide each App's efficacy, a successful outcome to appropriate clinical trials to obtain clinical evidence is necessary; this should encapsulate all the above issues. As mentioned earlier, the overall process of adopting and regulating m-Health Apps is a complex process. Recently, the UK NHS proposed the following multilayered stages for the development process of mobile health Apps:

- Stage 1: Predevelopment
- Stage 2: Design and development
- Stage 3: User testing
- Stage 4: Stakeholders review
- Stage 5: Medical device process

- Stage 6: External deployment
- Stage 7: Monetization of Apps
- Stage 8: Digital marketing and adoption considerations

These stages are described in detail in a NHS guidance document (NHS, 2014).

3.3.3 The Verdict on m-Health Apps

The current mobile health application sector is a popular and lucrative industry. It is estimated that the market for smart Apps will reach US$26 billion by 2017 (Research2guidance, 2014b). These developments represent a successful outcome on the translation of m-Health to commercial products and services. However, most of these market Apps and services have not yet reached large-scale adoption by patients and clinicians, compared to other smartphone applications, such as in the retail, finance, and entertainment sectors. This is mainly due to the lack of the clinical evidence to ensure their efficacy and effectiveness in different healthcare applications. Yet the current health App market covers a wide range of medical disciplines, including public health, lifestyle awareness, monitoring and management of chronic diseases, and support for care providers and specialists. It is also interesting to note that although the number of m-Health Apps is increasing, with tens of thousands currently in use, the overall long-term user adherence of these Apps is low (Butler, 2012; Helander et al., 2014). Moreover, a relevant usability survey revealed that of more than 3500 App users, 79% said they would retry an App only once or twice if it failed to work the first time, and only 16% said they would give it more than two attempts (Perez, 2013).

It is also important to note the increasing role of social networking in different chronic disease management areas, especially in diabetes management (McDarby et al., 2015). The combined popularity of smartphone Apps associated with online and social networking can perhaps contribute to the reshaping of future care delivery processes. This hybrid trend is compatible with the patient-centered m-Health model that has been advocated. As evidence of this trend, some recent examples are emerging of Apps that present patients with clinical information via online portals such as PatientKnowsBest (2016) and Health Fabric (2016). The former is basically a patient-owned healthcare record system and portal linked to an App; registered patients monitor their own vital signs, and connect to the PKB App via a hundred or more wearable sensors and devices linked to the system portal. The latter combines an online and App solution that allows patients to control their own health and social record, and to share this with their general practitioners for improved and integrated care planning. In 2015, Apple announced their "ReseachKit" (www.apple.com/ researchkit/) open source software platform. It allows medical researchers to create iPhone Apps for targeted clinical studies, such as for asthma, cardiovascular, and Parkinson's diseases, and to assist in patient participation and biometric data collection anonymously using these Apps. This approach combines crowd sourcing with open m-Health concepts for better understanding of these diseases. However, there is increasing concern about security and personal data privacy, in addition to the ethical quagmire associated with these open-source platforms.

Further empirical studies to validate the clinical efficacy and effectiveness of these new care models are as important as their proposed benefits. The absence of the "cognitive" element in current smart health Apps to understand patients' psychology and behavioral change is one such study. Another concerns the capability of providing feedback to sustain the use of an App on a long-term basis. The engagement and appropriate personalization of the technology in the App design process are important elements in the successful development of future m-Health Apps. There are also many challenges concerning the regulation of m-Health Apps for clinical uses as opposed to consumer usage.

Furthermore, in the absence of large-scale clinical evidence, it is likely that healthcare Apps will continue in the foreseeable future as market-driven products. The trend will be to offer advice to patients in the form of "Apps prescriptions" for health information and advice purposes only. A smart platform and diabetes App has been approved by the FDA as the first approved "Diabetes Prescription App" (Waltz, 2014).

Finally, since we are all living in the midst of a "smart devices culture," we have yet to see if a parallel "m-Health App" culture evolves in a future digital healthcare era.

3.4   CLOUD COMPUTING AND m-HEALTH

In a world of computers, data have to be stored somewhere. In the stone age of computers, when storage was measured in kilobytes or megabytes, data were stored on a single machine. If necessary, it could be transferred to a storage device, such as a "floppy disk," a compact disc, or a digital video disc. Nowadays, computers have enormous memory capacity measured in gigabytes or terabytes, and files can be transferred to portable devices such as memory sticks, which now also have memories of many gigabytes.

In m-Health applications, such as personal monitoring, the amount of data generated from a single patient can be very large and while it may be possible to store data from many patients on a single computer, it becomes impractical because of the risk of losing valuable, often irreplaceable, files. This could easily happen if the hard disk fails or if the computer is stolen. For a clinical practice such a loss could be catastrophic and would be the result of culpable bad management. The least that could be done to obviate this nightmare would be to share the data around a local network of several computers. The network may be accessible from outside the practice, perhaps by a hospital consultant, as long as suitable security measures are in place. This simple and straightforward network configuration was acceptable until recently, but the world has become a data-hungry place, not least the medical world, in which the records from countless numbers of patients, typically tens of millions in most countries, and hundreds of millions in others, may need to be stored. The solution is "cloud computing," which amounts to outsourcing of computer and storage resources with ubiquitous access. The physical location of specific data may not be known because they may be stored in one or more of thousands of computers, or servers. (As these words are being typed they are automatically being stored in a cloud somewhere out there in cyberspace!) The data for a given patient may not all be in the same location but may be spread among several, or even many, different computers. It is convenient to think of this huge collection of computers as a cloud and is usually drawn as such in illustrations.

Cloud computing is generally understood as a new approach for delivering computing resources. The cloud is playing an increasing role at the heart of developing new m-Health systems (Park, 2015). These days most of the popular Internet and entertainment services, such as NetFlix, Instagram, and Dropbox, make use of the commercial cloud in their core businesses. Similarly, global healthcare IT service providers are following suit. In this section we describe the basic features of cloud computing, and illustrate how the cloud is reshaping the future of m-Health services and care delivery processes.

3.4.1   What Is Cloud Computing?

The "cloud" is basically a pool of computer data storage provided by a host company, usually called a cloud service provider. There are several definitions of cloud computing (Buyya et al., 2011), but in a nutshell, cloud computing is an Internet-based computing platform allowing shared hardware and software resources that are provided to users on-demand in a scalable and simple way. It is important to understand that the cloud is a new approach to delivering computing resources, and not a new computing technology. Cloud computing can also be understood as the hardware, software, and applications delivered as services over the Internet (Antonopoulos and Gillam, 2010). However, in more generic terms, the cloud can also be defined as "data centers with hardware and software resources that provide services over the Internet" (Armbrust et al., 2009). These services include the availability of software delivered through the Internet on a browser without any installation, hosting an application on the Internet, or setting up remote file storage and database systems. The cloud was initially driven by IT business markets, and used as an umbrella term to describe a category of sophisticated on-demand IT and computing services. Cloud services are delivered in different classes, and are offered by companies and providers, such as Amazon, Google, Microsoft, IBM, and others.

3.4.2   Basics of Cloud Computing

In recent years, *cloud* and *cloud computing* have become buzzwords synonymous with most e-Health and m-Health services and applications. The cloud started with the simple notion of letting someone else take ownership of setting up an IT infrastructure and getting users to pay for storage. This notion leads to the question: Why is the cloud so important for mobile health compared to traditional web services? The answer is twofold.

The first factor is driven by business needs and financial incentives. For these we need to consider the following cloud principles in terms of services (Rosenberg and Mateos, 2011):

- *Pooled Computing Resources Available to Any Subscribing Users.* This cloud principle provides the benefit of efficiency, for example, in optimizing the computing resources for a healthcare organization, and commercial cloud providers charge a subscription on a "pay as you grow" basis.
- *Virtualized Computing Resources to Maximize Hardware Utilization.* This principle is particularly beneficial to the fast-growing small-to-medium sized mobile health businesses in utilizing, for example, their mobile hardware devices and resources to be compatible with their Internet-based software and computing infrastructure.
- *Elastic Scaling Up or Down According to Demands.* This principle allows, for example, different m-Health providers to minimize their up-front investment levels by knowing the level of their computing infrastructures and utilization costs.
- *Automated Creation of New Virtual Machines or Deletion of Existing Ones.* This principle allows m-Health service providers to deploy, configure, and alter all their computing facilities without too much manual intervention in their business provision and services.
- *Resource Usage Billed Only as Used.* This principle is particularly feasible for smaller m-Health companies and start-ups, which often cannot afford to spend large sums of money at the beginning of their businesses by developing a "per-usage," or pay only, for what their user business models consume.

The second factor on the importance of the cloud for m-Health is based on the tangible benefits that cloud computing and data storage brings to m-Health services, as compared to earlier noncloud computing and storage solutions. The cloud offers effective and efficient Internet-based solutions, whereby shared resources, software, and information are provided to computers and other devices on a demand basis. These solutions overcome many of the traditional server-based m-Health services and applications, such as their cost, lack of scalability, difficulty of replication, and their vulnerability to hardware outages, in addition to the underutilization of their processing powers. The emergence of the "big health data" from the billions of connected users, devices, and things also necessitate the importance of the cloud in storing, processing, and tailoring these data for personalized care.

3.4.3 Cloud Computing Deployment Models and Service Classes

Cloud computing technology can be understood from two perspectives: cloud deployment models and cloud service delivery classes. This is illustrated in Fig. 3.4:

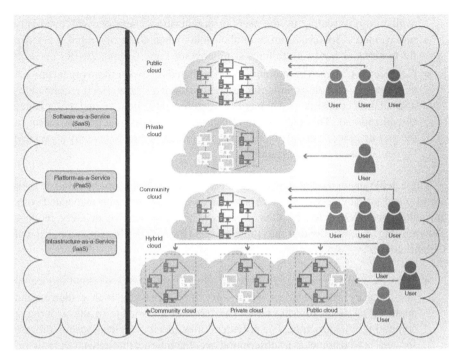

FIG. 3.4   Cloud computing deployment models and service delivery classes.

Figure 3.4 shows that the cloud can be classified into four main categories, depending on their accessibility restrictions and deployment models (Mell and Grance, 2011; Grace, 2010; Armbrust et al., 2009):

- *Public Cloud:* The public cloud is defined as a "cloud made available on a pay-as-you-go basis to the general public and users," and can be accessed by any subscriber with an Internet connection. This deployment model is based on the public availability of the cloud service and the public network that is used to communicate with it. Public cloud is suitable for public and government services and nonsensitive data applications.

- *Private Cloud:* A private cloud is defined as "internal data center or storage space of a business or user entity, not made available to the general public, and is established for a specific group of users and limited access to just that group, and not shared by other users." In other words, it makes use of the virtualization solutions in consolidating distributed IT services, often within the data centers belonging to the organization. This type is best suited for services for high privacy and security needs, such as in the healthcare sector.

- *Hybrid Cloud:* A hybrid cloud is a composition of two types (private and public), and takes shape when a private cloud is supplemented with computing capacity from public clouds. This type of cloud is able to maintain high service availability by scaling up the users or organization's services with externally provisioned resources from a public cloud when there are rapid workload fluctuations or hardware failures (Antonopoulos and Gillam, 2010).

- *Community Cloud:* A community cloud is shared by two or more organizations, and supports specific community services that have similar cloud requirements and shared concerns (e.g., security requirements, policy, and compliance considerations). This type of cloud can be understood as a generalization of the private cloud, a private cloud being the infrastructure that is only accessible by certain users or organizations.

These deployment models are appropriately selected by subscribers, based on their security and privacy needs, and access requirements, together with associated costs and allowances demanded by their services. The cloud service delivery models, shown in Fig. 3.4, are divided into three classes (Barnatt, 2010; Antonopoulos and Gillam, 2010):

- *Software as a Service (SaaS):* This is the most familiar type of cloud service. In this category, the providers allow subscribers access to both resources and applications. Basically, the customers have no knowledge of the underlying computing infrastructure. The providers offer specific applications or software services to a customer as a subscription over the Internet through a web browser. Customers simply use these web-based applications, or suite of applications, to handle the tasks required. The main advantage of using a SaaS is that the user does not need to worry about installation, storage space, data loss, and so on. It thus alleviates the burden of software maintenance and support. The SaaS customers can only use the provider's applications, and cannot have a physical copy of the software to install on their smart devices or other smart computing platforms. Thus, users relinquish control over software versions and require-ments. Commercial SaaS services examples include Dropbox, Apple's iWork, Googlemail, Microsoft's Office 265, Red Hat OpenShift, and Heroku.

- *Platform as a Service (PaaS):* In this category, the customers or businesses are allowed a platform on which they can create and deploy their custom Apps, databases, and line-of-business services integrated into one platform. In PaaS, the users have very little control over their software and programming environ-ment, as this service basically creates an environment for developing online applications that run on the cloud provider's equipment. The PaaS providers usually implement a software layer over the hardware they offer, so that users are obliged to work with the providers' own software layer. Customers can create their own applications using different application interfaces, software libraries, and other tools, such as databases and middleware operated by the virtual computers or machines, as required. Customers do not manage these virtual computing machines, but control the software deployment and configu-ration settings, whereas the PaaS cloud service operators provide the networks, servers, storage, and other services. Commercial examples of PaaS include

Windows Azure, Google's Apps Engine, SaleForce's Force.com, Red Hat OpenShift, and Heroku.

- *Infrastructure as a Service (IaaS):* This is the lowest-level model of the cloud services. It basically offers or outfits customers with one or more of the virtual machines running on the provider's physical equipment. Customers or businesses can purchase or lease the infrastructure from IaaS providers, such as virtual-machine image, operating systems that run on these machines, libraries, row (block) and file-based storage, firewalls, load balancers, IP addresses, virtual local area networks (VLAN), and so on. The customers are not required to manage or control the underlying cloud infrastructure but have control over tasks of the operating systems, storage capabilities, and deployment structures of their applications. Examples of commercial IaaS providers include Amazon Elastic Compute Cloud, Rackspace, Microsoft Azure, and Google Compute Engine.

In simple terms, the three different cloud computing services allow businesses or customers to "rent" the technology from cloud vendors, on a specific fee basis, to store data and run their applications. The differences between SaaS, PaaS, and IaaS are in the level of control that specific customers or businesses have over the applications they use, how these applications are created, and the type of hardware on which their specific applications are run (Barnatt, 2010; Curtis, 2014). Other popular cloud service providers not mentioned above include IBM, iCloud, and OpenDrive.

We will consider the health App development process described in Section 3.3.2 as an example to clarify the differences between these service classes. The SaaS cloud service simply means that the user rents the App required for the specific health application or service, and everything else relating to the management layers (networking, storage, servers, middleware, O/S, etc.) are managed by others. In the PaaS, the user rents everything except the relevant App and data services (which are user-managed), and all remaining layers are managed by others. This cloud service is often used by health App developer businesses. In the IaaS service, the user rents the hardware and relevant facilities to maintain the hardware (i.e., the user manages the App, data, run-time, middleware, and O/S functions), whereas the other layers (virtualization, servers, storage, and networking) are managed by others. This service option is usually popular with start-ups and small enterprises.

The appropriate partition of these cloud services and infrastructure requirements is usually driven by each healthcare or other service provider's needs, and their priority options from costs, scaling of the service for better management, maintenance of IT, and computing resources, security, and privacy.

In recent years, the healthcare sector has begun to embrace different cloud-based platforms for their health IT services, especially in hospital settings, despite the common concerns of compliance and security issues. In general, cloud computing offers flexible Internet-based healthcare service delivery formats with several benefits, such as better service delivery, efficiency, cost-savings in IT infrastructure, and leveraging new healthcare applications to support different mobility platforms and new workforce structures.

3.4.4   m-Health and Mobile Cloud Computing

As explained earlier, cloud computing technology is well-suited to meet the increasing demands of different m-Health and e-Health services, particularly since cloud computing provides reduction in capital costs, operational expenditures, and other infrastructure IT costs, and optimizes the resources required for these services.

One of the expectations of m-Health services is to provide better mechanisms to share and exchange medical information and improve the quality of care using different mobility platforms. Several secure e-Health cloud architectures and service models have been proposed (Lohr et al., 2010, Coats and Acharya, 2013), mostly targeting the following services:

- *PHR*, allowing patients to track, collect, and manage their own information.
- *EHR*, where patients' data or records are managed by their healthcare provider or specialist; these are usually secure in hospital settings.

The functional requirements between these two e-Health service models are different, due to the involvement of different stakeholders; and EHR cloud services are simpler than PHR cloud services (Lohr et al., 2010). In addition, the security and privacy requirements of EHR models need to be more stringent compared to the PHR service models. In terms of the various cloud service delivery classes listed earlier, SaaS acts as a gateway for e-Health services, patient registry, intelligent data analytics, and so on; PaaS is used for development and integration of tools for building, testing, and deploying new e-Health services; and IaaS is used for medical data storage purposes, e-prescriptions services, patient data processing, and so on.

There are in general various layered architectures available for cloud computing to provide the aforementioned services as a utility (Khan et al., 2012). Figure 3.5 shows a bird's-eye view of the cloud from an m-Health perspective. In simple terms, the cloud can be interpreted as the Internet equivalent of the "kitchen sink," acquiring different sources of medical data, for example, m-Health-monitoring sensors and patients' medical history data, which are pushed in turn to the cloud. Data are then stored and processed from any cloud access terminal or node for further analysis and feedback.

The m-Health cloud structures are usually built around the "mobile cloud computing or MCC" platforms. Mobile cloud computing is generally referred to as "an infrastructure where both the data storage and data processing happen outside the mobile device. Mobile cloud applications move the computing power and data storage away from mobile phones and into the cloud, bringing applications and mobile computing to not just smart phone users, but to a much broader range of mobile subscribers" (Dinh et al., 2013). As shown in Fig. 3.5, the architecture of the mobile cloud is based on the following blocks: mobile client and mobile network operators; Internet service providers; and application and cloud service providers.

Generally, mobile clients use mobile networks to communicate with the cloud through Internet connectivity. Communication between the client and the cloud provider is achieved via a native mobile App or embedded browser application developed, for example, to access EHRs located on the cloud server. The service

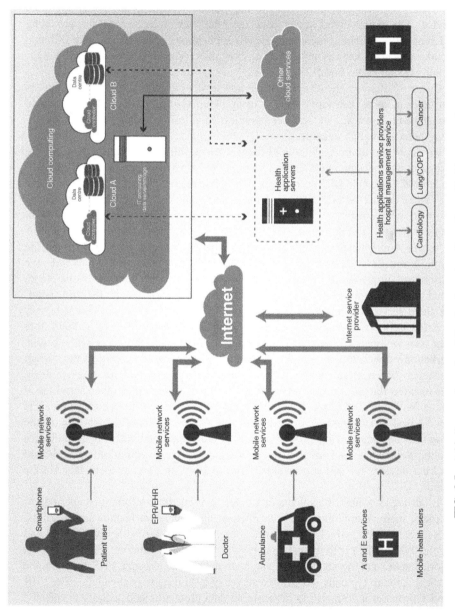

FIG. 3.5 A bird's-eye view of mobile cloud services from the m-Health perspective.

provider manages the cloud resources, and operates and maintains the cloud servers (public or private). There are typically two types of cloud servers: portal cloud servers and back-end cloud servers. The portal cloud server (sometimes referred as a cloud controller) receives and processes the mobile client requests for using the cloud service. The portal cloud server uses the back-end cloud servers for providing different services. These include m-Health applications and services, with different security functions such as authentication, key management, encryption, and authorizations (Khan et al., 2012).

Currently many m-Health applications and services leverage different data sensors and other medical data input sources using cloud services. These include hospital management systems, care services, wellness monitoring, mobile disease management, home-care, and ageing.

However, the future scaling-up of these services will depend on different drivers, such as costs, legislation, regulatory, security, and clinical validation. In a typical m-Health MCC, mobile healthcare users (e.g., patients, healthcare providers, doctors, nurses) are connected to the cloud via their smart mobile devices. Their requests and information are transmitted to central processors that are in turn connected to network operator servers. The mobile network operators provide the services to mobile users as AAA (Authentication, Authorization, and Accounting) requests, based on the Home Agent (HA) and subscribers' data stored in their databases. Afterwards, the subscribers' requests are delivered to the cloud through the Internet. In the cloud, cloud controllers process these requests to provide mobile users with the corresponding cloud services (Dinh et al., 2013).

An example of a simple MCC model is an access service where patients store their own wellness-related data on their Apps and synchronize these with specific market-based web servers (vaults). This model is applicable to some of the current consumer-driven cloud health services. In these systems patients track, collect, and manage the information about their health, wellness, and other activities at online Web sites or cloud providers such as the ones mentioned earlier. This cloud-based m-Health-monitoring model can also be applied, for example, to the *quantified self-monitoring* process in which individual patients collect their own health, well-being, and other daily data using different technological tools and tracking devices for self-improvement and to store and access their data via the cloud.

In all these scenarios, patients can enter their medical information, periods of sickness, medical appointments with doctors, and any other health- and wellness-related data. Alternatively, their wearable devices can seamlessly acquire and wirelessly send their activity and wellness data (e.g., sleep patterns, walking activity, calorie intake, and moods) to these vaults. In the case of a simple PHR scenario, the medical data are typically stored on a server of a third party in the cloud. The PHR service provider is responsible for ensuring data protection and security. Typically, patients can define role-based access rights for individual health professionals. For example, they can define full access to their family doctor, but only restricted access to some data to their fitness trainer or health coach. The advantages of such an approach are that the PHR is accessible from everywhere because of the centralized management (IT outsourcing). Patients can also, for example, allow one doctor access to data and test results that were determined by another doctor, when the data are stored in their PHR.

Another typical, and more complex, example is the cloud-based EHR access model. EHR access models are usually created, maintained, and managed by healthcare providers, and can be shared (via the central EHR server in the cloud) with other health professionals in the hospital or specialist centre. This model usually involves different stakeholders in the care service delivery chain, for example, clinicians, healthcare providers, and health insurance companies. It also embeds robust technical measures to enforce the medical data security and privacy of these EHRs. In addition to the storing and processing functions of EHRs, other services can also be outsourced to the cloud, such as financial and management services.

One of the critical challenges of mobile health cloud computing are the security, privacy, and threats challenging these cloud services, which are explained next.

3.4.5 Security Features and Threats for Cloud Computing

There are numerous challenges for MCC, such as data replication, consistency, limited scalability, unreliability, unreliable availability of cloud resources, portability (due to lack in cloud provider standards), trust, security, and privacy (Khan et al., 2012). Some of major concerns of using cloud computing for healthcare services in general, and for m-Health services in particular, are the security and privacy risks involved. These concerns are still subject to further research and scrutiny. Although this is a specialist topic, cloud computing provides several security features such as (ENISA, 2009; Balding, 2008):

- *Economy of Scale:* Cloud computing allows pooling and utilization of security measures, which are more cost-effective, and provide better resource allocations when implemented on a larger scale.
- *Data Centralization:* Using centralized data models provides better and more secure data management.
- *Ease of Monitoring Benefits:* Cloud computing, with its centralized storage facilities, provides easier control and monitoring processes, with the potential of better implementation of security measures compared to sparse data storage locations, especially when these are attacked.
- *Efficient Audit and Evidence Gathering:* Cloud computing provides a dedicated virtual forensic server image in the same cloud that can be implemented and placed offline for added security.
- *Ease of Management and Automatic Logging:* Cloud computing allows better logging mechanisms, required for security investigative procedures, and provides ease of management by the deployment of gold images (a unit or an instance that has been fully subjected to proper stability and vulnerability tests, and is ready for public deployment).

However, there are also general security risk factors attributed to cloud computing and identified as follows (Brodkin, 2008; Samson, 2013; Khan et al., 2012):

- Privileged user access and account traffic hijacking
- Data protection and breaches
- Data isolation and shared technology vulnerabilities
- Data sanitization and malicious insiders
- Data location and data loss and recovery
- Investigative support
- Insecure interfaces and APIs
- Denial of service

Furthermore, specific threats in the privacy-sensitive context of cloud services are summarized as follows (Horowitz, 2011; Lohr et al., 2010):

- *Data Storage and Processing:* Threats usually result from storing sensitive data, and the risk of data leakage to unauthorized entities; and also from client platforms and healthcare professional computing access platforms that use these services.
- *Management of e-Health Infrastructure:* Medical and administrative data of patients are usually processed at different e-Health cloud locations. In these scenarios, usage of smartcards and access control mechanisms, for example, does not provide effective and robust enough protection. Usually, the following issues need to be considered:
  - *Cryptographic Key Management:* Complex infrastructures must be managed with additional security and privacy measures. These include management of certificates and hardware/software components of the service; and access to mobile storage devices, which is particularly critical when access to patient medical data is carried out from mobile devices or smartphones.
  - *Usability and User Experience:* These threats are concerned with the people who use these services, namely, patients and healthcare professionals. Challenges such as remembering complicated personal identification numbers (PIN) can be compromised due to handwritten notes, or a breach of password information.

These are only some of the usability security risks. Others, such as necessary usability knowledge of the IT security set-up information potentially operated by doctors and other healthcare professionals to operate such systems, can also lead to different security breaches.

An example to illustrate typical security measures adopted in mobile cloud-based monitoring applications, such as sensing body temperature, heart rate, oxygen saturation, blood pressure, or blood glucose levels, is to add a cloud security layer with an encryption algorithm to the smartphone application software. When the application is activated, and the encrypted data acquired from medical sensors are connected to the smartphone, the encrypted sensor data are transmitted via mobile Internet connectivity, and pushed for storage in the cloud for further analysis and processing. At the healthcare provider's remote cloud access point (e.g., in a hospital, healthcare center), the received encrypted data stream is downloaded from any cloud web-accessed terminal, and de-encrypted by the same encryption key shared with the sender. Only then can the medical data be downloaded, processed, and viewed for further analysis and feedback.

Security threats and vulnerabilities are inherent in all cloud deployment (Public/ Private/Hybrid) models and service classes (IaaS, PaaS, SaaS). These usually appear once the cloud service technology is deployed. The timing and method of detecting these threats becomes more demanding, especially in healthcare services of a critical and sensitive nature. Securing the mobile health cloud is still an emerging topic, and subject to further research. More recently, the concept of Peer-to-Peer (P2P) cloud computing has been proposed as an alternative approach to business provider-based cloud platforms. In P2P, the cloud structure can be formulated by connecting geographically distributed cloud computing infrastructures that are made up of numerous individual computers and devices around the globe, connected through the Internet.

This P2P cloud structure can be built around different users' computing and storage facilities, for example, from routers, laptops, desktop computers, and broad-band modems, thus potentially alleviating the burden of control of the commercial cloud provider's model, and providing cost savings (Buyya et al., 2011). Although P2P cloud models provide less quality of service guarantees compared to the existing commercial cloud services, they can possibly offer an alternative source of new community-based m-Health2.0 cloud services, especially where competitive cloud services and providers are unavailable.

Further details on this topic and other cloud computing architectures, implemen-tation, security, and standardization issues can be found in specialist references (Antonopoulos and Gillam, 2010; Barnatt, 2010; Erl et al., 2013).

Finally, the "hype cycle" for emerging technologies has mapped both the *mobile cloud* and *mobile health monitoring* in the *trough of disillusionment* region of the cycle (Gartner, 2014). This predication parallels with the ongoing debate that cloud computing offers m-Health better opportunities for more effective and efficient mobile health services, especially in the areas of prevention, public health, and treatment. These are yet to be seen, considering the absence of large-scale clinical evidence.

3.5 m-HEALTH AND "BIG DATA"

Big data in the context of healthcare could be one of the transformative drivers of future m-Health systems. However, the key challenges of "Big Data and mobile health" remain largely untackled. The former has become increasingly a buzzword in the healthcare sphere, but without proper correlation of the appropriate theoretical computing and analytical framework that best matches the challenges of the latter. Increasingly, the health data are becoming more complex, voluminous, and multi-dimensional. The volume of digital data being collected and stored globally is exploding. This game-changing increase is being propelled by the unprecedented global usage of Internet-connected devices, and massive amounts of smartphone data generated by services and applications linked to these devices. As a consequence, there is major push to manage and analyze this volume of data, and to convert it into meaningful information that can benefit both users and stakeholders.

The term "big data" was introduced to describe an evolving concept, whose aim is to examine and interpret these data using different intelligent computing methods. In parallel to this explosion of big data in general, there is a similar increase in the healthcare context, driven by the increasing level of digital healthcare data from the m-Health and e-Health services and applications. This is becoming a major focus for all healthcare stakeholders, from the medical and pharmaceutical industries, clinicians and healthcare providers, to biological and computing scientists, and other experts. From the m-Health context, big health data are fuelled by recent advances in smartphone monitoring technologies, and the increasing number of health sensors embedded in or connected to these phones that generate the data. This is also complemented by the parallel increase in biomedical or genomic big data generated from recent developments in personalized medicine, diagnostic systems, and genome sequencing technologies. These rapid developments are accumulating an abundance of medical and personalized health data that can offer new and transformative opportunities for developing intelligent healthcare delivery systems by analyzing these data.

What is Big Data? In general terms, there are many definitions of big data, but most widely it is understood as the massive volume of both structured and unstructured data that exceed the capacity or capability of current or conventional analytic methods or systems to be mined for meaningful patterns and trends. It is also the correlation of many variables with costs and outcomes, and usually associated with four dimensions: volume, velocity, variety, and veracity (Laney, 2001). Others add two further dimensions of value and variability (Andreu-Perez et al., 2015).

There is ongoing debate about the sources of big health data. From the healthcare perspective, the input of big data is usually generated from multiple sources. These include sources of patient medical data, mobile device and sensor data, and social networking data that reflect personal behavior. Other sources include data generated from healthcare providers, for example, structured EHR, EPR, unstructured clinical notes, medical imaging, and pharmaceutical and relevant laboratory data. In addition, there are biomedical, epidemiological, behavioral, and genetic data. The output generated from big health data goes to EHRs, clinical decision support, health information exchange systems, and to payers' and government data bases. Hence, big health data are not only about the size of the captured data but also about storing, searching, sharing, and analyzing the data, and finding the appropriate insights from the different heterogeneous, dimensional data sources to provide meaningful and useful trends (Hansen et al., 2014).

As discussed earlier, we are witnessing a major increase in the number of smart health Apps linked to wireless, wearable sensors that are able to monitor all sorts of medical symptoms and conditions. These range from users' sleep patterns, heart rates, and wellness activity levels to self-management of different chronic diseases such as diabetes, heart failure, and chronic obstructive pulmonary disease (COPD). However, the empowerment of m-Health technologies with big data generated from these sources, and leveraging it for better patient care and outcomes, is still a challenging task. This is mainly due to the complex nature of such data, as reflected in the heterogeneity, multidimensionality, and volume of such data.

In addition, smart mobile phones are getting smarter, not only because of their embedded smart sensors, but also due to their ability to use more intelligent data processing capabilities. An example of this is the simple comparison of two smartphone models, iPhone6-Plus and Samsung Galaxy-Note4, revealing that in addition to their existing embedded sensors (gyroscope, compass, accelerometer, proximity, ambient light, cameras, microphone, GPS, WiFi, and Bluetooth radio connectivity), these models have additional sensors with extra functionalities to generate further data. These include, for example, security access capabilities (for fingerprint detection), a health and well-being sensor (for heart rate measurement), and Near Field Communications (NFC) detection (for secure payments).

The number of embedded sensors is expected to increase to more than 20 in future models, for example, in the Apple watch. These incredible technological advances and similar developments will generate a further deluge of digital data, including health-related data, that will increasingly demand better harnessing and utilization, especially for clinical and medical purposes.

Overall, many challenges in this context remain open for further research. These include, for example, how to control and rectify any potential errors that occur from the heterogeneity and uncontrolled sampling processes underlying the massive m-Health data sets generated from the millions of connected monitoring devices. Another problem is associated with time constraints of these types of data, where time-critical data processing with clinically accurate outcomes can pose clinical challenges, especially in many of current m-Health applications from smart diagnostic Apps to real-time monitoring systems. Furthermore, many of the existing theoretical principles of computation and statistics, such as dimension reduction, distributed optimization, Monte Carlo sampling, compressed sensing, low-rank matrix factorization, streaming, and hardness of approximation, can be relevant to m-Health big data analytical approaches, but major challenges of their application and accuracy remain open for future research.

3.5.1 Big Health Data and Big Biomedical Data

Big data are simply sets of data that are too large and complex to manipulate by standard methods. However, there are different definitions used for big data. One such definition is "referring to things one can do on a larger scale that cannot be done on a smaller one, to extract new insights, or create new forms of value, in ways that change markets, organizations, the relationship between citizens and governments, and more" (Mayer-Schönberger and Cukier, 2013). Another source defines big data as "large volumes of high-velocity, complex, and variable data that require advanced techniques and technologies to enable the capture, storage, distribution, management, and analysis of the information" (IHTT, 2012).

Big data have been successfully used in such diverse areas as health, retail, web engine searches, astronomy, politics, advertisements, and others (Mayer-Schönberger and Cukier, 2013). According to a market study by McKinsey & Company, it is estimated that the business opportunities inherent in making sense of big data is potentially between $300 and $450 billion a year, with major companies such as Apple, Qualcomm, and IBM driving the investment in these technologies (Byrnes, 2014). Furthermore, from the healthcare perspective, there is currently an increasing volume of activity on different applications of "Big Health Data" (Kayyali et al., 2013; Weber et al., 2014). Some sources summarize the importance of big health data in terms of the following benefits (Murdoch and Detsky, 2014):

- Big data may greatly expand the capacity to generate new medical knowledge. These include analyses of the unstructured data embedded in EHRs and EPRs for better healthcare outcomes. Also, these methods provide the potential to create an observational evidence base for clinical issues that are otherwise unavailable without such analyses.
- Big data may help with knowledge dissemination to clinicians, by providing better knowledge access to areas such as understanding data-driven clinical methods and decision-support systems, and a better understanding of new, diagnostic treatment methods and guidelines. Big data can also be harnessed and linked to enable physicians and researchers to test new hypotheses, and to identify new areas of possible intervention. Some interesting examples of these scenarios include understanding grocery shopping patterns obtained from stores in various areas to predict rates of obesity and type 2 diabetes linked with public health databases, or correlating levels of exercise recorded by home-monitoring devices, with response rates of cholesterol-lowering drugs, as measured by continued refills at pharmacies (Weber et al., 2014).
- Big data may be able to help translate advances in personalized medicine into clinical practices.
- Empowering patients by delivering information directly and enabling them to play a more active role in their health and disease management.

Others argue the additional benefits in diverse healthcare areas, including public health, evidence-based medicine, patient profiles, genomic analytics, and population health management (IHTT, 2012; Raghupathi and Raghupathi, 2014). These and other benefits crystallize the importance of big data in different healthcare sectors, and particularly for m-Health applications.

Big health data can be generated from the proliferation of e-Health technologies in different clinical settings, notably the widespread use of different e-Health services such as EHRs and EPRs, individual EMRs, and laboratory and imaging systems. These sources are referred to as "electronic clinical data" (Jain et al., 2014). For example, recent academic and industrial collaborations have focused on the use of electronic health data to study the value and comparative effectiveness, safety, and efficacy of medications, vaccines, and healthcare delivery models (Selby et al., 2012). Other examples of academic and pharmaceutical industry partnerships for big data solutions include conducting validation work in the following specific themes of clinical data (Jain et al., 2014):

- New methodologies for observational research
- Analysis of clinical data from practice settings
- Novel applications of health information technology
- Clinical interventions
- Education

Big data generation leads to big data overload. In life sciences they are effected by large-volume data sets, specifically by the "omics" information (genomes, transcriptomes, epigenomes, and other "omics" data from cells, tissues and organisms). The use of DNA sequencing machines, which are smaller in size, but capable of generating piles of data faster and at a lower cost, have also changed science and medicine in ways never seen before (Costa, 2012). The advances in microarray technologies and genomic signal processing are further sources of these data and these signal processing methods tailored for genomic data can also act as predictive analytical tools (Istepanian et al., 2011). These sources are sometimes termed as "Big Biomedical Data" (Weber et al., 2014) or "Genomic Big Data" (Costa, 2012). The increasing trend toward the predictive, preventive, participatory, personalized medicine is fuelling the big data trends in the different omics-data domains from genomics, epigenomics, metagenomics, proteomics, metabolomics, lipidomics, transcriptomics, epigenetics, microbiomics, fluxomics, phenomics, and so on.

There is also growing volume of Patient-Generated Health Data (PGHD) from diverse sources, from mobile devices, social networks, and other media channels generated by patients. This volume of data is creating an ever-greater stream that is difficult to analyze within the current healthcare services context. In general, PGHD are health-related data created, recorded, or gathered by or from patients (or family members or other caregivers) to help address a health concern (National Learning Consortium, 2014). These data can include health history, treatment history, symptoms, biometric data, and patient-reported outcome measures (Shapiro et al., 2012). In general, PGHD differs from data generated in clinical settings and through encounters with healthcare providers in two important ways (National Learning Consortium, 2014). Patients, not providers, are primarily responsible for capturing or recording these data, and patients decide how to share or distribute the data to providers and others. From the m-Health perspective, PGHD opens new opportunities and challenges.

The benefits of big data include the potential for better clinical evidence for m-Health from patient monitoring, especially for chronic disease management. These data can also open up new opportunities to better understand the causes and onset of some diseases that relate to environmental and lifestyle behaviors. The risks include the challenges of ownership, together with risks of security and data privacy. The ethical issues relating to the use of these data for clinical research also need to be considered carefully.

This tapestry of health and genomic data can provide transformative developments in m-Health systems by translating these to more intelligent systems, typically by combining advances in personalized medicine with similar ones in computing technologies. Examples are the development of new real-time smart diagnostic tests and fast smart mobile systems that are able to perform DNA assays, identify genetic predispositions to certain diseases, and gene mutations associated with the onset of different diseases such as cancer and diabetes.

Big data may be categorized into two general types:

- *Structured Data:* These refer to big data that have a defined length and format (numbers, dates, strings, etc.), and are generated by sources such as computers, mobile phone sensors, and web logs. These types of data represent the minority (generally around 20% of the total data generated). Examples include data extracted from EPRs, EHRs, home monitoring and treatment, and medical prescriptions.
- *Unstructured Data:* These refer to big data that do not follow a specified format, and represent the majority of data generated from different sources, such as general data from social media, mobile data, video, and web content. Examples include social health data (e.g., tweets and Facebook blogs), digital health and clinical notes, and medication diaries and instructions.

The trend for more evidence-based medicine pathways of healthcare services from medical and pharmaceutical industries is driven by the increasing demands of healthcare costs, wider care access, and the need for better patient outcomes. Several of these factors will drive this new era of care pathways, where big data will formulate the catalyst for such transformation. These include cost, reimbursement, new pharmaceutical and therapeutic drug discoveries, patient behavior, and social patterns, which are summarized as follows (Groves et al., 2014):

- *Right Living:* Informed lifestyle choices that promote well-being and active engagement
- *Right Care:* Evidence-based care with improved patient outcomes and safety
- *Right Provider:* Care providers (nurse, physician, etc.) and the settings for best healthcare delivery and impact
- *Right Value:* Sustainable approaches that enhance healthcare values with reduced cost and better quality
- *Right Innovation:* Innovation that advances the frontiers of medicine and boosts research and development capabilities in new therapeutics and care delivery

m-Health can contribute significantly in achieving these outcomes, for example, by enabling appropriate big data computing platforms required for the analytical framework, and thus accelerating such a transformation process. Also, m-Health can play an important role in accessing the appropriate healthcare data, which can lead to the acquisition of new medical knowledge, understanding the effectiveness of medical treatments, or for the better prediction of healthcare outcomes.

However, the appropriate analysis and processing of these big data, generated from these sources, requires the development of compatible analytical tools that need to be clinically effective, and applicable for different patient populations (Schneeweiss, 2014). Some examples of the challenges of big health data from the mobile health perspective can be summarized as follows:

- Better understanding of the structured and unstructured data generated from the variety of mobile users' data and information sources.
- Adapting and translating big health data generated from mobile health users to intelligent individualized or personalized behavioral change tools that can motivate patients to better improve their health or well-being.
- Enhanced capturing and effective correlation between individualized behavioral patterns obtained from mobile devices (e.g., wearable trackers and smart-phones) with their own health/wellness data patterns toward personalized healthy goals and better care outcomes.
- Improved correlation between the individualized genomic data sequencing and other factors such as environment, diet, lifestyle, and so on with accurate analytical prediction for potential predisposition to genetically related diseases.
- Robust analysis of the medical imaging and diagnostic data from mobile imaging devices and correlating these with patients' mobile EHR systems.

These challenges are subject to ongoing studies. It is expected that the benefits associated with big health data will include patient involvement and engagement, the ability to allocate the right care to the right patient at the right time, enhancing the quality of care, and reducing the costs, together with the knowledge gained from big data analytics on population health and genomics. All these can outweigh the risk factors associated with big data, including data security and privacy, clinician–patient interaction, physician judgment, selectivity of the data to be monitored, and who will monitor it.

3.5.2 Big Health Data Analytics

Data analytics are referred to as the science of examining raw data, with the purpose of drawing conclusions about that information. Traditionally, advances in different analytic techniques, especially in machine learning techniques, were important for analyzing large information sets. However, these techniques represent the necessary analysis framework for big health data that are in contrast to traditional statistical methods, and are largely not useful for the analysis of unstructured data (where the majority of the big health data falls). A general conceptual architecture of big data analytics can be represented by the following components (Raghupathi and Raghupathi, 2014):

- *Big Data Sources:* These represent the different sources of data and their formats.
- *Big Data Transformation:* In this component, the data are in a "raw" state and need to be processed, or transformed, at which point several options are available depending on whether the data are structured or unstructured. Several data formats can be input to the big data analytics platform component.
- *Big Data Analytical Tools:* In this component, several decisions are made regarding the data input approach, distributed design, tool selection, and analytic models. These are accomplished using different available analytical tools, for example, Hadoop Distributed File System (HDFS), MapReduce, and Hive.
- *Big Data Analytical Applications:* These big data analytic tools are then applied to different health data applications, for example, visualization, queries, reports, data mining, and online analytical processing (OLAP).

The basic health analytical platform structures of information extraction, feature extraction, and predictive modeling have been successfully applied in many big health data analytics examples. These include optimal processing of EMRs, timely management of medications of critically injured patients, and diagnosis of complications in stroke patients and others (Raghupathi and Raghupathi, 2014). A further example is the application of data analytics to pharmacies and medication knowledge. In this domain, recent studies from Express Scripts, which manages pharmacy benefits for 90 million members in the United States, and processes 1.4 billion prescriptions a year, has scoured its data from doctors' offices, pharmacies, and laboratories to detect patterns that might alert doctors to potential adverse drug interactions, and other prescription issues (Byrnes, 2014). From this data analytical approach, doctors were able to know 12 months in advance, with an accuracy rate of 98%, which of their patients may fail to take their medicine. Considering that the United States spends $317 billion each year on unnecessary emergency visits and other treatments, big data analytics can provide major savings (Byrnes, 2014). Other examples are research and development on genomic data, which are cited in different biological and genomic analytical areas (Costa, 2012).

The recent trend of digital self-tracking using wearable fitness trackers and watches, and intelligent processing of the logged data, will be a new area of research. It will be important in the future to develop a potential framework for big health data analytics for personalized m-Health applications. Examples include the usage of real-time activity tracking and recognition in smart home environments (Rashidi et al., 2011). These should seamlessly capture mobile patient data from their wearables and smartphone sensors, process the information in the cloud, and provide new and effective individualized care and coaching feedback with improved outcomes. This health "data mine phoning" approach is a possible candidate in the framework formulation process.

IBM has introduced the "Watson" system as an example of implementing smart big data analytics in healthcare, particularly in assisting different U.S. healthcare institutions in their clinical trials and genomic-based studies of cancer predictions and heart conditions (Rhodin, 2014). A recent review cited many examples of health-related big data projects related to social media and the quantified-self movement (Hansen et al., 2014).

In summary, big data analytics constitute a promising trend, which is in its infancy for the mobile health domain. The benefits and potential impact of big health data on various segments of healthcare services are evident, but there are still many unanswered questions and m-Health challenges, which include the following:

- Privacy and security concerns, including the right to privacy of patients' data and the potential risks involved in unauthorized access of such data from smart mobile phones, especially of sensitive data; for example, any mental health or biological markers/genetic data (Gibbs, 2015).
- The unclear role of big data in current m-Health ecosystems. These include the role of various stakeholders and the relevant spectrum of such ecosystems from healthcare providers to telecommunications industries, pharmaceutical companies, and health insurers.
- The lack of a clear regulatory framework in addition to any ethical and legal challenges associated with it.
- The personalization of m-Health and the ethical dimension of the ownership of personal data from self-tracking users.

Although the role of big health data and analytics from the mobile health perspective is in its early stages, the capabilities of m-Health technologies and platforms in using big data remain both an opportunity and a challenge for further research and development.

3.6 SUMMARY

At the inception of the second decade of m-Health, its future from the perspective of computing and Internet technologies is in flux. This is driven by the rapid changes and technological advances shaping their evolution. These changes are mainly centered around the next generation of the Internet (Web 2.0 and beyond), the pervasiveness of mobile cloud computing, big health data, the increasing role of social networking and media technologies for health, together with the unprecedented proliferation of mobile health Apps. All these and other future developments are transforming m-Health to a new era of m-Health 2.0 that encapsulates this evolutionary process. We are already witnessing examples of such technology, such as wearable health data hubs (smart watches and personal fitness monitors), self-learning and intelligent software that effectively turns smartphones into *de facto* personal diet advisors, intelligent healthy food companions, and others.

There remains much work to be done to translate this vision into reality. For example, more effort is required for a global regulatory framework for m-Health Apps and their usage. The ongoing debate on net neutrality, with a specific focus on current and future m-Health services, is also an important topic in the process. The development of simple web tools can enable ordinary Web.2.0 users (e.g., patients, healthcare providers, and doctors) to better utilize the different telecom services to improve healthcare, without the need for complex computing knowledge and dependency on expert programmers and developers. The power of mobile computing is increasing massively, but this power has not yet been harnessed adequately for m-Health systems. Unraveling the mobile big health data complexities can provide many potential benefits about the accurate decisions at the right time for patients and other users.

This harnessing will also be linked to future developments in the IoT and new smart wireless device connectivity. These and other developments described in this chapter will determine the future of m-Health and how it will evolve in its second decade.

REFERENCES

AHIMA (American Health Information Management Association) (2013) *Best practices for mobile health? There's an App guide for that—AHIMA develops a best practice primer for consumers.* Available at http://www.myphr.com/HealthLiteracy/MX7644_myPHRbrochure .final7-3-13.pdf (accessed October 2014).

Aitken M, Altmann T, and Rosen D (2014) *Engaging patients through social media: is healthcare ready for empowered and digitally demanding patients?* IMS Institute for Health Informatics, New Jersey. Available at http://www.imshealth.com/deployedfiles/imshealth/ Global/Content/Corporate/IMS%20Health%20Institute/Reports/Secure/IIHI_Social _Media_Report_2014.pdf (accessed October 2014).

Al Anzi T, Istepanian, RSH, and Philip N (2014a) Usability study of mobile social networking system among Saudi type 2 diabetes patients (SANAD). In: *2nd Middle East Conference on Biomedical Engineering (MECBME)*, February 17–20, 2014, Doha, Qatar, pp. 297–300.

Al Anzi T, Istepanian RSH, Philip N, and Sungoor A (2014b) A study on perception of managing diabetes mellitus through social networking in the kingdom of Saudi Arabia. In: *XIII Mediterranean Conference on Medical and Biological Engineering and Computing 2013*, vol. 4, Springer International Publishing.

Alvaoro S, Istepanian, RSH, and Garcia J (2006) Enhanced real-time ECG coder for packetised telecardiology applications. *IEEE Transactions on Information Technology in Biomedicine* 10(2):229–236.

Andreu-Perez J, Poon CY, Merrifield RD, Wong CST, and Yang GZ (2015) Big data for health. *IEEE Journal of Biomedical and Health Informatics* 19(4):1193–1208.

Antonopoulos N and Gillam L (Eds.) (2010) *Cloud Computing: Principles, Systems and Applications.* London: Springer.

Armbrust M, Fox A, Griffith R, Joseph AD, and Katz R (2009) *Above the clouds: a Berkeley view of cloud computing*, UC Berkeley Reliable Adaptive Distributed Systems Laboratory White Paper.

Balding C (2008) *Assessing the security benefits of cloud computing.* Available at http:// cloudsecurity.org/blog/2008/07/21/assessing-the-security-benefits-of-cloud-computing.html (accessed October 2014).

Barnatt C (2010) *A Brief Guide to Cloud Computing: An Essential Guide to the Next Computing Revolution.* London: Constable and Robinson Ltd.

Baron J, McBain H, and Newman S (2012) The impact of mobile monitoring technologies on glycosylated hemoglobin in diabetes: a systematic review. *Journal of Diabetes Science and Technology* 6(5):1185–1196.

Bernhardt JM (2015) *The future of health is mobile and social.* Available at http://smhs.gwu .edu/mhealth/sites/mhealth/files/GWU%20Future%20of%20Health%20is%20Mobile% 20and%20Social.pdf (accessed January 2016).

Boyd DM and Ellison NB (2007) Social network sites: definition, history, and scholarship. *Journal of Computer-Mediated Communication* 13(1):210–230.

Brodkin J (2008) *Gartner: seven cloud-computing security risks*, CIO. Available at http://www .cio.com/article/2435262/enterprise-software/gartner--seven-cloud-computing-security-risks.html (accessed October 2014).

Busch J, Barbaras L, Wei J, Nishino M, Yam C, and Hatabu, H (2004) A mobile solution: PDA-based platform for radiology information management. *American Journal of Roentgenology* 183(1):237–242.

Butler C (2012) How to pick useful health apps for mobile devices. *The Washington Post (Health and Science).* Available at http://www.washingtonpost.com/national/health-science/how-to-pick-useful-health-apps-for-mobile-devices/2012/07/16/gJQAQ1uFpW _story.html (accessed October 2014).

Buyya R, Broberg J, and Goscinski A (Eds.) (2011) *Cloud Computing: Principles and Paradigms.* John Wiley & Sons, Inc.

Byrnes N (2014) Can technology fix medicine? *MIT Technology Review*, July 21. Available at http://www.technologyreview.com/news/529011/can-technology-fix-medicine/ (accessed October 2014).

Centola D (2013) Social media and the science of health behaviour. *Circulation* 127: 2135–2144.

Chomutare T, Fernandez-Luque L, Arsand E, and Hartvigsen G (2011) Features of mobile diabetes applications: review of the literature and analysis of current applications compared against evidence-based guidelines. *Journal of Medical Internet Research* 13(3):e65.

Coats B and Acharya S (2013) Achieving electronic health record access from the cloud. In: Kurosu M (Ed.), *Human-Computer Interaction: Applications and Services*, Lecture Notes in Computer Science 8005, Springer, pp. 26–35.

Conn J (2012) Most healthful apps. *Modern Health Care* 42(50):30–32.

Cortez NG, Cohen GI, and Kesselheim AS (2014) FDA regulation of mobile health technologies. *The New England Journal of Medicine* 371(4):372–379.

Costa FF (2012) Big data in genomics: challenges and solutions. *G. I. T. Laboratory Journal* 11–12/2012, 1–4.

Curtis J (2014) *10 top cloud computing providers for 2014*, Computer Business Review. Available at http://www.cbronline.com/news/cloud/cloud-saas/10-top-cloud-computing-providers-for-2014-4401618 (accessed October 2014).

Dehling T, Gao F, Schneider S, and Sunyaev A (2015) Exploring the far side of mobile health: information security and privacy of mobile health apps on iOS and Android. *Journal of Medical Internet Research* 3(1):1–17.

Dinh HT, Lee C, Niyato D, and Wang P (2013) A survey of mobile cloud computing: architecture, applications, and approaches. *Wireless Communications and Mobile Computing* 13(18):1587–1611.

Duggan M, Ellison B, Lampe C, Lenhart A, and Madden M (2014) *Demographics of key social networking platforms*, Pew Research Centre. Available at www.pewinternet.org/2015/01/09/demographics-of-key-social-networking-platforms-2 (accessed October 2014).

Eddabbeh N and Drion B (2006) Mobile access to electronic health records: DOCMEM. In: Istepanian R, Laxmnaryan RSH, and Pattichis C (Eds.), *m-Health: Emerging Mobile Health Systems*, London: Springer, pp. 187–194.

El-Gayar O, Timsina BE, Nawar N, and Eid W (2013) Mobile applications for diabetes self-management: status and potential. *Journal of Diabetes Science and Technology* 7(1):247–262.

Eng DS and Lee JM (2013) The promise and peril of mobile health applications for diabetes and endocrinology. *Paediatric Diabetes Research* 14: 231–228.

ENISA (European Network and Information Security Agency) (2009) *Cloud computing, benefits, risks and recommendations for information security.* Available at http://www.enisa.europa.eu/activities/risk-management/files/deliverables/cloud-computing-risk-assessment) (accessed October 2014).

Erl T, Mahmood Z, and Puttini R (2013) *Cloud Computing: Concepts, Technology & Architecture.* Prentice-Hall.

European Commission (2015a) *Revisions of Medical Device Directives.* Available at http://ec.europa.eu/growth/sectors/medical-devices/regulatory-framework/revision/index_en.htm (accessed June 2015).

European Commission (2015b) *Summary report on the public consultation on the green paper on mobile health.* Available at ec.europa.eu/newsroom/dae/document.cfm?doc_id=8382 (accessed June 2015).

FDA (U.S. Food and Drug Administration) (2015) *Mobile medical applications: guidance for Industry and Food and Drug Administration Staff*, February 9. Available at http://www.fda.gov/downloads/MedicalDevices/./UCM263366.pdf (accessed June 2015).

FDA News Release (U.S. Food and Drug Administration) (2015) *FDA permits marketing of first system of mobile medical apps for continuous glucose monitoring.* Available at http://www.fda.gov/NewsEvents/Newsroom/PressAnnouncements/ucm431385.htm (accessed June 2015).

Foh KL (2012) *Integrating healthcare: the role and value of mobile operators in e-Health*, GSMA m-Health. Available at http://www.gsma.com/mobilefordevelopment/wp-content/uploads/2012/05/Role-and-Value-of-MNOs-in-eHealth1.pdf) (accessed October 2014).

Fox S and Duggan M (2012) *Mobile health 2012*, Pew Internet and American Life Project, Pew Research Centre, USA. Available at http://www.pewinternet.org/2012/11/08/mobile-health-2012/ (accessed October 2014).

Free C, Phillips G, Watson L, Galli L, Felix L, Edwards P, Patel V, and Haines A (2013) The effectiveness of mobile health technologies to improve health care service delivery processes: a systematic review and meta-analysis. *PLoS Medicine* 10(1):1–26.

Gartner (2014) *Gartner's 2014 hype cycle for emerging technologies maps the journey to digital business.* Available at http://www.gartner.com/newsroom/id/2819918 (accessed October 2014).

Gibbs S (2015) Mobile phones hacked: can the NSA and GCHQ listen to all our phone calls? *The Guardian*, February 20. Available at http://www.theguardian.com/technology/2015/

feb/20/mobile-phones-hacked-can-nsa-gchq-listen-to-our-phone-calls (accessed January 2016).

Grace L (2010) *Basics about cloud computing*. Available at http://www.sei.cmu.edu/library/abstracts/whitepapers/cloudcomputingbasics.cfm (accessed October 2010).

Green A (2011) When HIPAA applies to mobile applications. *Mobile Health News*. Available at http://mobihealthnews.com/11261/when-hipaa-applies-to-mobile-applications/ (accessed October 2014).

Groves P, Kayyali, B, Knott D, and Van Kuiken B (2014) *The big data revolution, accelerating value and innovation*, Report of the Centre for U.S. Health System Reform Business Technology Office, McKinesy & Company. Available at www.mckinsey.com (accessed October 2014).

GSMA (2012) *m-Health and the EU regulatory framework for medical devices: GSMA connected living report*. Available at www.gsma.com/connectedliving/wp-2015 (accessed October 2014).

Hansen, M, Miron-Shatz T, Lau AY, and Paton C (2014) Big data in science and healthcare: a review of recent literature and perspectives. *IMIA Yearbook of Medical Informatics*, Contribution of the IMIA Social Media Working Group, 21–26. Available at http://dx.doi.org/10.15265/IY-2014-0004.

Health Fabric. Available at ww.healthfabric.co.uk (accessed January 2016).

Helander E, Kaipainen K, Korhonen I, and Wansink B (2014) Factors related to sustained use of a free mobile app for dietary self-monitoring with photography and peer feedback: retrospective Cohort Study. *Journal of Medical Internet Research* 16(4):e109.

HIMSS (Healthcare Information and Management Systems Society) (2009) *Defining and testing EMR usability: principles and proposed methods of EMR usability evaluation and rating*, EHR Usability Task Force. Available at www.himss.org/files/HIMSSorg/content/files/himss_definingandtestingemrusability.pdf (accessed October 2014).

Holtz B and Lauckner C (2012) Diabetes management via mobile phones: a systematic review. *Telemedicine Journal and e-Health* 18(3):175–184.

Horowitz BT (2011) *Cloud computing brings challenges for healthcare data storage, privacy*. Available at http://www.eweek.com/c/a/Health-Care-IT/Cloud-Computing-Brings-Challenges-for-Health-Care-Data-Storage-Privacy-851608 (accessed October 2014).

IHTT (Institute for Health Technology Transformation) (2012) *Transforming health care through big data: strategies for leveraging big data in the health care industry*. Available at http://ihealthtran.com/wordpress/2013/03/iht%C2%B2-releases-big-data-research-reportdownload-today/ (accessed October 2014).

IMS Institute of Healthcare Informatics (2013) *Patient apps for improved healthcare: from novelty to mainstream*. Available at http://www.theimsinstitute.org/ (accessed June 2014).

Inspire: Health and Wellness Support Groups. Available at www.inspire.com (accessed January 2016).

Internetmedicine (2014) *Best free med apps*. Available at http://internetmedicine.com/best-free-medical-apps (accessed October 2014).

Istepanian RSH, Laxminarayan S, and Pattichis C (Eds.) (2006) *m-Health: Emerging Mobile Health Systems*. London: Springer.

Istepanian RSH, Sungoor A, and Nebel JC (2011) Comparative analysis of genomic signal processing for microarray data clustering. *IEEE Transactions on NanoBioscience* 10(4):225–238.

Istepanian RSH (2014) m-Health: a decade of evolution and impact on services and global health. *British Journal of Healthcare Management* 20(7):334–337.

Istepanian RSH (2015) Mobile applications (Apps) for diabetes management: efficacy issues and regulatory challenges. *The Lancet: Diabetes and Endocrinology* 3(12):921–923.

Iverson J and Eierman M (2014). *Learning mobile app development: a hands-on guide to building apps with iOS and Android*. Addison-Wesley.

Jain SH, Rosenblatt M, and Duke J (2014) Is big data the new frontier for academic-industry collaboration? *Journal of American Medical Association* 311(2):2171–2172.

Kaplan A and Haenlein M (2010) Users of the world, unite! The challenges and opportunities of social media. *Business Horizons* 53(1):59–68.

Kayyali B, Knott D, and Kuiken SV (2013) *The big-data revolution in US health care: accelerating value and innovation*. Chicago, IL: McKinsey & Co.

Keckley PH and Hoffmann M (2010) *Social networks in health care: communication, collaboration and insights*, Deloitte Centre for Health Solutions. Available at https://www.ucsf.edu/sites/default/files/legacy_files/US_CHS_2010SocialNetworks_070710.pdf (accessed October 2014).

Khan AN, Mat Kiah ML, Khanb SU, and Madanic SA (2012) Towards secure mobile cloud computing: a survey. *Future Generation Computer Systems* 29(5):1–22.

Kumar S, Nilsen W, Pavel M, and Srivastava M (2013) Mobile health: revolutionizing healthcare through trans-disciplinary research. *Computer* 28–35.

Laney D (2001) *3D data management: controlling data volume, velocity and variety*, META Group Inc., Publication File 949. Available at http://blogs.gartner.com/doug-laney/files/2012/01/ad949-3D-Data-Management-Controlling-Data-Volume-Velocity-and-Variety.pdf (accessed October 2014).

Liu C, Zhu Q, Holroydb K, and Seng E (2011) Status and trends of mobile-health applications for iOS devices: a developer's perspective. *Journal of System and Software* 84: 2022–2033.

Lohr H, Reza- Sadeghi A, and Winandy M (2010) Securing the e-Health cloud. *Proceedings of the 1st ACM International Health Informatics Symposium*, New York, pp. 220–229.

Malvey D and Slovensky DJ (2014) *m-Health: Transforming Healthcare*. New York: Springer.

Martínez-Pérez B, Torre-Díez I, and López-Coronado M (2013) Mobile health applications for the most prevalent conditions by the World Health Organization: review and analysis. *Journal of Medical Internet Research* 15(6):e120.

Marvel FA, Chase J, and Madruga M (2014) Ideas to IPhones: a 10 step framework for creating mobile medical applications with case report from Madruga and Marvel's black book app. *Journal of Mobile Technology in Medicine* 3(2):55–61.

Mayer-Schönberger V and Cukier K (2013) *Big Data: A Revolution That Will Transform How We Live, Work, and Think*. Boston, MA: Houghton Mifflin Harcourt, p. 242.

McDarby V, Hevey D, and Cody D (2015) An overview of the role of social network sites in the treatment of adolescent diabetes, *Diabetes Technology & Therapeutics* 17(4):291–294.

McGowan BS, Wasko M, Vartabedian BS, Miller RS, Freiherr DD, and Abdolrasulnia M (2012) Understanding the factors that influence the adoption and meaningful use of social media by physicians to share medical information. *Journal of Medical Internet Research* 14(5):e117.

MedData Group (2014) *Physician adoption of social media: physician opinions and challenges of using social media in the workplace*. MedData Point Report: Q2. Available at http://www

.meddatagroup.com/wp-content/uploads/MedDataGroup-Physician-Adoption-of-Social-Media-Q22014.pdf (accessed May 2015).

Mell P and Grance T (2011) *The NIST definition of cloud computing: recommendations of the National Institute of Standards and Technology (NIST)*, U.S. Department of Commerce Publication 800-145. Available at http://nvlpubs.nist.gov/nistpubs/Legacy/SP/nistspecial publication800-145.pdf (accessed May 2015).

Morris ME and Aguilera A (2012) Mobile, social, and wearable computing and the evolution of psychological practice. *Professional Psychology: Research and Practice* 43(6):622–626.

Murdoch TB and Detsky AS (2014) The inevitable application of big data to health care. *Journal of American Medical Association* 309(13):1351–1352.

National Learning Consortium (2014) *Patient-generated health data: fact sheet*, March 1–2. Available at www.healthit.gov/sites/default/files/patient_generated_data_factsheet.pdf (accessed October 2014).

NHS (National Health Service) (2014) *App development: an NHS guide for developing mobile healthcare apps*, NHS Innovations South East, UK. Available at http://innovationssoutheast.nhs.uk/files/4214/0075/4193/98533_NHS_INN_AppDevRoad.pdf (accessed May 2015).

O'Reilly T (2005) *What is Web 2.0?* O'Reilly Media. Available at http://oreilly.com/web2/archive/what-is-web-20.html (accessed September 2014).

Park A (2015) *The 25 best inventions of 2015*. Available at http://time.com/4115398/best-inventions-2015/ (accessed January 2016).

PatientKnowsBest. Available at www.patientsknowbest.com (accessed January 2016).

PatientsLikeMe. Available at www.patientslikeme.com (accessed January 2014).

Pearson JF, Brownstein CA, and Brownstein JS (2011) Potential for electronic health records and online social networking to redefine medical research. *Clinical Chemistry* 57(2): 196–204.

Perez S (2013) *Users have low tolerance for buggy apps: only 16% will try a failing app more than twice*. Available at http://techcrunch.com/2013/03/12/users-have-low-tolerance-for-buggy-apps-only-16-will-try-a-failing-app-more-than-twice/ (accessed October 2014).

PwC (PricewaterhouseCoopers) (2013) *How supportive is the regulatory framework for mobile health applications? m-Health Insight Report*. Available at http://www.pwc.com/en_GX/gx/healthcare/mhealth/mhealth-insights/assets/pwc-mhealth-how-supportive-is-the-regulatory-framework-for-mobile-health-applications.pdf) (accessed October 2014).

PwC (PricewaterhouseCoopers) (2014) *Top health industry issues for 2015: outline of a market emerges*. Available at http://www.pwc.com/en_US/us/health-industries/top-health-industry-issues/assets/pwc-hri-top-healthcare-issues-2015.pdf (accessed May 2015).

RAE (Royal Academy of Engineering and Academy of Medical Sciences) (2014) *Health apps: regulation and quality control: summary of a joint meeting*, London, November 19. Available at http://www.raeng.org.uk/publications/reports/health-apps-regulation-and-quality-control (accessed June 2015).

Raghupathi W and Raghupathi V (2014) Big data analytics in healthcare: promise and potential. *Health Information Science and Systems* 2(3):1–10. Available at http://www.hissjournal.com/content/2/1/3 (accessed October 2014).

Rajan RD (2013) *Wireless-enabled remote patient monitoring solutions*, Medical Design Technology, Qualcomm Life. Available at http://www.mdtmag.com/articles/2013/05/wireless-enabled-remote-patient-monitoring-solutions (accessed October 2014).

Rashidi P, Cook DJ, Holder LB, and Schmitter-Edgecombe M (2011) Discovering activities to recognize and track in a smart environment. *IEEE Transactions on Knowledge and Data Engineering* 23(4):527–539.

Research2guidance (2013) *Global mobile health market: trends and figures 2013–2017.* Available at http://research2guidance.com/the-market-for-mhealth-app-services-will-reach-26-billion-by-2017 (accessed October 2014).

Research2guidance (2014a) *Diabetes app market: how to leverage the full potential of diabetes app market.* Available at http://www.research2guidance.com/shop/index.php/downloadable/download/sample/sample_id/305/ (accessed October 2014).

Research2guidance (2014b) *m-Health app developer economics 2014: the state of the art of mHealth app publishing.* Available at http://mhealtheconomics.com/mhealth-developer-economics-report/ (accessed October 2014).

Rhodin M (2014) *IBM Watson takes on global challenges in healthcare, business and finance.* Available at Forbes.com,ww.forbes.com/sites/ibm/2014/10/09/ibm-watson-takes-on-global-challenges-in-healthcare-business-and-finance (accessed November 2014).

Rosenberg J and Mateos A (2011) *The Cloud at Your Service: The When, How and Why of Enterprise Cloud Computing.* Manning Publication.

Samson T (2013) *Nine top threats to cloud computing security*, Info World. Available at http://www.infoworld.com/article/2613560/cloud-security/9-top-threats-to-cloud-computing-security.html (accessed October 2014).

Scher DL (2013) *The five pitfalls of designing a medical app*, Medical Practice Insider. Available at http://www.medicalpracticeinsider.com/blog/technology/5-pitfalls-designing-medical-app (accessed Oct 2014).

Schneeweiss S (2014) Learning from big health care data. *New England Journal of Medicine* 370(23):2161–2163.

Selby JV, Beal AC, and Frank L (2012) The Patient-Centered Outcomes Research Institute (PCORI) national priorities for research and initial research agenda. *Journal of American Medical Association* 307(15):1583–1584.

Shapiro M, Johnston D, Wald J, and Mon D (2012) *Patient-generated health data*, White Paper, Office of Policy and Planning Office of the National Coordinator for Health Information Technology, RTI International, April. Available at www.healthit.gov/sites/default/files/rti_pghd_whitepaper_april_2012.pdf (accessed October 2014).

Skillings J (2013) *CNET: Apple reveals details of 50 billionth App Store download.* Available at http://www.cnet.com/news/apple-reveals-details-of-50-billionth-app-store-download (accessed October 2015).

Steinhubl SR, Muse ED, and Topol EJ (2013) Can mobile health technologies transform healthcare? *Journal of the American Medical Association* 310(22):2395–2396.

Swan M (2009) Emerging patient-driven health care models: an examination of health social networks, consumer personalized medicine and quantified self-tracking. *International Journal of Environment Research and Public Health* 6(2):492–525.

Van De Belt T, Engelen L, Berben S, and Schoonhoven L (2010) Definition of Health 2.0 and Medicine 2.0: a systematic review. *Journal of Medical Internet Research* 12(2):e18.

Waltz R (2014) *BlueStar, the first prescription-only app*, IEEE Spectrum. Available at http://spectrum.ieee.org/biomedical/devices/bluestar-the-first-prescriptiononly-app (accessed October 2014).

Wang X and Istepanian RSH (2005) A feasibility study of a personalised, Internet-based compliance system for chronic disease management. *Telemedicine and e-Health Journal* 11(5):559–566.

Weber GM, Mandl KD, and Kohane IS (2014) Finding the missing link for big biomedical data. *Journal of American Medical Association* 311(14):2479–2480.

WebMD. Available at WebMD.com (accessed January 2016).

WEGO Health. Available at WEGOhealth.com. (accessed January 2016).

Wicks P, Massagli M, Frost J, Brownstein C, Okun S, and Vaughan T (2010) Sharing health data for better outcomes on PatientsLikeMe. *Journal of Medical Internet Research* 12(2):e19.

World of Marketing. Available at http://worldofdtcmarketing.com/millennials-as-consumers-of-healthcare/health-information-online (accessed January 2016).

Zhang C, Sun J, Zhu X, and Fang Y (2010) Privacy and security for online social networks: challenges and opportunities. *IEEE Network* 24: 13–18.

4

m-HEALTH AND MOBILE COMMUNICATION SYSTEMS

> In the case of the cell phone, there's a chance to go beyond that and actually be there with the patient, there in the clinic, which might not be staffed with a fully trained doctor.
>
> Bill Gates, Bill and Melinda Gates Foundation, 2010

4.1 INTRODUCTION

Along with sensors and computing, wireless communication technologies with different configurations are essential for the realization of any m-Health model. Today, billions of people worldwide stare virtually every minute at their mobile phones, but this predominant view of "cellular communication" technologies as the central hub of m-Health is inadequately justified.

Most of the clinical evidence from smartphone-centric m-Health applications and services is either uncertain or ambiguous at best, as we will illustrate in Chapter 5. This status strengthens the argument for an alternative framework for m-Health outside this communications box.

Although the proliferation of smartphones cannot be underestimated, with nearly 85% of people living within range of a cellular base station, there are many other wireless communication and networking technologies used for various m-Health models that are either unrealized or not fully exploited. Each of these technologies has different configurations, wireless capabilities, mobility functions, and operating

m-Health: Fundamentals and Applications, First Edition. Robert S. H. Istepanian and Bryan Woodward.
© 2017 by The Institute of Electrical and Electronics Engineers, Inc. Published 2017 by John Wiley & Sons, Inc.

modes. Typically, m-Health systems are centered on one of these wireless communications standards.

Current developments of wireless technologies are now focused on the 5th Generation (5G) of mobile networks. These advances, together with different smart architectures, will potentially reshape the communications and networking "building blocks" of future m-Health.

To illustrate this notion, Table 4.1 shows categories of mobile networking technologies and standards applicable for current m-Health systems and applications. This table also shows the technical specifications in terms of operational range, connectivity, data rates, and so on.

In Chapter 2, we looked at some of the relevant wireless networking technologies associated with wearable sensing for m-Health (WPANs, WBANs). In this chapter, we expand this discussion and describe the wider communications and networking aspects associated with m-Health systems. The detailed operational modalities of each of these networks are defined by a set of associated standards that govern their core functional configurations. The technical descriptions presented here are intended for general readership. For more technical information, particularly on communications and networking aspects that are beyond the scope of this book, readers are referred to the specialized texts referenced in each section. As shown in Table 4.1, the current wireless communication and networking systems applicable for any m-Health system are generally categorized into one of the following:

Wide Area Networks (WAN): These are also referred to as Wireless Wide Area Networks (WWAN). They mainly consist of digital cellular networks and represent the first and most widely used networks for m-Health communications. Historically, the early beginnings of m-Health applications were developed around the second generation of cellular systems (2G) in the 1990s. This was followed by the third generation (3G), leading to the current fourth generation (4G). Efforts are now underway to develop the fifth generation (5G) of these networks. The numerical terms 1G, 2G, 3G, 4G, and now 5G, are the names given for the successive generations of the global standards introduced since the 1980s, governing communications protocols, core architectures, radio network functionalities, layers, and operational modalities of these networks.

Wireless Metropolitan Area Networks (WMAN): These networks generally refer to a combination of IEEE networking standards that govern their operating functionalities. They include IEEE 802.16, or as it is widely known, the Worldwide Interoperability for Microwave Access (WiMAX), and the lesser known IEEE 802.20 (Mobile Broadband Wireless Access), IEEE 802.21, and IEEE 802.22 standards. The main characteristics of these networks are the provision of broadband wireless data access in metropolitan areas, with peak data rates of up to 70 Mbps and distances of up to 50 km, with practical data rates of at least 1 Mbps. These networks operate on both licensed and license-exempt spectrums. The most widely known standard, the IEEE 802.16, is developed with different formats: the IEEE 802.16d for fixed access, 802.16e-2005 for

TABLE 4.1 Mobile Networks and Wireless Technologies for m-Health

Mobile network	Wide area networks (WAN) Wireless Cellular Networks	Metropolitan area networks (MAN) WiMAX	Local area networks (WLAN) Wi-Fi	Personal area networks (PAN)
Technology/standard	GSM (2G), GPRS, (2.5G), UMTS (3G), HXSPA (3.5G) LTE, LTE-A (4G), 5G	IEEE 802.16, IEEE802.16a, IEEE802.16e, WiMAX-2	802.11b, 802.11g, 802.11n, 802.11ac	Bluetooth (802.15.1) and Bluetooth Low Energy (BLE) ZigBee (802.15.4) WBAN (802.15.6) and others
Data rate	From low ~10 kbps (2G, 2.5G) to high data rate (~100 Mbps (4G) Expected above 1.2 Gbps (5G)	Very high data rates (up to 300 Mbps with the new standards)	High data rates (~11 to 54 Mbps) with much higher rates (~max 1 Gbps) for 80211ac	BT: (~1–2 Mbps)/Operating Frequency: 2.4 GHz. ZigBee: 20–250 kps Operating Frequency: 868 MHz, 915 MHz, and 2.4G Hz WBAN: ~10 kbps to 10 Mbps
Range	Long (global) range Global connectivity and Internet of things access (5G)	Medium or metropolitan range (few to up to 50 km)	Short range (10–100 m)	Very short range BT: (~1–100 m) BT- Radio: Class 3: 1 m; class 2: 10 m; class 1: 100 m ZigBee: (~75 m) WBAN: (~1–2 m)
Examples of m-Health applications	Smartphone Apps. health and wellness monitoring, mobile disease management, real-time diagnostics, global health	Ambulance and emergency care scenarios, remote real-time diagnostics, and healthcare services in rural areas	Home care, assisted living, hospital management and communications	Wellness monitoring, wearables, patient and disease monitoring, in-body and on-body medical application (WBAN)

mobile stations and access, and IEEE 802.16m (Mobile WiMAX-2). These WiMAX standards have been designed to operate at frequencies between 2 and 66 GHz, both in licensed and unlicensed ranges. The 2.3, 2.5, and 3.5 GHz bands are the most common. This standard, together with the Third Generation Partnership Program Long Term Evolution (3GPPLTE), is widely known as the fourth generation (4G) system. The fundamental communications principles between WiMAX and cellular networks are basically similar. However, packet-switched WiMAX networks are more applicable to Internet applications, which are suitable for data access services that require communication over metropolitan ranges. For m-Health, WiMAX networks can provide a useful communications infrastructure for services and applications that demand Internet Protocol (IP) data access, with higher bandwidths and quality of service. These include, for example, real-time remote medical diagnostics in rural areas, critical traveling ambulance data, and interactive medical, educational, and consultation referrals.

Wireless Local Area Networks (WLAN): WLAN, or Wi-Fi networks as they are known, have become widely used everywhere, mainly because of the rapid spread of this technology for smartphones and tablets. These networks are defined by IEEE 802.11 (Wi-Fi) standards with its different variations (a, b, g, and n). They usually operate in the unlicensed 2.4 and 5 GHz radio frequency bands. These networks are the most popular wireless technology for any broadband Internet access, and their operating range (approximately 20–60 m) depends on the specific 802.11 version and frequency used. Their theoretical peak data rates range from 54 Mbps (defined by earlier 802.11a standards) up to 600 Mbps for the latest standard (802.11n). For m-Health, these networks are mainly used for any indoor wireless services and applications, such as in-hospital wards and home-monitoring care services. These networks can also be ubiquitously coupled with WPANs for medical sensing and monitoring applications.

Wireless Personal Area Networks (WPAN): These networking technologies operate over short ranges, typically in the vicinity of the body (Kim and Cho, 2009). They are closely related to a variety of medical and nonmedical applications and operate in the unlicensed 2.4 GHz spectrum. Examples include Bluetooth (IEEE 802.15.1), ZigBee (IEEE 802.15.4), WBAN (IEEE 802.15.6, 2012), and ultra-wide band (UWB). The ZigBee standard can also operate at reduced data rates in the 868 and 915 MHz bands. Bluetooth is the most popular option, and is usually designed to operate over approximately 10 m range; the latest Bluetooth low-energy (BLE) version has a maximum range of 100 m, while ZigBee radio devices can achieve a maximum range of 75 m. A WPAN for m-Health is suitable for short-range, personalized monitoring, tagging, and other medical sensing applications.

Most of these technologies are evolving rapidly, and are continuously reshaping different mobile communication architectures with increasing ubiquity, and operating across heterogeneous networks and seamless connectivity links.

In this chapter, we look at each of these technologies and describe their role in shaping the future m-Health. We also describe the emerging paradigm of Internet of things (IoT) connectivity and discuss the enabling machine-to-machine (M2M) communications concept.

4.2 WIRELESS COMMUNICATIONS FOR m-HEALTH: FROM "UNWIRED HEALTH" TO "4G-HEALTH"

The early development of simple prototypes of mobile health systems for wireless biomedical data transmission started in the early 1990s with the introduction of the Global System for Mobile Communications (GSM) (Istepanian et al., 1997, 1999; Istepanian, 1998; Woodward and Vyas, 2000). This was the era of "pre-m-Health" (unwired health) based on 2G/2.5G cellular systems. Systems developed after about 2003 in the era of 3G/3.5G cellular systems (Istepanian et al., 2006). This evolution is continuing with the current "4G-Health" systems associated with the introduction of 4G mobile communication networks (Istepanian and Zhang, 2012). Today, mobile telecommunications offer unprecedented opportunities for mobile services to billions of people worldwide; This is an essential part of the global economy (GSMA Intelligence, 2015).

Figure 4.1 shows the evolution of m-Health corresponding to the generational evolution of mobile communication systems. The figure also illustrates an

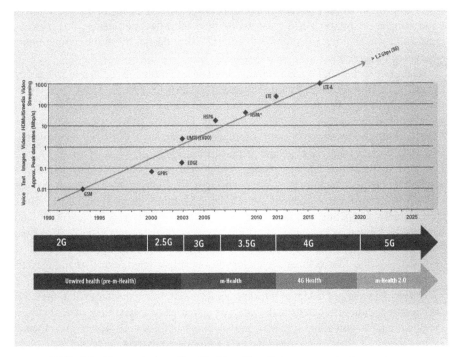

FIG. 4.1   The evolution of m-Health with cellular communication systems.

approximate 10-fold increase in peak data rates every 5 years, leading to expected peak rates above 1 Gbps (in high-mobility scenarios) for the forthcoming 5G systems by 2020. This would benefit developing countries immeasurably, but it is unlikely on the same timescale (Gatherer, 2015).

The massive increase in peak data rates has enabled the fast evolution of m-Health over the last decade (Fiordelli, 2013). This evolved from the transmission of text and single-channel wireless telemedical data using GSM and GPRS to the present wide range of smart healthcare applications and mobile-connected services. This process is expected to continue with the introduction of the next generation of m-Health (m-Health 2.0), leading to a new and uncharted era.

Next we present, in historical sequence, the evolution of these wireless cellular network technologies since the early 1990s, and chart their development from their first digital (2G) systems (1G corresponds to analogue networks) to the future 5G. Figure 4.2 shows this evolution, together with the associated technologies and operational modalities, and we discuss it in terms of its role in shaping the development of mobile health. This evolution is ongoing, with the earlier generation of cellular networks (e.g., GSM) still in worldwide use today.

4.2.1   Global System for Mobile Communications

Figure 4.2 shows that GSM is the standard-bearer of 2G that was developed in the 1980s and completed in the late 1990s (Eberspächer et al., 2009). It is estimated that over 80% of the world population still uses GSM when placing wireless calls, with approximately 4 billion GSM users worldwide (GSMA Intelligence, 2014a). This generation was the first to combine the digital voice capability of the mobile phone with data transmission.

2G mobile technologies are generally based on either time-division multiple access (TDMA), or code-division multiple access (CDMA). In general, TDMA allows for the division of signals into time slots, while CDMA allocates each user a special code to communicate over a multiplexed physical channel. For example, GSM used TDMA initially to support eight users per carrier in a 200 kHz band with a 900 MHz carrier frequency. At the same time, CDMA, also known as IS-95, was launched in the United States. Along with GSM, a number of other 2G digital standards emerged, such as Pan-American Digital Advanced Mobile Phone System (D-AMPS), and Direct Sequence Code-Division Multiple Access (DSCDMA). These were usually competing and incompatible standards, collectively representing what came to be known as 2G wireless cellular network technologies. The IS-95 system had an evolved path with the Pan-American CDMA2000 system.

GSM was the first cellular technology to establish international roaming. This enabled subscribers to use their mobile phones in many countries worldwide. The initial maximum data rate of GSM was 9.6 kbps per time slot, but with later evolution (2.5G) such as for circuit-switched data, the rate was increased to a maximum of 43.2 kbps in the downlink (downloading) channel, and 14.4 kbps in the uplink (uploading) channel.

In general, 2G had much lower data rates compared to today's 4G systems; these were insufficient for advanced video and wireless Internet applications, but provided

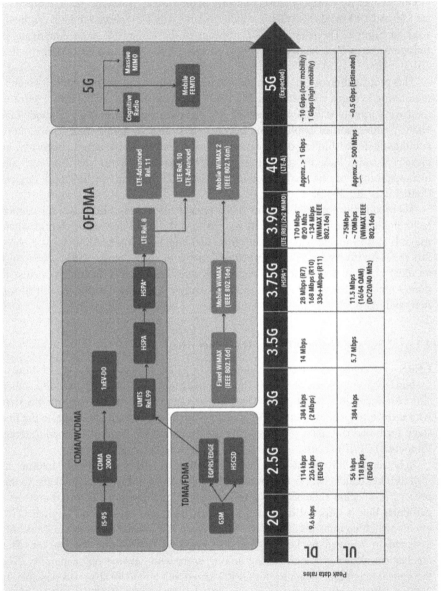

FIG. 4.2 The evolution of cellular communication systems and their access technologies.

the first facility for sending text messages using the short message service (SMS). This later facilitated the first m-Health studies on the use of SMS for clinical applications. In Europe, the GSM carrier frequencies are 900 and 1800 MHz (1.8 GHz), enabling mobile phones to be used in most countries. In the United States, the usual frequencies are 850 and 1900 MHz (1.9 GHz), which explains why U.S. phones did not originally work in Europe. The solution for Europeans in the United States or Americans in Europe was to use a triband phone, which works at 900, 1800, and 1900 MHz (Sauter, 2010).

This proliferation of GSM technologies triggered the early work of "pre-m-Health" monitoring systems (Istepanian et al., 1997, 1999; Istepanian, 1998, 2000). Most of these early prototypes suffered from the obvious shortcomings that they precluded the wider proliferation of applications beyond physiological monitoring. These short-comings included limited data rates, high end-to-end network delays, security and privacy issues, lack of ubiquitous wireless connectivity to the Internet, and incompatibility of the early mobile phones with medical sensors and devices (Pattichis et al., 2002).

There are still potential benefits to be gained from using GSM and 2G cellular technologies for m-Health applications, particularly in the developing world where usage is still available with cheaper tariffs and wider coverage (Ahmed and Kohno, 2013). Cost-effective tariffs combined with network access makes GSM technology an ideal platform for m-Health applications and services that are much needed in poorer regions. These include, for example, simple health monitoring, health educa-tion, and alerts via text messaging.

4.2.2 2.5G and General Packet Radio Services (GPRS)

GPRS was enhanced, or 2.5G extension, for GSM networks. This generation of mobile technologies was obviously the intermediate stage between 2G and 3G, which was introduced to support packet-data wireless communications. This interim step, with its bizarre label, was the GSM operator's first stage toward 3G, and also the first stage for broadband data services using the Transmission Control Protocol/Internet Protocol (TCP/IP) access, enabling the "always-on" Internet access.

In lay terms, GPRS is based on packet switching technology used in the Internet, as well as other data networks that bundle data into blocks of bits called packets. In this process, a specific time slot in a time-division multiplexing (TDM) frame on a particular link is thus not dedicated to a particular user, as with the original GSM circuit-switching transmission. In the case of packet-switched CDMA, a code is only assigned to a given packet for the time required to transmit that packet. This mechanism allows efficient sharing among users with "always-on" capability. The theoretical maximum data rates for GPRS varied, but up to 140 kbps was possible. In 2003, a better performance technology was developed with the introduction of the "Enhanced Data rates for GSM Evolution" (EDGE), designed to deliver multimedia applications such as streaming video and audio to mobile phones, with a peak data rate of 300 kbps using 59.2 kbps per time slot (Schwartz, 2005; Muzic and Opatic, 2009; Sauter, 2010). The CDMA2001x (or 1xRTT) technology was the 2.5G extension, or evolution of CDMA. As shown in Fig. 4.2, GPRS mobile technology supported peak download data rates of up to 115 kbps, with average rates of 40–50 kbps, which was comparable to other 2.5G technologies (Sauter, 2010; Schwartz, 2005).

GPRS was considered the "best effort" service technology, implying that the network's data rates and latency (delay) depended on the number of users sharing the network service concurrently. EDGE, with a bandwidth of up to 236 kbps, could handle four times as much traffic as GPRS with reduced latency. These technologies are still widely used in different regions worldwide (Halonen et al., 2003; Sauter, 2010).

For m-Health, 2.5G technologies provided the capability of sending and receiving larger amounts of data in reasonably short times and with adequate rates using the Multimedia Messaging Service (MMS) and web access facilities. Applications of 2.5G-based m-Health systems were limited and remain largely in the health data monitoring domain (Woodward et al., 2006; Jones et al., 2006; Verulkar and Limkar, 2012).

4.2.3 Universal Mobile Telecommunications Service (UMTS)

UMTS represents the third generation of mobile communication systems. Its core network elements represent the successor to the GPRS and EDGE technologies, although UMTS uses a different air interface, with less complex development and installation tasks for the migration process from 2G to 3G.

3G introduced the era of mobile broadband services, with faster speeds and Internet access capabilities. Figure 4.2 shows that the evolution of UMTS benefited from the use of Wideband CDMA (W-CDMA) technology that evolved from the earlier IS-95 (CDMA One) standard for radio transmission. As also shown, the initial UMTS network releases provided significant improvement in their capabilities compared to the 2G and 2.5G services, with peak data rates of up to 2 Mbps in the downlink and 384 kbps in the uplink directions. These relatively high data rates allowed for faster Internet surfing, video downloads, and other multimedia services by a larger number of cell users. This resulted in the development of the High Speed Packet Access (HSPA) technology that optimized the data channel for W-CDMA, and provided increased data transfer rates for mobile broadband users in both the downlink and the uplink channels. The HSPA was introduced by the Third Generation Partnership Project (3GPP) and comprises a combination of two technologies, High Speed Downlink Packet Access (HSDPA) and High Speed Uplink Packet Access (HSUPA).

Technically, the 3G HSPA specification is an enhancement of W-CDMA technology, with peak data rates up to 14 Mbps for the downlink and 5.7 Mbps for the uplink. In parallel to this process, the Evolution-Data Optimized (EV-DO) standard was developed for optimizing the data channel for the CDMA2000 standard, with peak data rates of 3.1–14.7 Mbps. Further releases in 2008 and 2010 introduced the HSPA Plus (HSPA+), which represents the protocols and technology thresholds between 3G and 4G. HSPA+, or Evolved High Speed Packet Access, is an enhanced version of the HSUPA and HSDPA 3G standards.

In 2014, the Global Spéciale Mobile Association (GSMA) reported that all W-CDMA operators had launched HSPA, with 547 networks providing access to more than 1.5 billion subscribers worldwide. Two-thirds of these networks had upgraded to HSPA+, and 159 of them supported Dual Cell HSPA+, enabling downlink peak data rates of more than 168 Mbps (Brydon, 2014). GSMA has also reported on mobile health devices (GSMA, 2011, 2012)

The introduction of 3G (UMTS) technologies represented a turning point for m-Health, leading to the development of diverse new mobile health systems and applications that were not possible with earlier generations. These systems allowed seamless and ubiquitous Internet access and high-speed medical data transmission for many simultaneous users of 3G terminals with mobile Internet access capabilities. Some early examples of mobile broadband m-Health applications included remote real-time ultrasound diagnostics and 3G-based medical video streaming applications (Garawi et al., 2006).

Until the introduction of smartphones in 2007, most m-Health systems had limitations that may be summarized as follows (Istepanian et al., 2004):

- There was a lack of integrated "m-Health-on-demand" connectivity, with different mobile telecommunication standards aimed at different healthcare services. Thus, there was operational incompatibility between mobile technologies, mobile terminals, and their relevant protocols.
- The complexity of the healthcare sector and the organizational changes required to convince decision makers in industry took no account of the benefits of m-Health services, especially for improved patient health delivery.
- Reimbursement methods and payment models for m-Health services were not yet fully developed and understood.
- There were no interoperability standards between existing m-Health systems and e-Health information systems, for example, health records, and referral and ordering systems.

Subsequent developments, in parallel with advances in medical sensing and web access technologies, signaled a transformation of m-Health to new levels of adoption and acceptability that were not previously feasible. During the period 2003–2007, m-Health was mainly in the "awareness" era, to be transformed to the "mainstream and business market" era after 2007. This "smartphone-centric" m-Health era mitigated some of the earlier technical challenges, but obstacles remained, especially in terms of clinical effectiveness, security, and privacy concerns.

This unprecedented interest in mobile health was mainly driven by its potential as a major "vertical market" opportunity. Global telecommunication and healthcare device industries therefore embraced m-Health. The mainstream healthcare and medical services sectors were also beginning to embrace the concept, albeit in a less cautionary approach. These and other relevant issues are discussed in more detail in Chapters 5 and 6.

4.2.4   4G: Long Term Evolution (LTE), LTE Advanced (LTE-A), and 4G-Health

The main drivers behind the development of 4G communications systems can be summarized as follows:

- There was increasing demand for mobile technologies, coupled with an unprecedented exponential increase in mobile broadband data traffic, which led to affordable high-speed data with cost-effective access.
- There was a demand for seamless, ubiquitous, and heterogeneous network access across different wireless network technologies, not only via cellular links, in every scenario, for example, travel, home, workplace, airports, and hotspots. This push was also driven by other competing standards, for example, WiMAX and Wi-Fi, and the increasing demand for seamless broadband wireless access using noncellular (3GPP) technologies.
- There was increasing demand for more smart mobile devices with more powerful capabilities. These include more user-friendliness, more video streaming, better handling, access, and large-volume storage of Internet-based data, for example, the cloud, with multifunctional capabilities for mobile users.

These drivers, combined with more mobile business opportunities, including the mobile healthcare sector, led to global efforts to develop the LTE mobile system; this is widely known as the pre-4G generation, or (3.9G), as shown in Fig. 4.2. In general, LTE is widely considered as the first truly global standard for mobile broadband services. It was developed by the global alliance of the 3GPP, under the auspices of the International Telecommunications Union (ITU), the international body responsible for these standards. The LTE network standard was introduced with 3GPP Release 8 (LTE-R8) in 2008, followed by the launch of the first LTE commercial network in Sweden in 2009. As shown in Fig. 4.2, LTE is marketed as one of the initial 4G candidates. This was based on the GSM/EDGE and UMTS/HSPA network evolution with a peak data rate of 170 Mbps in the downlink and 75 Mbps in the uplink channels. The other 4G standards candidate is the mobile version of the WiMAX standard (IEEE 802.16e).

Table 4.2 shows a simple comparison of the main technical features of LTE and Mobile WiMAX (Korhonen, 2014; Riegel et al., 2009). In general, LTE has the following characteristics that differentiate it from the earlier 3GPP legacy standards (Korhonen, 2014):

- High data rates of over 300 Mbps in the downlink, and over 50 Mbps in the uplink.
- Low latency, the time required to connect to the network, which is as low as 10 ms compared to 100 ms or more for 3G.
- Seamless services in higher mobility environments, allowing LTE networks to connect seamlessly with other networks, such as GSM and W-CDMA.

TABLE 4.2   Comparison of the Technical Features of LTE and Mobile WiMAX (IEEE 802.16e) Communication Technologies

Network Specification	LTE	Mobile WiMAX (IEEE 802.16e)
Access technologies (Uplink-UL and Downlink-DL)	OFDMA (DL), SC-FDMA(UL)	OFDMA (DL and UL)
Frequency bands	Existing 800, 900, 1800, 1900 MHz and also new frequency bands within the range of 800 MHz to 2.62 GHz	2–11 GHz
Antenna scheme	MIMO	MIMO
Peak data rates (UL and DL)	100–326.4 Mbps (DL) 50–86.4 Mbps (UL)	75 Mbps (DL) 25 Mbps(UL)
Range (cell radius)	5 km	~8.4–20.7 km (depending on the BW used (5,10 MHz–3.5, 7 MHz)
Users (cell capacity)	>200 users @ 5 MHz and >400 for larger BW	100–200 users
Mobility (speed handover)	Up to 350 km/h	Up to 120 km/h

- Higher quality of service, especially for video and other content-rich applications.
- Simple network architecture with higher capacity, thus providing cost-effective and affordable mobile access.

Besides, the following are some of the main technical features that distinguished LTE and WiMAX from earlier 3GPP technologies (Korhonen, 2014):

- LTE and WiMAX both use the Orthogonal Frequency-Division Multiple Access (OFDMA) for their access technology. In lay terms, this is a modulation technique used for transmitting large amounts of digital data over a channel by splitting the signal into multiple smaller subsignals that are then transmitted simultaneously at different frequencies. Thus, the amount of cross-talk attributed to the signal transmissions is reduced. This method is now widely used in most of the latest wireless technology standards. LTE networks use the OFDMA technology in the downlink channel for efficient multiple access, and for countering the multipath frequency-selective fading problems that are usually encountered in indoor environments.
- Both LTE and WiMAX use advanced multiple-input multiple-output (MIMO) antenna technology, which allows multiple antennas at both the source (transmitter) and the destination (receiver) ends to minimize errors and optimize data speed.

- LTE also uses the hybrid automatic repeat request (HARQ) method. This is a processing technique used in modern data communication systems that ensures minimum packet loss and better reliability and transmission performance.
- LTE also uses a flat-IP architecture for the core network structure to provide for better access and seamless Internet services.

In general, LTE network technology provides much improved coverage, download speeds and browsing capabilities, better video streaming, and faster functionalities. It also provides better mobile connection at higher speeds (e.g., in trains and cars), with improved access in indoor as well as outdoor environments, for example, in hospitals and clinics, thereby embracing m-Health applications.

The introduction of 4G has been in response to the major proliferation of mobile devices and wireless connectivity, with an estimated projection of 7 trillion network-connected wireless devices to serve 7 billion people by 2017 (Sørensen and Skouby, 2009). This prediction is in line with similar predictions of the increase in mobile data and video content by 2016 at about five times the current demand rates, with a doubling of mobile traffic each year afterward (Cisco VNI, 2012).

These statistics fuel further demands for much higher data rates, with more efficient access and smarter network capabilities than are already available. Hence, the launch of the LTE-A mobile networks that was announced by the 3GPP as Release 10 (R10) (Wannstrom, 2013). This release brought a much improved network capacity, faster data rates, and better coverage. LTE-A cellular networks are termed as a "true 4G," offering an increased theoretical peak data rate of up to 3 Gbps on the downlink and 1.5 Gbps on the uplink, combined with higher spectral efficiency and support available to a larger number of simultaneously active users (Wannstrom, 2013). More realistically, practical data rates are currently around 1 Gbps for the downlink and 0.5 Gbps for the uplink. By comparison with LTE networks, with peak downlink rates of around 170 Mbps and uplink rates of 75 Mbps, this constitutes a major leap for many smart m-Health applications.

Figure 4.2 shows that LTE-A technology also includes new transmission protocol capabilities, with multiple antenna schemes that enable smoother transfer between the cells at higher mobility speeds, reaching up to 350 kph with increased throughput at network cell edges. This is also complemented by more bits per second per hertz of the spectrum, resulting in possible lower tariffs and higher network capacity, that is, the amount of mobile traffic the network can handle at one time with different users (Shen et al., 2012). Furthermore, LTE-A, unlike LTE, meets the requirements of the ITU's IMT-Advanced Standard (ITU, 2013).

In summary, current 4G provides all Internet access capabilities, facilitating voice data with fast, mobile multimedia data streaming services "anytime, anywhere." These combine the capabilities of the 4G networking technologies, namely, the LTE-A and mobile WiMAX. Most global commercial network operators initiated the deployment phases of 4G worldwide in late 2014, commensurate with the launch of newer smartphone models operating with LTE-A communications.

Figure 4.2 shows further that the introduction of 4G launched the era of smart "4G-Health." The concept of 4G-Health was defined as "the evolution of m-Health towards targeted, personalized medical systems with adaptable functionalities and compatibility with future 4G communications and network technologies" (Istepanian and Zhang, 2012). The introduction of 4G mobile broadband technologies has significantly benefited developments of 4G-Health systems. These developments are exemplified by the following:

- The proliferation of a variety of personalized smart healthcare services and applications (Apps) and their seamless integration with wearable devices, especially for wellness tracking and health monitoring.
- The introduction of interactive mobile broadband services and applications, such as health-focused social networking, real-time mobile remote diagnostics, and mobile health educational services.
- The support of "big mobile health data" and cloud processing capabilities.

These developments can potentially bring major benefits to different services such as emergency care, patient monitoring, home care, follow-up services, and intra-hospital and medical information management systems. However, these developments also bring additional technical challenges:

- *The Importance of the Quality of Service (QoS) and Quality of Experience (QoE):* These services typically define the "best effort" requirements of 4G mobile systems to provide acceptable mobile user services and experiences obtained, or expected to be obtained, from mobile health networks. However, these issues are not fully utilized, or robustly clinically tested from the m-Health perspective. Technically, 4G-Health systems that utilize the full IP packet switching capacity of the current LTE-A networks need also to satisfy the clinical service model requirements of health applications. Although 4G can provide higher data rates, larger network capacity, and much smaller network delays suitable for different seamless mobile broadband services, these specifications do not, for example, include the specific clinical requirements of more demanding health care services and applications (Debono et al., 2013; Panayides et al., 2013). These and other applications, in order to provide reliable and clinically acceptable 4G-Health services, must comply with specific clinical quality and efficacy issues. These are typically more stringent than the typical quality and user experience requirements specified by the network telecommunications operators. These requirements must also take account of all the users and contributing stakeholders involved in these healthcare services, for example, patients, specialists, and healthcare settings. They also need to comply with a set of both clinical and nonclinical targeted measures defined by acceptable medical quality of service (m-QoS) and medical quality of experience (m-QoE) metrics that can satisfy these requirements (Istepanian et al., 2009a, 2013).
- *The Development of New Security and Privacy Models:* This is particularly important in the personalized wellness domain. The phenomenal growth of big health data and secure cloud access connectivity are not yet fully realized,

particularly for 4G-Health. The current cybersecurity concerns add to the importance of this challenge, especially from the healthcare data privacy perspective.

- *The Relevant Challenges of 4G-Health Systems:* These are applicable to systems being developed around the emerging machine-to-machine communications, particularly in terms of their interoperability, standards, and conformity with future mobile health applications. These challenges are also not yet fully understood, or properly defined.

- *The Development of Smarter m-Health Apps:* The Apps are those that utilize both 4G smartphone functionalities and wearable sensing devices. There is thus a need for a balance between the provision of better health outcomes and Apps-related issues such as effectiveness, efficacy, and user compliance. These issues, explained in Chapters 3 and 5, are open to debate by clinicians.

4.2.5 5G Cellular Communication Systems and m-Health 2.0

Growth of mobile data traffic was about 69% during 2014, reaching 2.5 exabytes (10^{18}) per month by the end of the year, compared with 1.5 exabytes per month at the end of 2013. This massive increase in mobile data traffic in a year was nearly 30 times the size of the entire global Internet in 2000, with 497 million mobile devices and connections added in 2014 (Cisco VNI, 2015). These staggering statistics of mobile data usage can only be accommodated with new advances in mobile communications and networking technologies that can cope with this global demand. The ongoing developments in 5G, once deployed in post-2020, will be capable of providing very high mobile broadband speeds and information access, with data sharing anywhere, anytime, for anyone. Its capability will go far beyond that of 4G.

Ongoing 5G development is currently focused on introducing efficient new wireless transmission technologies and network solutions, such as Heterogeneous Networks (HetNet), multiple-input multiple-output, and millimeter wave (mmWave) technologies that can meet the 5G requirements of very high data rates, ultralow latency, large numbers of connected devices, high levels of energy efficiency, and guaranteed quality of service and experience. According to the GSMA, there are two views of 5G (GSMA Intelligence, 2014b):

- The first view is the "hyperconnected vision," where mobile operators would create a blend of preexisting technologies covering 2G, 3G, 4G, Wi-Fi, and others to allow higher coverage and availability and higher network density in terms of cells and devices, with the key differentiator being greater connectivity as an enabler for M2M services and the IoT.

- The second view is the "next-generation radio access technology," which represents the more traditional "generation-defining" view, with specific targets for data rates and latency.

From the European perspective, the vision of 5G, as presented by their 5G public–private partnership (5GPP) of leading European telecommunications and other mobile industries in partnership with the European Union, was summarized as follows (European Commission 5GPPP, 2015):

- *Key Drivers for 5G:* These include not only the evolution of mobile broadband networks but also completely new heterogeneous networks and service capabilities.
- *Design Principles of 5G:* Infrastructure should be flexible and rapidly adapt to a broad range of requirements, and designed to be a sustainable and scalable technology.
- *Key Technological Components of 5G:* Networks should encompass optical, cellular, and satellite solutions, which could rely on emerging technologies such as software-defined networking (SDN), network functions virtualization (NFV), mobile edge computing (MEC), and fog computing (FC).
- *Spectrum Considerations of 5G:* Access networks will require "hundreds of MHz up to several GHz" to be provided at a very high overall system capacity, and higher carrier frequencies above 6 GHz need to be considered. Maintaining a stable and predictable regulatory and spectrum management environment is critical for long-term investments.

Within the ongoing 5G activities, several leading global and European research projects and initiatives have been undertaken. These include, for example, the following European 5G flagship projects: METIS (2014), 5GNOW (2014), MiWEBA (2014), and CREW (2014). The detailed technical aims and objectives of these are beyond the scope of this book and can be viewed on the projects' web links listed in the references. However, for completeness, according to the METIS (Mobile and Wireless2 Communications Enablers for the 2020 Information Society) project, the main challenges for 5G systems are driven by three influential factors (Osseiran et al., 2013):

- *An Anticipated Avalanche of Mobile Traffic:* Volume expansion of mobile broadband services and usage is predicted to increase 1000-fold in the next 10 years. This includes video streaming, with high quality of service and user experience requirements, which are important driving factors for future m-Health services.
- *Massive Growth in "Communicating Machines":* An estimated 50 billion wireless-connected devices are expected by 2020. Further outcomes will be 10–100 times higher data rates and 10 times longer battery lifetimes (5 years or more) for these devices, minimum signaling overheads, low device cost, and support for efficient transmission of small payloads with fast setup and low latency (Osseiran, 2014; Fallgren and Timus, 2013).
- *Large Diversity of User Requirements:* These will be due to, for example, an increasing level of communicating machines, such as smart vehicles.

The general technical requirements for 5G mobile communications are shown in Table 4.3 (Osseiran, 2014; Osseiran et al., 2014). From the 3GPP projections, the

TABLE 4.3   A General Selection of 5G Technical Requirements

Requirements	Desired Technical Parameters
Data rates	1–10 Gbps
Capacity and data volume	9 Gbyte (in busy periods) and 500 Gbyte/month/subscriber
Spectrum	Higher frequencies and flexibility
Energy	~10% of today's consumption
Latency reduction	Less than 5 ms
Coverage	>20 dB of LTE (e.g., sensors)
Battery life	~10 years
Connected devices	300,000 devices per access point (AP)

Adapted from Osseiran et al. (2014) and METIS (2014).

most likely technology trends for 5G are extensive capacity and mobility needs in dense areas, ubiquitous coverage and access quality, and ever-increasing cost pressure and future mobile markets (Bertenyi, 2014).

At this stage, there is a dearth of detailed technical information, as well as fuzziness on the telecommunications and engineering aspects of the 5G technologies currently being implemented and tested. As for the outcome, we can only infer this based on what has been published so far, and speculate from the telecommunications industry's reports. Global standard bodies, including 3GPP, have not yet fully defined the specific parameters required to meet the anticipated 5G network performance levels (Bertenyi, 2014).

An ITU working party was established and has worked on the necessary 5G deliverables and recommendations for International Mobile Telecommunications (IMT). In September 2015, ITU-R finalized its "Vision" of the 5G mobile broadband connected society, detailing the schedule for setting the scene for future IMT-2020 specifications (ITU, 2015). There is near consensus that 5G mobile systems should achieve a theoretical peak data rate of 10 Gbps (in low mobility) for indoor and campus environments, and a peak data rate of 1 Gbps for high mobility and for urban and suburban scenarios. Where 4G theoretically achieves a maximum rate of 1 Gbps, this 10-fold increase to 10 Gbps can allow the download of standard high definition movies in about 6 s (Thompson et al., 2014). These rates are also predicted to support latency of less than 1 ms with a network capacity of more than 10,000 times the capacity of current 4G networks; 4G has typical latency of about 50 ms. These systems will also adapt new communication techniques in their development process. They will include, for example, the development of heterogeneous network architectures that will have a multitude of communication technologies and network configurations (Mavrakis, 2014) with the following characteristics:

- Multi-network association, that is, 5G networks to coordinate different wireless domains, such as Wi-Fi, cellular, and WiMAX.
- Virtualization, software control, and ubiquitous mobile cloud connectivity.
- Cognitive radio technologies and terminals.

- Mobile Femto cells.
- Smart antennas and massive MIMO configurations.
- Full duplex configurations, so that the same mobile device both transmits and receives at the same time, thus achieving almost double the capacity of a frequency-division duplex (FDD) or time-division duplex (TDD) system.
- Energy efficiency and very low power devices.
- Advanced traffic management and off-loading.

The possible time frame and availability for the first 5G commercial systems according to leading telecommunications operators and industries is expected to be 2020–2025 (European Commission 5GPPP, 2015). It is more likely that the research will be completed by the end of 2016–2017, followed by the standardization process during the following 2 years, with early system deployment after 2020. Currently, several 5G global telecommunications and university consortia and research partnerships from the United States, Europe, and the United Kingdom are leading the development of 5G testbeds for different application areas required for 5G mobile coverage and data demands (CREW, 2014; 5GNOW, 2014; METIS, 2014; MiWEBA, 2014).

Figure 4.3 shows a future m-Health architecture scenario based on the 5G heterogeneous multitier network structures that can offer different cellular

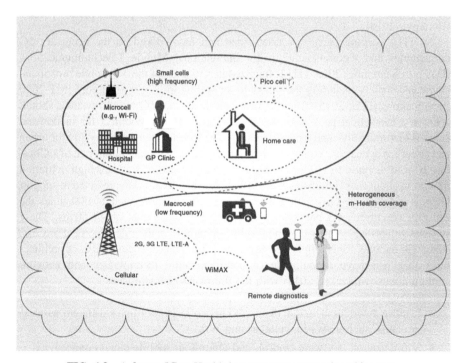

FIG. 4.3   A future 5G m-Health heterogeneous network architecture.

configurations and multiple frequency ranges. The smaller cells operate at higher frequencies to provide guaranteed high data rates in home and hospital environments for diagnostic, large medical data, and image transfer applications. Existing cellular networks (3G, 4G, WiMAX) use macrocells connectivity to provide wider coverage at lower data rates applicable, for example, for smart wearable monitoring devices, health, social interactions, and wellness monitoring applications.

As shown in Figs 4.2 and 4.3, these and other options of future wireless architectures will provide unprecedented opportunities for developing new m-Health systems (m-Health 2.0). These will provide seamless global connectivity paradigms, connecting people to people, people to machines, and machines to machine, anywhere, anytime, whenever such connectivity is needed. For example, these solutions will enable the development of future smarter personalized m-Health systems, integrating new wearable devices with M2M communications at very low power consumption rates that are linked seamlessly with intelligent mobile cloud computing hubs for personalized health monitoring and connected care management.

We might also witness new mobile 3D multimedia diagnostic m-Health terminals, for example, for ultrahigh definition ultrasound and medical imaging with acceptable clinical QoS and QoE. Other examples such as developments in home care and robotic assistive living services will also be forthcoming with 5G. The expected energy-efficient, enhanced coverage, and cost-effective tariffs should enable similarly enhanced m-Health solutions, particularly for global outreach, including for use in disaster areas. More open m-Health architectures and services might also benefit from the enhanced communications architectures and functionalities. At this stage, these issues only reflect the general vision of 5G and the expected direction of the ongoing research and development.

5G Challenges for m-Health For the foreseeable future, there are two main challenges: clinical and engineering. The first can be summarized as follows:

- Rapprochement between the requirements of clinical and healthcare delivery and those of the telecommunication, mobile phone, and medical device industries. This can be achieved by the development of a new m-Health framework to bridge the existing divergence.
- Introduction of unified global m-Health interoperability standards whose functionalities are compatible with 5G.
- Development of m-Health security and privacy standards to overcome cybersecurity and privacy threats, especially for big health data analytics, IoT, and cloud computing.

In terms of the engineering challenge, the overarching aim of 5G is to overcome the current limitation of 4G to meet the future demands of an efficient scaled-up healthcare service. The challenges can be summarized as follows:

- *To accommodate an exponential increase in mobile connections, users of 5G mobile health applications, and data estimated in exabytes/month.* This will be augmented by tens of billions of IoT and M2M connected devices for health-care. Special m-Health-related access control and scheduling will be required to enable the associated increase in data capacity.

- *To reduce the cost of future m-Health services, regardless of the expected increase in the user capacity and volume of data.* This important factor is pushing forward the development of new mobile networking methods, such as the separation of the control and user data planes. This can be carried out by providing different signaling methods in macrocell coverage.

- *To adapt 5G for "mission critical" m-Health applications and services.* These must cater for ultrahigh-speed real-time data combined with a high level of clinical reliability and accuracy. Examples include robotics in remote surgery and diagnostics and high-resolution emergency services, among many others. They will require quality of service and experience at elevated levels, which we may call "ultra-m-QoS" and "ultra-m-QoE." The design of new reconfigurable and ad hoc backhaul networking architectures and adaptive M2M direct communications are ingredients for consideration.

- *To make 5G networking and air interfaces compatible with future m-Health (m-Health 2.0) systems.* The associated ongoing research includes compatibility with MIMO and advanced beam forming, and the design of new m-Health-tailored resourcing and scheduling, coverage optimization, network radio resource management, and device mobility methods.

In summary, the pace of developments in 5G has been growing rapidly but standards are still pending and are yet to be ratified and finalized. These standards need to embrace the impact of mobile health and the potential for global markets. The introduction of 5G will have a profound impact on the future of m-Health 2.0 systems and their ability to provide more efficient and cost-effective global healthcare services and connectivity.

4.2.6 Medical Quality of Service and Medical Quality of Experience

To illustrate the complexity of the challenges for future m-Health systems outlined earlier, we present here an exemplar from the wireless communication and network-ing perspective, reflecting on specific m-Health requirements for future services and applications. In particular, we will explain the importance and relevance of m-QoS and m-QoE.

Although this topic is important, even critical, for mobile health, it has been largely overlooked. It has rarely been discussed by the telecom proponents of m-Health and not generally understood by clinicians. The anticipated increase in availability, miniaturization, and enhanced performance, together with the expected convergence of future wireless communication and networking technologies, will accelerate the deployment of future generations of m-Health systems with smarter and more enhanced capabilities. This includes the cognitive understanding of different users' needs, and how these systems will function according to requirements. The timely consideration of any critical issues is therefore becoming increasingly important, especially in the design and development of future m-Health systems from the communications perspective.

It is well known that in mobile networks, each application may have particular QoS requirements that are typically defined as the capability of a specific mobile network to provide a satisfactory service to the user. This includes provision of voice and data quality, signal strength, low call blocking, signal dropping probability, and high data rates. These qualities are typically needed by smartphone users in their everyday tasks, such as Internet browsing or quality video streaming. These quality requirements are generally translated into the following parameters (Lloyd-Evans, 2002):

- *Throughput:* This represents the rate at which data packets are transmitted through the network, and should be maximized.
- *Delay:* This is the time taken by a data packet to travel from one end of the network to the other end, and should be minimized.
- *Packet Loss Rate:* This is the rate at which data packets are lost in the network, and should be minimized, especially in m-Health applications with high clinical data requirements.
- *Packet Error Rate:* The error that is present in a data packet due to corrupted bits, and should also be minimized.
- *Reliability:* The availability of a network connection and its reliability in certain terrains and environments. It is particularly important in remote areas where the accuracy of vital signs monitoring and data quality is critical. These QoS metrics can also be translated into two levels from the network and users' perspectives (Jamalipour, 2003):
    - *Network QoS:* The requirement that specific network technology can offer users, such as bandwidth, time delay, and reliability.
    - *Application QoS:* The quality of the perceived services by users, where different users have different translations for QoS.

This differentiation is important for m-Health because varying levels of service quality are offered by network operators and service providers. To illustrate these issues further for m-Health, Table 4.4 is a summary of different m-Health applications and services in real-time and non-real-time scenarios, together with their associated data rates, corresponding QoS requirements, and metrics (Istepanian et al., 2013).

To reflect further on these issues, the concept of m-QoS was tailored for m-Health applications as a subcategory of QoS. It was defined as "an augmented requirement of critical mobile healthcare applications and traditional wireless quality of service requirements" (Istepanian et al., 2009a). The m-QoS can be categorized as a set of specific parametric requirements for particular m-Health applications and services provided by a specific network.

These issues can be important, especially in certain m-Health-specific applications with large numbers of users, and also for higher data rates or high level of clinical

TABLE 4.4   Different m-Health Services and Monitoring Parameters and Their QoS Requirements

m-Health Services and Monitoring Parameters	Real-Time Application	Non-Real-Time Application	Data Rate	QoS Indices
Electrocardiography (ECG) monitoring	✓		24 kbps/12 channels	Delay
Blood pressure monitoring (sphygmomanometer)	✓		<10 kbps	Delay
Digital audio stethoscope (heart sound)	✓		~120 kbps	Packet loss, delay
Region of interest JPEG Image	✓		15–19 MBytes	PSNR, frame size, packet loss,
Radiology		✓	~6 MBytes	PSNR, frame size, packet loss
Magnetic resonance imaging (MRI)		✓	<1 MBytes	PSNR, frame size, packet loss
Ultrasound video streaming and diagnostics	✓		250 kbps to 1.2 Mbps (WMV2)	PSNR, frame rate, frame size, packet loss, delay
Mobile medical consultation (e.g., accessing to patient records)		✓	~10 Mbps	Packet loss
Video/audio conferencing	✓		~1 Mbps	Packet loss, delay
Remote control applications (e.g., robotic control)	✓		~1 kbps	Packet loss, delay

Adapted from Istepanian et al. (2013).

accuracy, such as in real-time mobile medical video streaming and diagnostic applications (Istepanian et al., 2009b).

To demonstrate this, Table 4.5 summarizes the m-QoS requirements and the clinically acceptable bounds required for one such example of remote mobile ultrasound video streaming service. These are calculated from earlier clinical evaluations and mobile communication studies (Istepanian et al., 2013; Garawi et al., 2006). For this particular application example, three important metrics were identified as ultrasound image quality, frame rate of the received images, and end-to-end delay.

TABLE 4.5   Example of m-QoS for Ultrasound Video Streaming m-Health Application

m-QoS Index	Acceptable Value
Ultrasound frame size	QCIF (176×144), CIF (352×288), 4CIF (704×576)
PSNR	>35 dB
SSIM	>0.959
MSE	<14.07
Frames per second:	>5
End-to-end delay	<350 ms

Adapted from Istepanian et al. (2013).

Other examples of m-QoS include wellness and fitness monitoring applications, accident and emergency services, and home care applications; these may be for remote cardiac monitoring and diagnostics, or monitoring wearable and implantable devices from a large number of patients and data users.

More recently, the 3GPP defined the multiple QoS requirements for LTE-A (4G) and all IP networks, with different QoS Class Identifier (QCI) priority levels specified for different m-Health applications (3GPP, 2014a; Huang and Xie, 2015).

As shown in Table 4.6, each QCI metric is characterized by a specific priority level (ranked in terms of their packet loss rates and packet delays) that decides how these data packets are prioritized for scheduling across the LTE-A air interface network for each of the health application examples shown. This service model needs to be clinically validated and rigorously tested further for different medical settings for future m-Health 2.0 or 5G-Health applications. Future full-IP mobile networks need to adapt and prioritize their functionalities for different m-Health applications to provide "optimal effort" QoS levels. Similarly, the increasing demands for high bandwidth m-Health applications with specific QoS requirements also demand better

TABLE 4.6   QoS Class Identifiers for Different m-Health Applications for 4G and All IP Mobile Networks

QCI Index	Priority Level	Packet Delay (ms)	Packet Loss	Examples of m-Health Applications
1	2	100	10^{-2}	Emergency VoIP call
2	4	150	10^{-3}	Video consultations
3	3	50	10^{-3}	Patient video tracking
4	5	300	10^{-6}	Health monitoring
5	1	100	10^{-6}	Telemedical video consultation
6	6	300	10^{-6}	Medical data transmission with TCP
7	7	100	10^{-3}	Healthcare educational systems
8	8	300	10^{-6}	Daily health condition alerts
9	9	300	10^{-6}	Medical image transmissions

Adapted from 3GPP (2014a) and Huang and Xie (2015).

QoE requirements. For mobile communications, QoE can be defined as the "overall acceptability of an application or service, as perceived subjectively by end users" (ITU, 2008). The notion of QoE covers a broad spectrum of aspects representing factors that are important both to end users of an m-Health system and to stakeholders involved in the m-Health service provisioning chain.

To illustrate this further, smart mobile devices are expected to provide high-quality data connectivity and performance at all times. For example, in entertainment, the time required to download a specific video clip on to a smartphone, the stalling period, video content resolution, and responsiveness of a particular mobile App can determine the user's QoE.

A similar principle can apply for different m-Health applications. Today, different network operators and service providers consider QoE as the ultimate measure of the services tendered, as the overall satisfaction level of any service or application is perceived subjectively by the end users. Thus, although QoS is the kernel of technical quality, the QoE embraces much more than the performance of the mobile network itself; it usually concerns the overall experience of the user when accessing the specific network service being offered.

For different m-Health services that use 4G mobile networks, as shown in Table 4.6, m-QoE must take account of everyone concerned, usually patients, doctors, and nurses, all with their individual perception of the specific network services being offered by a targeted mobile health application.

In general, QoE has many contributing factors, among which some are subjective and not controllable, while others are objective and can be controlled. Some of the specific subjective factors include a patient's emotion, experience, and expectations and the doctor's acceptability level of the specific service. To illustrate this with the mobile medical video streaming application mentioned earlier, the mean opinion score (MOS) metric is used to quantify the quality of a particular multimedia data content that represents the received ultrasound video clip or images. This is typically given a numerical level between 1 and 5, with 5 representing the highest quality. This metric is a well-known method used to calculate and compare the quality of different multimedia clips in different mobility environments (Istepanian et al., 2013).

Table 4.7 shows the results from a study of the perceptions of end users, in this case junior doctors and specialist clinicians, on their experience with different remote diagnostic images of video-streamed ultrasound data samples transmitted wirelessly over both WiMAX and HSPA+ mobile networks (Alinejad et al., 2011, 2012; Istepanian et al., 2013). The table shows the MOS score for the tested medical images and data, and the relevant ratings allocated to the m-QoE metrics, as perceived by the participating clinicians. However, specifying more comprehensive QoS and QoE metrics for different m-Health services, especially in uncertain wireless and mobility environments, remains an open and challenging issue, and subject to further research.

4.2.7　Mobile Health and Future Mobile Communication Challenges

It is interesting to note that ongoing discussions cite some of the most likely future mobile communication trends as follows (Neira, 2015): 5G evolution; fiber networks

TABLE 4.7   Illustration Example of m-QoE Evaluation Using the MOS Metric for an m-Health Application

m-QoE Score	Overall Rating	Subjective Quality Indices
5	Excellent	Resolution: same as original, smooth and no jitters
4	Good	Resolution: good, almost same as original, smooth. Very few jitters
3	Fair	Resolution: good but occasionally bad; image jitters and breaks at periphery, but is tolerable as long as region of interest (ROI) not affected; obvious flow discontinuity of video due to image obstruction
2	Poor	Resolution: poor, image jitters throughout the clip. ROI was minimally affected
1	Bad	Resolution: bad, image jitters and breaks for longer intervals in various areas affecting ROI; not acceptable

Adapted from Alinejad et al. (2012) and Istepanian et al. (2013)

everywhere; virtualization; everywhere connectivity for IoT and IoE; cognitive networks; cybersecurity; green communications; smarter phones and connected sensors; network neutrality and Internet governance; and molecular communications.

These likely scenarios warrant some consideration of the impact on future m-Health systems. More global effort is required to identify specific healthcare and mission-critical medical scenarios and their clinical requirements, and how the associated technologies can best be used for future m-Health services. This needs to be carried out in a timely manner that is compatible with ongoing technical developments. Relevant work on the standardization of m-Health applications, particularly for IoT and M2M systems and services, also needs to be in place to avoid any gap between the current 5G development and the mobile healthcare applications most likely to adopt these technologies after 2020.

Future developments in emerging concepts such as the "Tactile Internet" (ITU, 2014a; Fettweis, 2014) will also result in interesting m-Health developments. These were originally derived from remote robotic–human sensing research and combine the expected extremely low latency of high-speed mobile networks with Internet computing and the cloud access of IoT devices. These emerging Internet and mobile communication paradigms will ultimately lead to developments of new "tactile m-Health" systems that will emulate the Tactile Internet characteristics with ultrafast reactions applicable to many future m-Health 2.0 applications, such as for real-time remote diagnostics and robotic surgical applications or assisted living environments. Future research and development also need to address the following issues:

- *User and Device Planes:* Future architectures are expected to focus on enhancing user interaction capabilities, especially for "context-aware" applications. For m-Health, this requires a clear definition of the users, for example, patients, clinicians, healthcare providers, the "worried well" and the "health savvy" or the millennial generation, together with their different expectations of these smart m-Health devices and how they will be designed, adapted, and used in different settings.

- *Security:* m-Health services are considered one of the main applications that will require more robust, secure, and reliable access, and exchange of future digital health data and information, together with challenges relating to cybersecurity.
- *Network and Services Planes:* These need to identify and provide independently the variety of future "asymmetric and symmetric" big health data traffic requirements expected to be generated from the high volume of users. These must address the seamless and intelligent delivery of "on-demand m-Health" services and how to differentiate these from other 5G services.
- *Socioeconomic and Business Perspective:* These need to focus on the integration of both the technological advances with the future demand of the m-Health-driven market and business opportunities that are expected to evolve from these developments.
- *Global Connectivity and Policy:* This important challenge represents the paradigm shift that 5G systems will provide to global m-Health connectivity and user interactivity. These will be reflected in the global spread of mobile users and potential access to services to be provided by these systems. A new focus on mobile health policies and on global mobile health standards will also be required.

In conclusion, the existing gap between the clinical landscape and the market-driven potential of current m-Health systems is increasing. This trend necessitates a fresh vision and new outlook on how future m-Health systems will be developed, designed, and implemented from the mobile communications perspective.

4.3 WIRELESS METROPOLITAN AREA NETWORKS FOR m-HEALTH

As explained earlier and as illustrated in Table 4.1, WMANs are associated with WiMAX networks. This wireless broadband technology currently coexists with the LTE-A standard as part of 4G. Globally, some mobile vendors and operators are opting for the WiMAX technology as flexible, all-IP network architecture, better suited than the HSPA+ technologies to provide all Internet-based services. Other vendors and operators, such as AT&T and T-Mobile in the United States, have opted for LTE technology for their 4G networks deployments (Stair and Reynolds, 2014). Increasingly, the LTE standard is gaining momentum, and the hype of WiMAX technology is gradually dissipating, especially with the increasing adoption of LTE-A technology as the dominating 4G system. Historically, WiMAX technology was based on a set of IEEE 802.16 standards that support different types of communications access. In 2004, the original IEEE 802.16d standard was drawn up for fixed installations to provide mobile broadband services, hence it was referred to as the

"Fixed WiMAX." Following optimization to overcome mobility and coverage limitations, "Mobile WiMAX" was introduced in 2006; this conformed to the IEEE 802.16e standard, and subsequent revisions have reached the IEEE 802.16m standard (Nuaymi, 2007).

Typically, Mobile WiMAX can provide very high data rates of around 70 Mbps over short ranges of 1 km or a little more, but much lower data rates over the maximum range of about 50 km because of increased bit error rates. Other area networks provide highly mobile long-range coverage but low data rates by providing a specification that supports mobile broadband access, including connectivity between base stations. An improved specification of 802.16m, called WiMAX Release-2 (WiMAX-2), has been released with peak data rates of up to 300 Mbps (Stair and Reynolds, 2014). The WiMAX network typically operates at frequencies between 2 and 11 GHz for non-line-of-sight (N-LOS) operation, and between 10 and 66 GHz for line-of-sight (LOS) operation. There is no uniform global licensed spectrum for WiMAX, but for standardization and interoperability purposes, the WiMAX Forum has published three licensed spectrum profiles, which are 2.3, 2.5, and 3.5 GHz (Ergen, 2009).

Technically, as shown in Table 4.1, both LTE and WiMAX share similar communication features (Etemad, 2008; Ergen, 2009). Both standards are all-IP access network technologies; support advanced MIMO antenna technology for their mobile terminals; and use similar OFDMA as their main modulation technology. However, with the introduction of the LTE-A standard, there are several communications and functional features that differentiate WiMAX from LTE-A:

LTE-A can generally operate with mobile speeds of up to 450 kph (280 mph), compared to WiMAX, which has an operational speed of up to 120 kph (75 mph).

WiMAX does not support the legacy systems of earlier mobile technologies, that is, 2G and 3G communications, while LTE-A allows the coexistence and roaming between them.

LTE-A offers better power consumption for their mobile terminals and devices compared to WiMAX terminals. For example, LTE-A mobile terminals are equipped with modulation technology called single-carrier frequency-division multiple access (SC-FDMA) in the uplink direction. This modulation technology provides much better battery power savings and consumption for mobile terminals compared to the scalable orthogonal frequency-division multiple access (SO-FDMA) technology used in WiMAX terminals in both link directions.

These technical differences do not preclude the fact that WiMAX technology can offer alternative advantages over LTE technology, particularly for m-Health services and applications. The deployment of WiMAX networks rather than LTE networks can be more cost-effective for some m-Health applications, especially in remote areas and metropolitan areas, and require less engineering and IT maintenance. Furthermore,

WiMAX technology can provide better market competition for mobile broadband services compared to LTE networks, which can be monopolized by specific operators and vendors.

Since both WiMAX-2 and LTE-A provide real-world quality performances, the decisions of global operators and vendors may determine the fate of WiMAX technology. Since WiMAX networks were initially designed to deliver wireless broadband services to homes and businesses, these can still be easily adapted for providing broadband services for hospitals and healthcare clinics in remote towns and cities.

There is increasing work on the feasibility of using WiMAX technology for real-time m-Health applications that require high data content (Alinejad et al., 2011, 2012; Markarian et al., 2012). To illustrate some of the relevant technical issues, Fig. 4.4 shows a WiMAX-based mobile system configuration used for remote real-time video streaming and ultrasound data scanning (Alinejad et al., 2012).

4.3.1   Ultrasound Video Data Streaming and Diagnostics

In this m-Health example, a mobile WiMAX-based station is deployed in a remote healthcare center in an area where there is neither specialist expertise nor cellular connectivity. The mobile healthcare worker or nurse, with some basic level of training in ultrasound scanning, is equipped with a portable ultrasound scanner connected to a

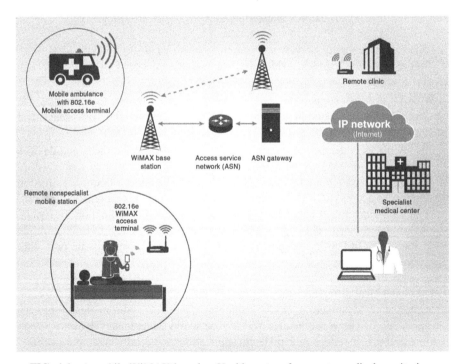

FIG. 4.4   A mobile WiMAX-based m-Health system for remote medical monitoring.

mobile WiMAX terminal. The healthcare worker can roam to different remote areas within the operational range of this network for any emergency cases. Table 4.1 shows that due to the limited range of WiMAX transmissions, the operational network range, that is, the distance between the remote area and a specialist center, can be increased by deploying additional base stations acting as communication cells. In the scanning process, the healthcare worker can be guided by a remote specialist via a real-time video link used by the broadband voice and data capabilities of the WiMAX network. The scanned ultrasound images and data streams can then be interactively transmitted in real time to the specialist for viewing and feedback.

Another more advanced m-Health application is the use of a robotic ultrasound probe that can be guided remotely by an expert clinician for real-time ultrasound scanning and feedback to the remote patient station (Istepanian, 2007). Other mobile healthcare scenarios that can benefit from hybrid WiMAX/WLAN network connectivity include applications such as the transmission of critical medical data from moving ambulances to hospitals equipped with WiMAX mobile terminals, especially in urban zones and inner cities, and medical care in disaster or disease-outbreak areas where the local cellular network infrastructure may be out of action. These and other scenarios of ad hoc WiMAX-based m-Health networks are feasible alternatives, especially in remote and isolated areas that require urgent healthcare services and attention. Most of the telecommunications operators embrace LTE-A technology for their 4G network implementation compared to the WiMAX solutions. However, new WiMAX solutions are being considered for the 5G options that we discussed earlier.

4.4 WIRELESS LOCAL AREA NETWORKS (WLAN) FOR m-HEALTH

As is widely known, WLANs, or Wi-Fi, are today located globally, with hotspots in every feasible location from homes, cafes, and hospitals to shopping malls, and almost any indoor public place, and even on buses and trains. This popular mobile network technology is widely considered to be the most preferred option for indoor broadband Internet access by users and developers. The expected proliferation of this technology is reflected in a market analysis, predicting that more than 725 million households worldwide would have Wi-Fi connection by the end of 2015, with more than 6 million Wi-Fi hotspots globally (Informa Telecom & Media, 2012).

These WLANs were originally based on the IEEE 802.11 standard released in 1997. Since then it has adopted different versions, namely, IEEE 802.11a, b, g and n that operate in the unlicensed radio frequency bands of 2.4 and 5 GHz. The high-speed data rates were the initial drivers for the development of these standards. The commercially available IEEE 802.11n version, with peak data rates of up to 600 Mbps (more than 10 times the initial 54 Mbps of the earlier IEEE 802.11a version) was introduced in 2009. Table 4.8 shows a comparison of most available and future IEEE 802.11 standards and their communication characteristics. The table shows that the operating range varies approximately between 30 and 50 m for

TABLE 4.8 Comparison of the Existing and Forthcoming Wi-Fi (802.11) Technologies

	Existing IEEE802.11 Technologies				Forthcoming IEEE 802.11 Wi-Fi Technologies				
	802.11a	802.11b	802.11g	802.11n	802.11ac	802.11ad	802.11af	802.11ah	802.11ax
Operating frequency	5.3–5.8 GHz	2.4 GHz	2.4 GHz	2.4 and 5 GHz	5 GHz	60 GHz	54–790 MHz	<1 GHz	2.4/5 GHz
Nominal range	30–35 m	30–50 m	30–50 m	60–70 m	~100 m	~10 m	~1 km	~1 km	~100 m
Channel BW	20/22 MHz	20/22 MHz	20/22 MHz	20/40 MHz	20/40/80/160/ 80+80 MHz	2.16 GHz	6/7/8/12/14/16/ 24/28 MHz	1/2/4/8/ 16 MHz	–
Peak data rate	54 Mbps	11 Mbps	54 Mbps	248 Mbps	6.933 Gbps	6.756 Gbps	586.9 Mbps	346.7 Mbps	–
Modulation technique/ features	OFDM	CCK/DSS	OFDM	OFDM using MIMO	Throughput enhancement	Long-range extension			

the earlier standards (a, b, g) and between 60 and 70 m for the 802.11n standard. This standard not only offers better performance in terms of capacity, range, and reliability but also provides better QoS for differentiated voice, video, and data services that are compatible with the earlier standards (a, b, g). It also provided the first wireless speed devices comparable to the wired networks, such as Ethernet. However, more recent enhancements for this standard for both long-range extension and greater ease of use have been the key directions of the IEEE 802.11 working group (Sun et al., 2014).

The two most recently approved standards, IEEE 802.11ac and IEEE 802.11ad, have been designed to provide a theoretical maximum data speed of 1 Gbps (Verma et al., 2013). The nominal range of these two new standards is approximately 100 m for 802.11ac and 10 meters for 802.11ad. Two newer standards, IEEE 802.11af and IEEE 802.11ah, are being introduced to provide a long-range extension, with the latter operating at an approximate range of 1 km at a frequency band of 1 GHz and data rates of around 346 Mbps (Verma et al., 2013; Sun et al., 2014).

The large growth of mobile broadband traffic and applications markets, supported by embedded WLAN technologies in smartphones and tablets, necessitated the need for mobile operators to integrate WLANs with a 3GPP Evolved Packet Core (EPC) connectivity link (Roeland and Rommer, 2014). This advanced integration feature will allow future smart devices to set up a trusted WLAN IP connection that gets routed to the EPC link and hence to any 3GPP-operated network. This process ultimately links any 3GPP network with non-3GPP networks, enabling trusted and heterogeneous Internet access with flexible roaming capabilities.

Further work is also ongoing on the integration of WLANs with 5G systems. These advances will open up new m-Health wireless access and connectivity models that we discussed earlier. The earlier studies, on the use of Wi-Fi technologies for m-Health applications (MedLAN), focused on the feasibility of using earlier WLAN standards for medical applications inside hospitals (Banitsas et al., 2002). Nowadays, most modern hospitals and healthcare centers are equipped with network hotspots for applications in their vicinity; these include, among others, internal e-mail communications; wireless access to electronic health records; wireless patient monitoring and tracking in hospital wards; identification of hospital assets, medical devices, and location tracking systems; and wireless wards and care staff management (Ruckus, 2013).

Figure 4.5 shows future m-Health scenarios using Wi-Fi technologies applied for various medical applications and environments. The wireless connectivity is based on using the combination of existing and future 802.11 Wi-Fi technologies.

Most recent models of smartphones and tablets are equipped with the 802.11a/b/g/n WLAN-compatible chips, with later models expected to be equipped with the more robust 802.11ac and newer technologies shown in Table 4.8, allowing users larger range and better Internet access capabilities. WLAN or Wi-Fi technologies are still evolving in many aspects: speed enhancement, longer operating range, and greater flexibility of use. They will also provide better interworking with future cellular connectivity and coexistence with other unlicensed band-based wireless technologies. These capabilities will also provide ample opportunities for designing better, more

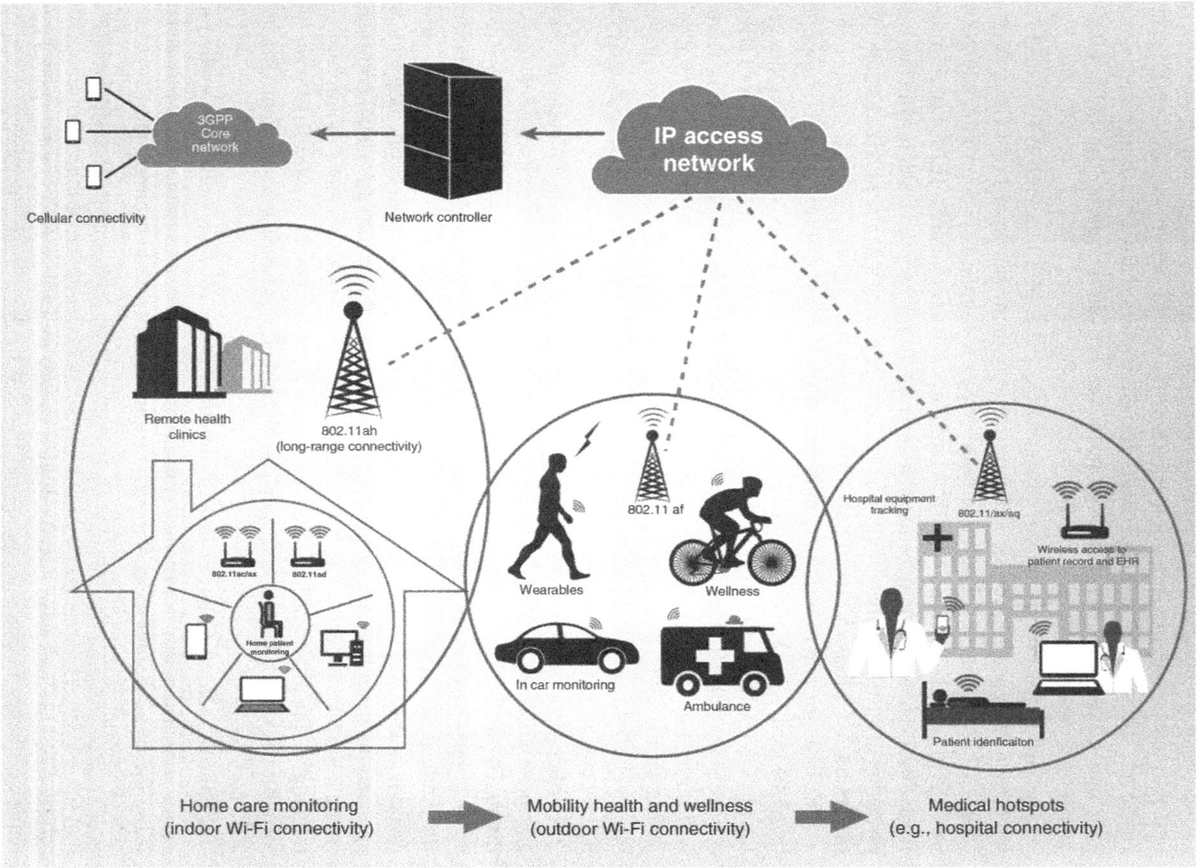

FIG. 4.5 Wi-Fi-based m-Health network architectures and medical care scenarios.

secure m-Health applications. A particular concern with their security and privacy is that most hospitals and healthcare centers use Wi-Fi (802.11n) devices for their hotspot access points. These are still equipped with fragile Wi-Fi secure channels that can be easily compromised. In the United States, enterprise-grade wireless security for healthcare applications is an absolute necessity to meet, for example, the Health Insurance Portability and Accountability Act (HIPPA) requirements (Ruckus, 2013). There are as yet unmet challenges in deploying and managing resilient and more secure WLAN systems dedicated to medical and data access requirements. These challenges are mainly attributed to the existing fragile security mechanisms of earlier WLAN models and devices deployed in these environments (Zhang et al., 2010). However, with the introduction of the latest technologies currently under development, which can provide better privacy and authenticated mechanisms, these issues can be mitigated. They will inevitably become more important, especially with Wi-Fi wireless connectivity to private hospital cloud systems for health data access and storage.

4.5 PERSONAL AREA NETWORKS (PAN) AND BODY AREA NETWORKS (BAN) FOR m-HEALTH

As explained in Chapter 2, the basic principle of wireless sensor networks (WSN) refers to the specific networking configuration of large numbers of sensors that are capable of computation, communication, and sensing tasks in different connectivity structures. In general, a WSN consists of miniaturized sensors, or nodes (sometimes referred to as motes), which typically have one or more base stations, or sinks. The WSN also includes an aggregation node and a management node. In the data transmission process, the nodes detect and transport the data through other nodes by "hops" and route these to the sink node. The data are then transported from the sink node to the management node via the Internet or a cellular network, where the information is collected and analyzed.

Examples of the use of WSNs come from industrial or automation applications, where these network configurations usually comprise stationary nodes in environments where access is difficult, or where monitoring requires a large number of nodes, or where they are physically unreachable after deployment. Among the numerous applications of WSNs are industrial and environmental monitoring, automation and control, and healthcare. The proliferation of WSNs in healthcare and wellness monitoring applications has led to the development of more tailored WSN configurations and standards that are more suited for personalized m-Health. In the following sections, we will briefly introduce these networking structures, and discuss mobile health applications.

4.5.1 WPANs and WBANs for m-Health

As discussed in Chapter 2, a wireless personal area network (WPAN) refers to a personal, short-distance wireless area network for interconnecting devices, or sensing

TABLE 4.9   IEEE 802.15.6 WBAN Specifications and Standard Requirements

Specifications	WBAN Technical Requirements
Data rate	~10 kbps to 10 Mbps
Approximate range	~1–2 m (<3 m)
Latency	<125 ms (medical); <250 ms (nonmedical)
Jitter	<50 ms
Reliability	<1 s for alarm; <10 ms for applications with feedback
Power consumption	>1 year (1% LDC+ 500 mAh battery; >9 h (always ON+ 50 mAh battery)
Topology	One-hop star, two-hop star, bidirectional link
Nodes (device numbers)	Typically 6, up to 256
Setup time	Insertion/removal time (<3 s)
Packet error rate (PER)	~<10% (with link success probability of 95% over all channel conditions
Sampling rate	~<100 Hz (for medical WBAN applications)

nodes, centered on or around the body, generally within a range of less than 10 m from the human body.

WPANs are generally defined by the IEEE 802.15 standard, as opposed to earlier networking systems that are associated with longer distances (outside the communications vicinity of the human body). WPANs are configured and applied for a multitude of wireless networking applications, such as in emergency rescue and disaster events, workforce safety, and health management (Alam and Hamida, 2014). Addressing the increasing demand of medically oriented applications, the IEEE 802.15.6 Task Group6 (TG6) approved and released in 2012 the Wireless Body Area Networks (WBAN) IEEE 802.15.6 standard for short-range wireless communications used on, or inside, the human body (Movassaghi et al., 2014). Table 4.9 is a summary of the general specifications of the WBAN standard (Zhen et al., 2008; Alam and Hamida, 2014).

WBANs are basically a number of heterogeneous nodes (sensors) that are connected to a single processing hub acting as the central aggregation controller and processing element. This standard supports communications in the vicinity of or inside a human body to serve a variety of medical and also nonmedical applications. However, WBANs are mostly associated with different mobile health and wellness monitoring applications (Latre et al., 2011; Alam and Hamida, 2014). The characteristics of medical WBANs include a low-duty cycle with a typical sampling rate of less than 100 Hz; low data rates, typically 500 kbps; low transmission power, typically less than 1 mW); and versatile latency, for example, less than 500 ms for ECG monitoring, but up to a few seconds for skin temperature.

In 2012, the U.S. Federal Communications Commission (FCC) allocated a specific radio spectrum (2360 and 2400 MHz) for Medical Body Area Network (MBAN) applications. These low-power wideband networks, consisting of body-worn sensors that transmit patient data wirelessly, are divided into two major types (Harbert, 2012; HIMSS, 2012):

- *In-Body MBAN Devices or Sensors:* These sensors can be ingested or implanted to track biological, chemical, pressure, and temperature changes that occur in the body.
- *On-Body MBAN Devices or Sensors:* These sensors can be attached to the body, or embedded in the fabric of garments and wearable equipment such as helmets, casts, or prosthetics. The sensors can be used to track different contextual information such as location, posture, moisture, and activity level, as well as biological changes to the body related to ECG, pressure, temperature, and chemical changes on the skin.

Figure 4.6 shows a typical WBAN m-Health monitoring architecture and connectivity system.

Although WBAN standards are designed for optimized, low-power consumption at relatively high data rates, the same enabling wireless technologies using the WPAN and WSN standards are also used for WBAN solutions, due to the current unavailability of market off-the-shelf IEEE 802.15.6-compliant products (Alam and Hamida, 2014). WBANs interact with currently available radio devices used

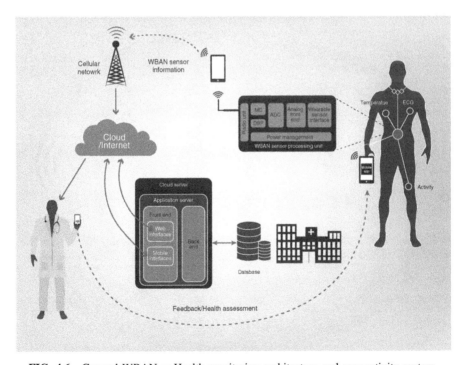

FIG. 4.6 General WBAN m-Health monitoring architecture and connectivity system.

for PAN wireless technologies, such as Bluetooth (IEEE 802.15.1) and Bluetooth low energy. There are also others used for WSNs, such as ZigBee (IEEE 802.15.4) and ultra-wide band (IEEE 802.15.4).

As described in Chapter 2 (Fig. 2.4), Fig. 4.6 illustrates the WBANs node architecture that consists of wearable medical sensors interface unit, an analogue-to-digital converter, a microcontroller or DSP unit, and a radio module, together with a power management module. As also mentioned in Chapter 2 for medical WBAN applications, the most common sensing tests for biomedical signal monitoring include heart rate and ECG, skin temperature, activity, body motion, blood pressure, respiration rate, electromyography (EMG), electroencephalography (EEG), and pulse oximetry (oxygen saturation). Figure 4.7 shows the communication tiers and range of WBAN in comparison with WPAN and WLAN networks.

The different communication tiers of WBAN (intra-WBAN, inter-WBAN, or beyond WBAN) can be configured with WPAN and WLAN and cellular communications networks, depending on the functionalities and range requirements of the specific m-Health application being targeted for specific medical scenarios, such as emergency care, in-hospital, or home care monitoring. Before looking at some of the popular and widely used short-range wireless communications standards and protocols for m-Health, it is worth mentioning that most of these are evolving continuously. Some never get into general acceptance, some become obsolete quickly, and some are upgraded frequently. It is therefore impossible to set in stone any protocol or standard

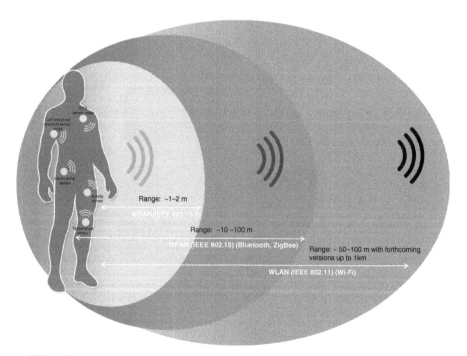

FIG. 4.7   Communication tiers and range of WBAN, WPAN, and WLAN networks.

that will be used indefinitely for mobile health. People strive for higher data rates, more data throughput, additional channels, and so on; what is presented in this section is a snapshot of the scene at the time of writing. Technology never stands still. Although we may think our communications systems are impressive—and they are, their specifications will inevitably seem modest half a century in the future. More technical details and information on this topic are explained extensively in specialist references and review articles (Cao et al., 2009; Adibi, 2012; Ullah et al., 2012; Movassaghi et al., 2014; Cavallari et al., 2014; Ha, 2015).

4.5.2 Short-Range Wireless Technologies for m-Health

In general, the classification of any short-range wireless communications system depends on the following diverse aspects (Kraemer and Katz, 2009):

- *Network topology*, which can be either centralized (e.g., cellular) or distributed (e.g., ad hoc, mesh, without infrastructure).
- *Air interface*, or physical layer, characteristics, which can be conventional (narrowband radio), ultra-wide band radio, mm-wave communications (60 GHz), or optical wireless communications (infrared or visible light).
- *Mobility*, which can be fixed, nomadic, and mobile.
- *Data rates*, which can range from low (below 100 kbps) to moderate (100 kbps to 10 Mbps) to high (10 Mbps to 10 Gbps).

These aspects apply to the m-Health domain, particularly for WPAN and WBAN applications. However, the choice of the best wireless technology options for WPAN or WBAN m-Health applications is not an easy task. The selection depends on several factors, such as the specific medical service requirements, range, power consumption, and security of the link.

Table 4.10 is a summary of some of these technologies and their general communications specifications, range, operating frequencies, and data rates for clarification. The wireless standards and their technologies described in this section constitute most of the short-range communication links required for different m-Health WPAN and WBAN applications.

Bluetooth Low Energy/Bluetooth Smart   This is one of the most popular and widely used data communication technologies for short-range, wireless applications and services, including m-Health. Bluetooth is already a *de facto* technology in most smart mobile handsets, and is expected to continue in the foreseeable future. The standard Bluetooth, or Bluetooth Classic, technology was originally proposed by Ericsson in 1994 as an alternative to cables to link mobile phone accessories (Bluetooth, 2014). The original standardized versions of Bluetooth 1.1 and 1.2 were based on the IEEE 802.15.1 standard and released in 2001 and 2003, respectively. This was followed by the Bluetooth-2 version in 2004. In 2009, Bluetooth 3.0 was introduced, adding the IEEE 802.11 standard as a high-speed channel to increase the peak data rate to 20 Mbps.

TABLE 4.10 Examples of Short-Range Wireless Technologies Used for m-Health

Technology/ Standard	Frequency Band/ BW	Range	Data Rates	Features	m-Health Applications
Classic Bluetooth (802.15.1) Bluetooth 2.0 Bluetooth 3.0	2.4 GHz/1 MHz	<10 m and	~1–3 Mbps	802.15.2: range 10–30 m) with power of 2.5 mW (4 dBm) 802.15.3: high data rate ~20 Mbps with range of few meters with 1 mW power	Different m-Health monitoring, wellness applications, home care, assistive living, and WBAN applications
Bluetooth low energy Bluetooth Smart	2.4 GHz/2 MHz	~10–100 m	~1 Mbps	Bluetooth Smart (Bluetooth 4.0/ 4.1) Supports Bluetooth Classic and Bluetooth low energy	Same application as above, but with very low-power requirements
ZigBee (802.15.4)	2.4 GHz/5 MHz and 868 and 915 MHz	<10 m	20–250 kbps (2.4 GHz)	Nodes per network (65,000) compared to only 7 for BT Classic	Home care, WPAN, WBAN and other health monitoring applications
Ultra-Wide Band –(UWB) (802.15.6)	3.1 to 10.6 GHz/ >500 MHz	<10 m	850 kbps –20 Mbps	Low power with high-speed data	Patient localization, hospital wards, short- range WBAN applications
Infrared (IrDA)	~800–1000 μm	<1 m	1.15–4 Mbps	High-speed short-range, point-to-point data transmission	Point-to-point wireless medical data transmission, for example, wireless intravenous (IV) therapy
RFID	125 kHz, 13.56 MHz, 902–928 MHz	<1 m	~26 kbps (HF range)	Low cost	Implantable RFID chips and tagging health monitoring Hospital tracking
NFC	13.56 MHz	<30 cm	<424 kbps	Security, tagging	EHR secure tagging, electronic cards, secure tagging of hospital assets and medical devices

Technically, Bluetooth Classic is considered as wireless short-range technology that enables any electrical device to communicate in the 2.4 GHz industrial, scientific, and medical (ISM) unlicensed frequency band. Bluetooth also allows real-time data transfer at rates of up to 1–3 Mbps, with a typical rate of around 400 kbps over ranges of 10–100 m depending on the version used. It also allows point-to-point wireless connections between devices such as body sensors and smartphones. It is configured with a master and up to seven slaves, forming a "piconet," and has been specifically designed as a low-cost, small-size, and low-power wireless technology, so it is particularly suited to the short ranges appropriate for WBAN applications.

Bluetooth low energy was introduced in 2010 and was previously known as Bluetooth low end extension, then Wibree, after incorporating more advanced features such as an ultralow-power wireless technology designed to connect small devices such as WBAN sensors (Bluetooth Low Energy, 2013). BLE resulted from a collaborative effort between the Medical Devices Working Group and the ISO/IEEE 11073 Personal Health Devices Working Group (Bluetooth Low Energy, 2013). BLE is a part of Bluetooth 4.0; in this specification, a dual mode device such as a smartphone is able to support Bluetooth Classic and Bluetooth low energy at the same time.

The main advantages of BLE technology is that it has a data rate of up to 1 Mbps, it uses only frequency hopping channels, and its synchronization between sensors (or other devices) is very fast. All of these parameters are superior to those for Bluetooth Classic, which employs frequency hopping across 79 channels, each with a channel bandwidth of 1 MHz and a raw symbol rate of 1 Msymbol/s. The modulation scheme is Gaussian frequency shift keying (GFSK), quadrature phase shift keying (4PSK), or 8PSK.

For BLE, the modulation scheme is GFSK, with a raw data rate of 1 Msymbols/s and a 2 MHz channel bandwidth, which is double that of Bluetooth Classic. While frequency hopping is still used in BLE to mitigate interference, the dwell time in BLE is longer than for Bluetooth Classic, so the timing requirements in BLE can be relaxed (Galeev, 2011).

Most of the current Bluetooth-enabled sports and wellness wearable monitoring devices include this standard. The latest updated specifications of the BLE standard, also known as Bluetooth 4.1, released in 2013, are aimed to improve data exchange rates and aid developers by allowing devices to support multiple roles simultaneously. This new release also lays the groundwork for an all-IP-based connectivity, extending Bluetooth technology's role as the essential wireless link for the IoT paradigm (Bluetooth Low Energy, 2013). Moreover, the latest BT version is marketed as Bluetooth Smart. This updated specification is designed to be used in power-constrained devices, such as wireless sensors and control applications, with flexible IP access and support for the dual functionalities of BLE and Bluetooth Classic. The selling point of these devices is that they can be powered for periods of months, or even years, on a single coin-size cell. The protocol used is a star network connection for single-hop communication over ranges of typically 50–100 m. This technology is also most likely to be powering the forthcoming IoT connectivity (Galeev, 2011).

For m-Health, there are numerous studies that exploit this technology (Mulvaney et al., 2012). Bluetooth technology is also embedded in most of the current commercial wearable, wellness, and sports activity monitoring devices. It is expected that Bluetooth Smart will continue to be the *de facto* short-range wireless technology for many consumer m-Health monitoring and sensing applications in the foreseeable future, due to its compatibility with most of the current smartphones.

Earlier versions of Bluetooth had inherent security flaws in their protocols that made their configurations vulnerable to privacy and security attacks. The latest versions have better security configurations, although not perfectly immune to such attacks.

Recent developments in the IoT and M2M communications are likely to have a major impact on network connectivity, as will be explained in the following sections. There are currently ongoing efforts to translate BT Smart as a viable solution for IoT connectivity features for various m-Health applications. This is ongoing work in the Bluetooth industry and vendor communities, reflecting its importance and market potential.

ZigBee   The original IEEE 802.15.4-2003 standard for ZigBee was superseded by IEEE 802.15.4-2006. After several iterations, an enhanced version called ZigBee PRO became available in 2007 (ZigBee Alliance, 2014). The name is supposed to describe the "dance" of a honeybee when arriving back at the hive; following the bizarre nomenclature of wireless protocols, the inventors presumably thought this had something to do with communications. ZigBee is intended for applications of PANs and WBANs. It is therefore suitable for medical monitoring, and considered a secure protocol using a mesh configuration.

As shown in Table 4.10, ZigBee operates in the ISM frequency bands, which are 868–870 MHz in Europe (one channel with a maximum data rate of 20 kbps), 902–928 MHz in the United States and Australia (10 channels, 40 kbps), and 2.4 GHz worldwide (16 channels, 250 kbps). It is intended for embedded applications such as biomedical monitoring, which are characterized by relatively low data rates, and very low power consumption, which means that these devices can have a battery life of 2 years or more, as required by the ZigBee certification. Ranges achievable indoors are typically less than 10 m at 2.4 GHz, depending on the transmitted power and the actual environment. For data coding, direct-sequence spread spectrum is used. Binary phase shift keying (BPSK) is used in the 868 and 915 MHz bands, and offset quadrature phase shift keying (OQPSK), with two bits per symbol, is used in the 2.4 GHz band (Elahi and Gschwender, 2009; Yang, 2014).

Among the m-Health applications adopting ZigBee is a wireless ECG sensor patch using a ZigBee-based BAN (Munshi et al., 2008); another is for various home or hospital environments (Shnayder et al., 2005; Gao, 2007). The ZigBee Health Care public application profile also meets the Continua Health Alliance specifications for remote health and fitness monitoring applications (ZigBee Alliance, 2010). The major advantage of ZigBee technology is that it can provide multihopping routing in various configurations, including a cluster tree or mesh topology, without the need for synchronization. These specifications make ZigBee a potentially strong competitor to Bluetooth for future m-Health short-range applications.

Ultra-Wide Band   As the name indicates, the ultra-wide band technology is a form of transmission that occupies a very wide bandwidth. UWB is defined by the Federal Communications Commission (FCC) as any wireless technology with a bandwidth greater than either 500 MHz, or 20% of the arithmetic center frequency, whichever is least. The FCC has approved UWB for indoor and short-range outdoor communication, with restrictions on the spread of the transmission frequencies and power limits. UWB may also be used in the unlicensed 3.1–10.6 GHz band. One advantage of UWB is its short range of less than 10 m (typically <2 m), which makes it suitable for BANs and for slightly longer range indoor applications, especially in a hospital, where radio frequency emissions tend to be discouraged. Further advantages of UWB for its use by BANs are its very high data rate (850 kbps to 20 Mbps), which allows precise localization for tracking a patient inside a building, effectively an "indoor GPS" (Hernandez and Mucchi, 2014).

In technical terms, there are generally two different UWB technologies developed for different applications (Hernandez and Mucchi, 2014):

- *Carrier-Free Direct Sequence UWB:* This form of UWB technology transmits a series of short-duration impulses, resulting in a signal spectrum occupying a very wide bandwidth. This technology is usually used in high data rate transmissions, such as short-range video transmissions and similar applications.
- *MBOFDM UWB:* In this form, UWB uses 500 MHz multiband orthogonal frequency-division multiplexing (MBOFDM) as a wide band method that is then "hopped" to enable it to occupy a sufficiently high bandwidth. This type of technology is adopted for common USB connections used in wireless devices.

The WBAN standard (IEEE 802.15.6) also defines the physical communications layer of UWB, together with other narrowband technologies (Yong et al., 2007). This technology has many industrial and consumer applications, but its adoption for m-Health applications is still limited. Many potential applications include, for example, mobile patient monitoring in intensive care units, smart mobile bedside care, portable medical data imaging, and mobile diagnostics. These can all be considered as candidates for using UWB for future short-range communications applications.

4.5.3   Other Short-Range Wireless Technologies for m-Health

In addition to these popular standards outlined above, there are many proprietary (technology owned by a company) and nonproprietary (not owned and open) short-range wireless technologies that are used for a range of medical and nonmedical purposes, for example, automation control and other industrial wireless sensing and networking applications. These market-driven technologies are available with their specific integrated circuit design linked to specific sensing configurations. Some of these are used for mobile health, including for hospital tracking and health status monitoring. We present here a short description of these technologies for completeness.

Radio Frequency Identification (RFID)　　This is a wireless protocol for transmitting data, but its main purpose is to identify or track so-called "tags" that are attached to objects, or carried by people. Typically, these tags store data that may be read at ranges of less than a meter to few meters. Some are battery-powered, and can operate over 100 m or more, while others are energized by the interrogating electromagnetic field, and then act as a passive transponder to emit microwaves, or UHF radio waves. Historically, the first RFID tag was demonstrated in 1971, and was a passive radio transponder with memory, powered by the interrogating signal. The wide range of applications includes transportation (vehicle identification, automatic toll payments, electronic license plate recognition, vehicle routing, vehicle performance monitoring, public transport access and payment), the pharmaceutical industry (tracking and labeling drugs), banking (electronic checks and electronic credit card), farming (identifying livestock), veterinary work (identifying pets), security (electronic passport recognition, personnel identification badges, automatic gates, surveillance), and medical (staff identification and location, patient history).

RFID tags can be passive, active, or battery-assisted passive. They contain a processor and a nonvolatile memory, an RF modulator and demodulator, and an antenna for receiving and transmitting signals. An RFID reader transmits an encoded radio signal to interrogate the tag. The tag receives the signal, then responds with its identification, a multi-bit Electronic Product Code (EPC), which can be used as a key into a global database to identify that particular tag. Unlike bar codes, RFID tags can be read anywhere within the range of a reader, even if it is not visible, and they can be read an indefinite number of times (Violino, 2005). The converse is that barcodes can be generated electronically, and sent by e-mail for printing or display; one well-known example is an airline boarding pass. On the RFID standardization issues, there is no fixed standard globally. As shown in Table 4.10, RFID tags use frequencies in the various license-free ISM bands according to the country where they are used. Passive RFID devices generally work at low frequency (125–134.2 kHz and 140–148.5 kHz), high-frequency (13.553–13.567 MHz), and ultrahigh-frequency (868–928 MHz), but there are variations. Active RFID devices use the 433.05–434.79 MHz and 2.4 GHz bands with wireless ranges up to 100 m, and data rates up to 54 Mbps. In North America, UHF can be used in the 902–928 MHz band (i.e., from 915 to ±13 MHz); in Europe, the 865–868 MHz band may be used, but with certain restrictions; and in Australia and New Zealand, the 918–926 MHz band may be used, but with power limitations.

There is increasing evidence that the use of RFID technologies, especially in hospital environments, indicates a cost-effective and economically efficient solution (Wang et al., 2006; Vilamovska, 2010). Other advances of RFID technologies include one of the smallest RFID chips, measuring only 0.05 mm by 0.05 mm, which can store a 38-digit number using 128-bit read-only memory (ROM) (Hitachi Global, 2006). These tags can be potentially used as "smart dust" in future m-Health applications, such as for miniaturized human implanted devices or in bioelectric medicine and other molecular tracking applications inside the human body. However, these are still possibilities and there are still many major technical and biological challenges for these to be applicable in healthcare. Some scenarios of human implantable RFID chips have been reported in potential applications such as digital patient identity information and medical record tracking (Dimov, 2014). Another is the digital RFID tattoo (BioStamp) that can be stamped directly onto the body to collect data on body temperature, hydration levels, and ultraviolet exposure (Wakefield, 2015). This technology can also be part of the current developments in ultrathin wearable devices. These are feasible scenarios, but they are difficult to implement, mainly because there are major personal privacy and security concerns that need to be addressed before such tagging applications become applicable in real medical settings (Kumar, 2007; Kitsos and Zhang, 2008).

Near-Field Communication (NFC) This technology allows contactless data exchange between devices like smartphones or tablets that are held close together, typically 10 cm or less. NFC uses magnetic induction between two loop antennas located in the near field of each device to form an air-core transformer (Near-Field Communications, 2014). NFC technology operates in the unlicensed ISM band at 13.56 MHz (ISO 14443), normally in a ±7 kHz bandwidth (which can be extended to 1.8 MHz) on the ISO/IEC 18000-3 air interface standard, at data rates of 106, 212, and 424 kbps. Communication is also possible between a smartphone and an unpowered NFC tag, of which there are four types, with different communication speeds, and between 96 and 4096 bytes of memory (Near-Field Communications, 2014).

NFC technology is used in some mobile and e-Health applications, such as secure medical tags, secure health cards for storing EHRs, and other tagging applications (Sethia et al., 2014). Although still limited, NFC technology has major potential for m-Health applications in the near future.

Infrared The use of infrared light, which is set by the Infrared Data Association (IrDA) standard, is familiar to anyone who has used a remote control to operate a television set. It normally works over a range of a few meters at frequencies below the human visible range, and was the early choice of the authors of this book for a physiological monitoring system using a GSM mobile phone before Bluetooth became widely available (Woodward et al., 2001).

It is evident that, unlike omnidirectional radio frequency communications, the disadvantage of infrared technology is that connectivity between one or more sensors and a mobile phone needs to be directional, that is, point-to-point in line of sight. In general, data are transmitted between devices using infrared light-emitting diodes (LEDs), with high-speed data rates of 1.15–4 Mbps over ranges of up to about 10 m. Due to these limitations, there is little scope for adopting this technology for most conceivable m-Health applications.

4.5.4 Proprietary Wireless Sensor Technologies

There are many proprietary wireless technologies designed and used for different automation and industrial monitoring applications. Increasingly, these technologies are used for m-Health wellness and activity monitoring. Table 4.11 shows some of the proprietary wireless technologies, with their basic specifications and potential

TABLE 4.11 Examples of Proprietary Wireless Technologies Used for m-Health

Technology	Frequency Band	Range	Data Rates	Features	Potential m-Health Applications
ANT/ANT+	868 MHz, 915 MHz 2.4 GHz	<10 m	<1 Mbps	Low power, short-range communications with low-power consumption	Health and fitness monitoring and other wellness devices
DASH7	433.92 MHz 2.4 GHz	<2 km	<200 kbps	Ultralow power consumption, with relatively longer distance	Patient monitoring in hospital wards Home care and patient tracking
EnOcean	868 MHz, 315 MHz	<30 m (indoor) <300 m (outdoor)	<125 kbps	Ultralow power	Home care, elderly care
INSTEON	902–924 MHz	<50 m	13- kbps	Low cost, improved security, power line connectivity option	Home care, elderly care, ambient assisted living
JenNet	2.4 GHz	200–450 m	<250 kps	802.15.4, ZigBee and IPV6 compliant, low power, low cost	Home care, and monitoring
Z-Wave	868.42 MHz, 908.42 MHz, 919.82 MHz, 921.42 MHz	<30 m	9.6 kbps, 40 kbps	Simpler Protocol, IPV6 and IOT compatible, improved security	Smart home care, patient care in hospitals

m-Health applications. These technologies can be used for WPAN and WBAN configurations and offer some of the following advantages over standardized wireless technologies:

- Stronger air transmission signaling and interference avoidance capabilities.
- Better security and encryption mechanisms; proprietary wireless technologies are not required to interact with standardized wireless technologies, hence can provide better security by comparison.
- Better quality of service, packet error corrections, and air interface protocols.
- Low power consumption and longer battery life.

What follows is a brief outline of these technologies. As technology is ever-changing, it is impossible to keep up with developments for more than a short period, so in a book it is inevitable that some of the details will be out-of-date soon after publication. Certain propriety technologies may well be omitted, so what we present here is not meant to be comprehensive, but it should give a flavor of the bewildering array of technologies available.

ANT and ANT+   These are proprietary wireless sensor network technologies marketed by Dynastream Innovations, which is based in Canada and owned by Garmin, the manufacturers of GPS monitors. It operates in the 2.4 GHz ISM band, and uses low-power transceiver chips, typically from Nordic Semiconductors and Texas Instruments (ANT Alliance, 2014). ANT is a flexible protocol, because each individual device, like a body sensor, can take on the role of transmitter, receiver, or bidirectional transceiver, and also as master or slave, depending on the network topology required, which can be point-to-point, star, tree, or mesh. An important characteristic of ANT is that it can spend long periods in a low-power "sleep" mode, "wake up" to transmit or receive data, and then return to sleep mode. Typical applications are characterized by the periodic transfer of small packets of sensor data every few seconds, so the protocol is well suited to health and fitness monitoring, particularly for cyclists and runners. It is also used in heart rate monitors, speed-ometers, calorimeters, blood pressure monitors, position tracking, homing devices, and thermometers. The usual arrangement has a transceiver embedded in a chest strap that is linked wirelessly to a watch-type display, forming a PAN, so that an athlete can monitor his or her performance in real time (ANT Alliance, 2014).

The *ANT* technology divides the 2.4 GHz ISM band into 1 MHz channels, and has a basic data rate of 1 Mbps. A TDM scheme accommodates multiple sensors. *ANT+* is an upgraded version of ANT, and is suitable for fitness and performance monitoring.

DASH7   This protocol is the ISO/IEC 18000-7 standard for wireless sensor net-works, and operates at 433.92 MHz in the unlicensed ISM band. It is distinguished from most other protocols by its extremely low-power dissipation, resulting in a battery life of up to 10 years. One of its major features is a nominal range of 1 km, and a maximum range of 2 km.

Another feature is a maximum data rate of up to 200 kbps, although the normal rate is typically less than 30 kbps. DASH7 Mode 2 is the latest version of the standard announced by a consortium called the DASH7 Alliance, which includes Texas Instruments and Analog Devices (Dash7 Alliance, 2014). One of the outstanding features of DASH7 is its capability of penetrating concrete and water; at first sight, this might seem an unlikely criterion in most biomedical monitoring scenarios, but it does mean that transmission through walls from one room to another, such as in hospitals, should be possible. It is worth noting that 433.92 MHz is a multiple of 13.56 MHz by 2^5 (32), which allows the use of antennas designed for 13.56 MHz (Swedberg, 2014). It also has good security features for medical applications.

Typical applications of DASH7 are in industry and commerce applications, for example, in access control to a building or room, tracking the location of things that can be "tagged," such as smart cards, key fobs, mobile advertising with "smart billboards," automotive innovation such as tire pressure monitoring, and "smart energy." This is relatively new technology, but it is easy to envisage its adoption for m-Health applications, notably in tracking patients or staff inside a hospital or care home (DASH7 Alliance, 2014).

EnOcean   The name is that of a German spin-off company of Siemens AG, which is a member of the EnOcean Alliance, a group of European and American companies that promote interoperability of products between partners (EnOcean Alliance, 2014). EnOcean is an "energy-harvesting" wireless technology, which converts small mechanical movements, or slight changes in lighting and temperature, into electrical energy. This is principally earmarked for applications in industry, transportation, logistics, and potentially for smart homes. EnOcean products and protocol conform to the ISO/IEC 14543-3-10 standard, published in 2012, which is optimized for wireless transmission at 315 MHz in North America and 868.3 MHz in Europe with ultralow power dissipation. It has a medium-range capability, typically 300 m outdoors and 30 m indoors. It uses ASK modulation, with data rates of approximately 125 kbps.

This technology has potential for m-Health applications, particularly in home care environments for elderly and disabled people. Its battery-free technology could be adopted in a smart home to allow, for example, wireless connectivity between switches and sensors for controlling lights and temperature, or for monitoring occupancy, carbon monoxide, and so on.

INSTEON   This technology is produced by SmartLabs, and it can use either AC power lines or radio frequency wireless links for connecting switches to a variety of devices, such as sensors, without the constraint of dedicated wiring. All devices in the protocol can be networked using mesh topology to send and receive data, and they all repeat each message in specific time slots synchronized by the AC power frequency (50 Hz in the United Kingdom, 60 Hz in the United States), which means there is no need for a master controller or hierarchical software. Automatic error detection and correction are included in all compatible products. The technology can support a large number of nodes (>1000). The Insteon hub mainly supports applications in home automation networking, so it is particularly appropriate for disabled or elderly people living in smart homes, although so far there appears to be little, if any, such exploitation of the technology. Its data rate of 13 kbps in the 902–924 MHz American unlicensed band is slow compared with other technologies, and at 50 m maximum range, requires a line of sight between source and receiver (Insteon Alliance, 2014).

JenNet   This wireless network protocol was developed to provide low-power, wireless connectivity and rated as a simpler alternative to the ZigBee Pro protocol. It is designed for monitoring and control applications, with up to 250 sensors or devices. One useful feature is that the battery-powered devices in the network can be omitted or others added without affecting its operation. Also, the protocol has a "listen-before-talk" facility, and can be used with other IEEE 802.15.4 wireless technologies, such as ZigBee and 6LoWPAN, in the same overall system (Jennic, 2014). Some of the applications of this technology include home automation, automated meter readings, and environmental control and security systems. More recently, the JenNet-IP version provides an IP-based networking solution, enabling an IoT IPv6 connectivity to smart devices with up to 500 nodes in the home and other buildings. This feature allows the technology to adapt to future IPv6-based m-Health applications and M2M connectivity architectures.

Z-Wave   Z-Wave was developed by a Danish company called Zen-Sys, which was taken over by Sigma Designs in 2008. It is the flagship of the Z-Wave Alliance, a consortium of over 250 companies worldwide that was established in 2005 to work to agreed-upon specifications for interoperable applications associated with wireless "smart" homes, remote control, and security applications (Z-Wave Alliance, 2014). It is a leading technology for the IoT, and is also a candidate for medical and healthcare applications, including use in care homes for the elderly and disabled people, as it uses sensor-based technology specifically for healthcare applications (Z-Wave Alliance, 2014). This protocol operates at a single frequency, using FSK coding, but at specific carrier frequencies in different countries: 868.42 MHz (Europe), 908.42 MHz (the United States), 919.82 MHz (Hong Kong), and 921.42 MHz (Australia/New Zealand). Thus, a BAN based on Z-Wave technology is not transportable between, say, the United States and the United Kingdom. There are also serious limitations that might inhibit its use. For example, in Europe, there is a 1% duty cycle limitation at the available frequency band, which ensures low-power dissipation but may make many monitoring applications impractical. Z-Wave uses a mesh network of up to 232 devices (nodes), which consist of controllers and slaves. The protocol is optimized for small data packets with data rates up to 100 kbps, although typical rates are between 9.6 and 40 kbps. There is no "sleep" mode, as devices have to be "awake" to retransmit messages to guarantee connectivity in a multipath environment, such as a home or hospital. A typical communication range between two devices is 30 m, which ensures good reliability in most homes, because of the in-built capability of "hopping" up to four times between devices.

In addition to these technologies mentioned above, there are many other proprietary wireless technologies and protocols. A complete list is beyond the scope of this book, and can be found in more specialist references (Kraemer and Katz, 2009). There is an increasing trend in the use of these technologies in consumer wellness and health monitoring applications (Baum and Abadie, 2013). However, so far there have been no comprehensive studies to validate the effectiveness of each of these wireless technologies for various m-Health applications in healthcare settings compared to other short-range technologies.

4.6   MACHINE-TO-MACHINE COMMUNICATIONS AND INTERNET OF THINGS

The IoT is a new and transformative paradigm that could have a profound impact on the future of m-Health, particularly for communications (ITU, 2005). Different market leaders and global businesses refer to the IoT concept with different terms and abbreviations such as "Internet of Everything" (Cisco), or the Industrial Internet (GE). This paradigm shift will transform current m-Health communications toward realignment with full Internet-based access modalities. The impact of IoT will straddle many domains and stakeholders, including retail, energy, transport, logistics, media, manufacturing, and of course healthcare. The IoT and its enabling M2M technology is rapidly crystallizing the Internet-based communications vision of billions of con-nected physical devices and objects, such as sensors, actuators, tags, smart meters, and home appliances, all connected to provide their associated functions and services via the Internet, and accessed via Internet-connected mobile network gateways (Verme-san and Friess, 2014). Furthermore, it is also projected that M2M mobile connections will reach more than a quarter (26%) of total devices and connections by 2020 (Cisco, 2016). In general, M2M communication broadly "encompasses a number of areas where devices communicate with each other without human involvement" (ITU, 2012a, 2012b). These communication technologies can potentially improve services in diverse industries and sectors, such as healthcare, transport and manufacturing, utilities, and other sectors. The M2M communications concept is effectively in the growing phase, or at least it is in its early stages of deployment in the healthcare sector. In this section, we describe some of the recent M2M developments in the mobile health domain. The reader can refer elsewhere to have in-depth details of communi-cation and networking aspects, together with standardization issues, that are beyond the scope of this book (Vermesan and Friess, 2014; Hersent et al., 2012).

4.6.1   IoT and M2M Communications for Healthcare

According to a recent report, the healthcare segment of the IoT market is poised to hit $117 billion by 2020 (McCue, 2015). It is predicted that IoT-based m-Health systems will constitute a substantial part of this market share, considering the projected growth in the consumer-driven systems and services such as wearable devices and smart wellness tracking devices. It is important to note that the IoT for healthcare is a complex concept and requires different implementation strategies and service archi-tectures. There are several interpretations of the basic idea behind the IoT, which reflect the concept of connectivity of anything at any time and from any place

(Jamoussi, 2010). According to the ITU-T Study Group 13, which leads the work of the ITU on the standards for next-generation networks (NGN), the IoT is defined as "a global infrastructure for the information society, enabling advanced services by interconnecting (physical and virtual) things based on existing and evolving inter-operable information and communication technologies" (ITU, 2012a, 2012b). IoT can also be envisioned as "a global network of connected devices having identities and virtual personalities operating in smart spaces, and using them with a future vision that represents a future where billions of everyday objects and surrounding environments will be connected and managed through a range of communication networks and cloud-based servers" (Ashton, 2009; Atzori et al., 2010; Li et al., 2011).

The major general trends of IoT systems are compatible with most of the future trends of m-Health systems discussed earlier. These include big data and analytics; cloud-based computing and storage systems; mobility and ubiquitous coverage; and universal and high-speed data access (Uckelmann et al., 2011).

IoT for healthcare is emerging as a major technology concept for m-Health. However, it represents important new opportunities whereby millions of IoT-enabled medical sensors, smart medical devices, and mobile health terminals may be connected to the Internet for access by patients and clinicians. There is currently extensive work in the developed world, including analysis using embedded Artificial Intelligence (AI), to enable this vision to come to fruition.

M2M communications act as an enabling technology for the practical implementation of the IoT concept. This emerging communications and networking paradigm envisions the interconnection of machines without the need of human intervention. The main concept lies in the ubiquitous connectivity between devices, along with their ability to communicate seamlessly in any autonomous and self-organizing heterogeneous wired or wireless communication network (Hammond, 2011).

The growth in the M2M sector is expected to increase exponentially in the near future. This will be fostered by the massive deployment of sensors, actuators, tags (RFIDs), smart metering, and other machine-type devices (MTDs) that will be used in M2M applications (ITU, 2012a, 2012b). Some of these applicable areas include security and public safety (surveillance systems, object/human tracking, alarms, and so on), smart grids (grid control, industrial metering, demand response), vehicular telematics (fleet management, enhanced navigation), manufacturing (production chain monitoring), and remote maintenance (industrial automation, vending machine control) (Wu et al., 2011). The healthcare sector will be one of the priority areas targeted by M2M services and industry.

M2M networks can be divided into two broad types: capillary M2M and cellular M2M networks (3GPP, 2010; ITU, 2012b). In capillary networks, M2M devices form a "device area network" wherein connectivity is provided through short-range communication technologies (e.g., ZigBee, 6LoWPAN, and Wi-Fi). The wider area connectivity is provided through an Internet gateway. These capillary M2M networks are generally characterized by large numbers of low-cost, low-complexity devices, such as sensors and RFID tags, with requirements of high-energy efficiency and reliability, low packet loss, and use of low-power link layer technologies. In cellular networks, M2M devices are each equipped with an embedded mobile subscriber identity module (SIM) card to communicate autonomously within the cellular network, and hence emulate a mobile device. Most of these cellular M2M devices have small volume data transmissions for uplink traffic, and minimum device mobility, together with specific service requirements of high-energy efficiency. On standardization efforts, there are currently working Standards Development Organizations (SDOs) whose aims are to create a unified co-operative M2M standards activity globally (3GPP, 2014a, 2014b; Vermesan and Friess, 2014; ISO, 2014). The future impact of IoT and its importance is best reflected in the different global IoT standardization and interoperability activities currently ongoing, which include the following:

- *ITU IoT Global Standards Initiative (IoT-GSI) (www.itu.int/en/ITU-T/gsi/iot/ Pages/default.aspx):* This is the global initiative that promotes a unified approach in ITU-T for development and recommendation of technical standards to enable the IoT on a global scale in collaboration with other SDOs.
- *AllSeen Alliance (*www.allseenalliance.org*):* This is a nonprofit consortium dedicated to driving the widespread adoption of products, systems, and services that support the Internet of Everything (IoE) with an open, universal development framework (AllJoyn). It has currently more than 150 members working on different aspects of IoT interoperability.
- *IEEE-IOT Association and Working Group (*standards.ieee.org/innovate/iot/*):* The IEEE P2413 is a working group that developed a standards draft for an architectural framework for the IoT Working Group and IoT ecosystem.
- *OneM2M (*www.onem2m.org/*):* This is a global initiative, established in 2012, of eight leading telecom SDOs for M2M communications that are developing cross-industry telecommunications standards for IoT. It released its specification (OneM2M release-1) in 2015.
- *Industrial Internet Consortium-IIC (*www.iiconsortium.org*):* This consortium was founded in 2014 by AT&T, CISCO, GE, Intel, and IBM, "to bring together the organizations and technologies necessary to accelerate growth of the Industrial Internet by identifying, assembling and promoting best practices and also to co-ordinate vast ecosystem initiatives to connect and integrate objects with people, processes and data using common architectures, interoperability and open standards."

From the mobile health perspective, there are ongoing standardization efforts, particularly from the ITU and other collaborating SDOs, to develop specific e-Health and m-Health architectures and specifications for M2M (ITU, 2014b; ISO, 2014). Similarly, the European Telecommunications Standard Institute (ETSI), acting as one of the leading OneM2M consortium SDOs, has developed generic M2M architecture proposals for devices and applications that are connected through the network domain (ETSI, 2011). This includes access and core networks with M2M servers to provide services and enable applications. The ITU-T telecommunication standardization sector, the ITU Focus Group FG M2M, has proposed a reference model (ITU-T

M2M reference model) tailored for m-Health and e-Health applications. This model supports the capabilities of the four layers identified in the ITU-IoT reference model (ITU, 2014a).

To illustrate these architectures and specification models for mobile health, Fig. 4.8 shows a proposal for an M2M m-Health monitoring architecture based on the ETSI-IoT reference model (Hersent et al., 2012; ETSI, 2011). In this proposal, medical devices, such as implantable or wearable sensors, are connected to the core M2M network in one of two ways. The first is to connect these devices through the cellular M2M network, as described earlier, that is, equipping each medical device with its own SIM chip (card) for its cellular network connectivity. The second is to connect to a capillary M2M network link, whereby these devices organize themselves locally, configuring as an M2M area network by exploiting the available connectivity options of any of the short-range networking technologies, such as a WBAN (IEEE 802.15.6), WSN (IEEE 802.15.4 or IEEE 802.15.4e), or Wi-Fi (IEEE 802.11). These capillary M2M networks are then connected to the core access networks through an M2M gateway that acts as proxy between the M2M sensors network and the access network that links these to any of the core network options (e.g., 3GPP, 3GPP2, and ETSI TISPAN). The core network then provides the IP connectivity link with the relevant M2M application and server (middleware layer), thus providing the required m-Health data services. This model also supports a service capabilities layer (SCL) that supports the different applications enabled by the M2M technologies (Hersent et al., 2012; ETSI, 2011). This SCL layer can be positioned with respect to the ITU-T's IoT reference model described next (ITU, 2014b).

A proposed m-Health M2M monitoring architecture, based on the ITU-T M2M and the IoT "reference model," is shown in Fig. 4.9. This model consists of the following main components (ITU, 2014b):

- *M2M Devices:* These are m-Health sensors that can be connected directly or via a suitable gateway to the network.
- *M2M Gateway:* This component hosts the M2M service layer capabilities and the specific application targeted (e.g., m-Health monitoring).
- *Network:* This is the component that connects the M2M devices (e.g., M2M-enabled sensors) with the M2M gateway and the network application server.
- *M2M Platform:* This hosts the M2M service layer capabilities, and can be used by one or more application servers; this is part of the network application server. These include, for example, the different functionalities shown in Fig. 4.8.
- *Application Server:* This component hosts the network applications and is also part of the network application server.
- *Network Application Server:* This component includes both the application server and M2M platform described above.

In this reference model, the M2M platform is part of the network application server. The service layer provides the different capabilities required for each service, such as communications management, device management, security management,

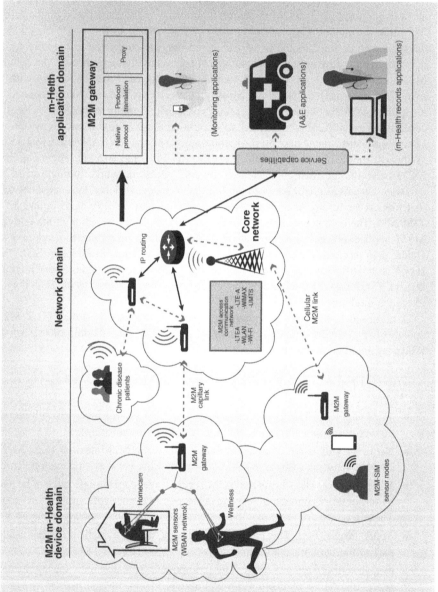

FIG. 4.8 Example of the ETSI M2M m-Health monitoring system architecture. (Adapted from ETSI (2011).)

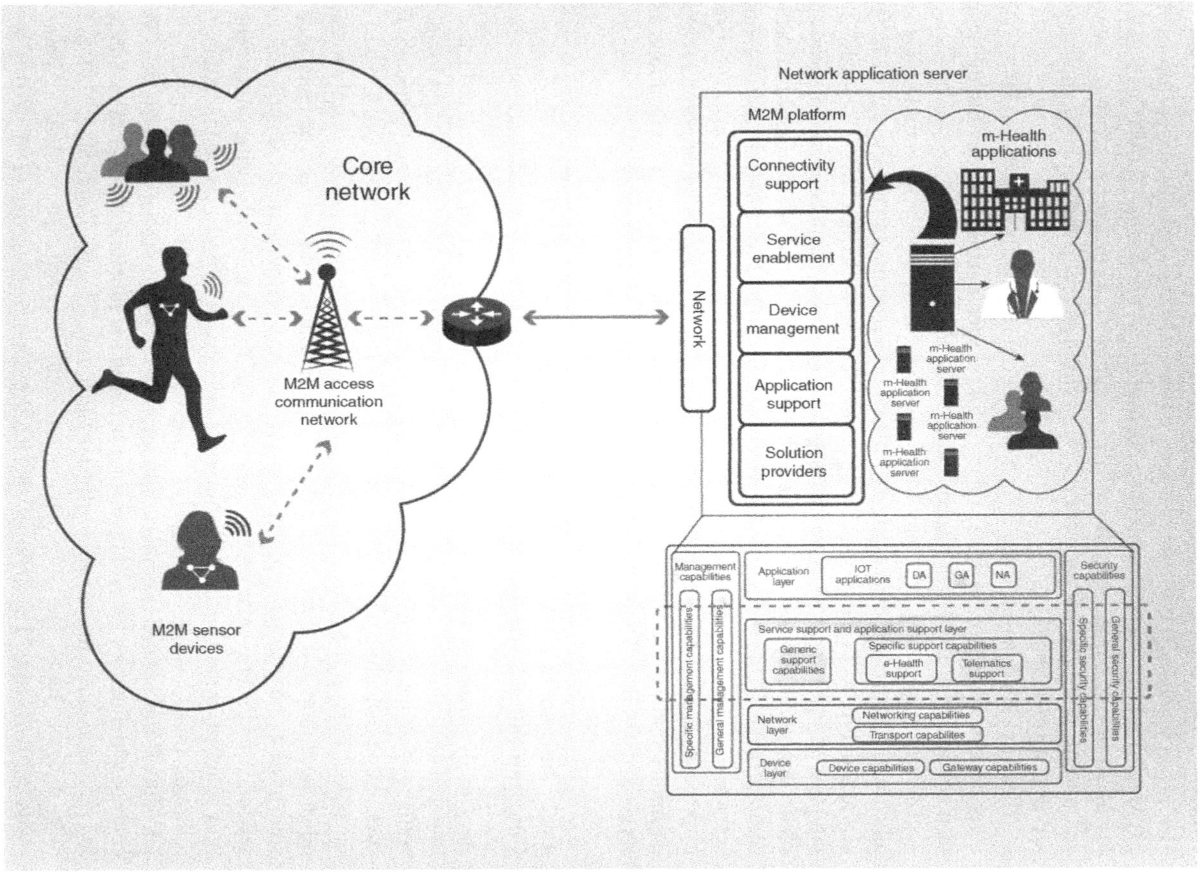

FIG. 4.9 Example of ITU-T M2M m-Health monitoring system architectural model. (Adapted from ITU (2014b).)

location provision, data processing, and many others required by the specific applications targeted.

Figure 4.9 shows that the service layer is linked to the application layer by reference points with specific support capabilities that include any e-Health and m-Health applications. Furthermore, three types of M2M applications on top of the M2M service layer are proposed (ITU, 2014b): *device application* (DA) that resides in the M2M device component; *gateway application* (GA) that resides in the gateway component; and *network application* (NA) that resides in the network application server described above.

These applications typically use the capabilities provided by the M2M service layer. Further details of these and other technical specifications are beyond the scope of this book, and may be accessed online (ITU, 2014b).

The main requirements of any M2M network for m-Health are complex. Generally, the aim is to connect securely to healthcare servers running the specific m-Health services that process the medical data acquired from networked M2M devices. It is vital that any final architecture addresses the specific m-Health ecosystem model and security requirements for acceptable implementations in real medical and healthcare settings (ITU, 2014b). This model is perhaps well-suited for future M2M m-Health services and applications. Analysis of this model reveals implementation mechanisms to provide more m-Health-specific support capabilities, supplemented with security capability options that are much needed for m-Health or e-Health applications and services.

4.6.2 Internet of m-Health Things (m-IoT) and 6LoWPAN Communications Architecture

As already mentioned, the global interest in M2M communications reflects the importance of the IoT concept in reshaping the future of mobile health and to adopt better global health connectivity options and effective care access services. The current understanding of the IoT concept from the healthcare perspective is based on the model of typical M2M-connected health sensor networks linked directly to each other to capture and share vital medical data through a secure service layer that is connected to the Internet or cloud, thus providing the relevant medical application or service. This generic view perhaps needs further refinement and clarification to allow it to better fit in the overall IoT concept, especially in terms of communications. To reinforce this view, the concept of "Internet of m-Health Things" (m-IoT) was introduced and defined, aiming to interpret these developments for m-Health. This concept is defined as "an amalgamation between the functionalities of m-Health and IoT, for innovative future (4G Health) applications" (Istepanian et al., 2011). Since then, different terms and abbreviations such as "The Internet of Medical Things" (Miles, 2014), and "The Internet of Healthy Things" (Kvedar et al., 2015) were introduced on the web and the relevant literature.

Figure 4.10 shows the M2M protocol stacks of M2M example links of ZigBee, 6LoWPAN, and 802.11 Wi-Fi standards for comparison. The figure shows the 6LowPAN adaptation layer for transporting the IPv6 packets over 802.15.4 links,

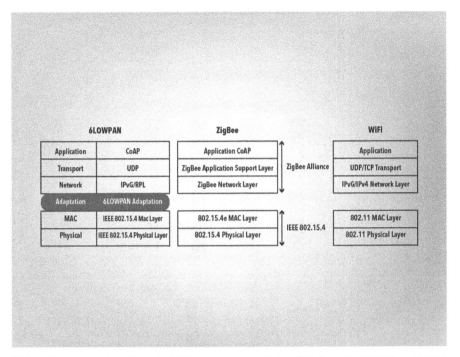

FIG. 4.10   Protocol stacks and M2M communication links of ZigBee, 6LoWPAN, and Wi-Fi technologies.

required to resolve the problems associated with sending the data over this wireless link (Shelby and Bornmann, 2009).

The figure shows that the adaptation layer lies between the 802.15.4 data link layer and the IP stack. This defines the necessary mechanisms required to compress the IPv6 headers, to segment their datagrams, and consequently to send these efficiently. This process makes the protocol suitable for any IPv6 connectivity applications, compared to the other protocols mentioned above. The basic functionalities of these protocol stacks are well known from the mobile networking perspective. In simple terms, they characterize and standardize the internal functions of any wireless communication and networking system by partitioning it into these abstraction layers to allow for the system to function. Further communication, networking, and technical details of IoT and M2M communications issues are beyond the scope of this book, but more specialized texts are available to interested readers (Perera et al., 2013; Shelby and Bornmann, 2009).

Figure 4.11 illustrates the communication protocols associated with the IP-based IoT layers and those corresponding to M2M communications, which are based on the protocol links below the network layer. These include, for example, 6LoWPAN, ZigBee, Bluetooth, and other wireless protocol technologies.

These new communication architectures will lead to a new generation of Internet-connected mobile health models that reflect the smarter and more personalized

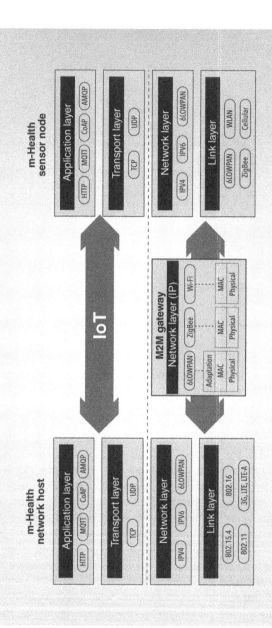

FIG. 4.11 Communication Protocols for the IoT and M2M Networks.

m-Health requirements and clinical options demanded by future healthcare services for more effective and efficient delivery. These new connectivity modes can play an important part of the future m-Health 2.0 vision described in Chapter 3. They can also improve or even eliminate the deficiencies of current m-Health architectures. Communications and computing advantages that these Internet-enabling technologies can benefit their relevant networking architecture include (Yang, 2014) the following:

- Facilitating interoperability by allowing the use of existing network infrastructures, based on all IP-based protocols.
- Allowing wireless devices to be connected directly to the Internet without the need for any further gateway connections.
- Establishing application protocols and data models that can be used in different applications.

These benefits will also bring more efficient and faster cloud access by using future m-IoT configurations, which could use IEEE 802.15.4 or 6LoWPAN wireless protocols to become part of future m-Health 2.0 ecosystems. With low-power and seamless Internet access functionalities, these will be potential candidates for future M2M-enabled wearable sensors used for wellness and sport and consumer health monitoring applications.

As M2M sensors for m-Health applications are typically very low-power devices or nodes, there are additional requirements for the gateways linking them to the Internet. The gateways need to be compatible with the level of data processing and the energy resources available in the devices. Typically, it is the gateways that perform data processing rather than the networked devices; this includes routing and facilitating communications between the different networks involved in the complete communication chain. Hence, there is need to have very low-power and efficient standard protocols that can adapt to these communications requirements and sustain the battery life of the sensors for as long as possible.

As shown in Figs 4.10 and 4.11, some of the wireless protocols and links for M2M connectivity include IEEE 802.11, IEEE 802.15.4 (Bluetooth/Bluetooth Smart, ZigBee/ZigBee Smart Energy 2.0), IEEE 802.15.6-2012 (Body Area Networking), Z-Wave, and ANT+.

For m-IoT applications, IEEE 802.15.4 is the most viable option because it satisfies the low-power communication requirements of the standardized and proprietary technologies. The 802.15.4 standard defines the network layers for low-duty cycles, low-throughput, and low-power devices, making it suitable for mobile and medical monitoring applications. The 6LoWPAN is well-suited for all IPv6 m-Health applications that adapt to future all-IP network connectivity. The Bluetooth/Bluetooth Smart protocol is more suitable for shorter range and relatively low data rate m-Health applications geared toward the current cellular networks.

Technically, 6LowPAN stands for IPv6 over 802.15.4 Low-Power Wireless Personal Area networks (L-WPAN). The 6LoWPAN standard enables the efficient use of IPv6 over low-power, low data rate wireless networks on simple embedded devices, and was initially developed by the Internet Engineering Task Force (IETF) in 2007 to allow the transport of IPV6 packets over IEEE 802.15.4 links (Shelby and Bornmann, 2009). This allows 6LoWPAN-based sensors to be seamlessly connected to IP-networked m-Health systems.

Most of the current work on m-IoT is still at an early stage and largely in experimental or prototyping phases, with more emphasis on the wireless communication and technical challenges. However, future m-Health connected sensors will be deployed using these M2M technologies. The need for further clinical studies using M2M technologies for m-Health applications in real patient settings is important. In addition, further work to evaluate the efficiency, user functionality, and reliability of these systems, compared to the more standardized wireless device configurations, is another important aspect that needs to be carried out before such systems can be used effectively in mainstream clinical healthcare services. The combined challenges of the big health data and IoT are another emerging area of work that will attract further attention in the mobile health domain.

There are other challenges for the m-IoT concept that need to be addressed carefully as part of its future evolution:

- Sensor power consumption, wireless range, coordination, and control. There are currently some proposals to address these challenges to provide robust long-range wireless signaling and control platforms (IBM, 2015).
- Healthcare service architectures and specifications required for m-IoT healthcare applications, including monitoring, diagnostics, and robotics.
- End-to-end attributes, including scaling, distribution, and other relevant factors.
- IoT security and privacy. The need to develop robust security and privacy protection for m-IoT systems and architectures in view of their potential vulnerability to future cybersecurity attacks is vital for successful m-Health applications. Current IoT devices have limited memory, processing capability, and power, and consequently any existing security mechanisms that might be adopted for m-IoT applications would be inadequate, given the sensitive nature of the data. There is ongoing global effort by the IETF and other standardization bodies to address various topics related to IoT security (Ishaq et al., 2013).

From the market perspective, recent predictions for the potential of IoT technologies in transforming healthcare services include (Isaacs, 2014) reducing device downtime through remote monitoring and support; proactive fulfillment by replenishing supplies before they are needed; and efficient scheduling by leveraging utilization to serve more patients.

The IoT market is growing fast and some countries, such as the United Kingdom, are now implementing these networks, such as in smart city infrastructure services (Onita, 2015). However, the expected potential IoT benefits for healthcare services are yet to be seen. There is increasing trend of IoT applications in some modern hospital IT workflow processes, and in computing and data management applications.

The success of the IoT and M2M communication technologies in healthcare will have to be clinically and rigorously validated as well as industrially led and marketed. The complexity of the medical devices and healthcare market poses additional challenges for this sector. The potential proliferation of technologies for other vertical market sectors, such as transport, entertainment, and smart cities, cannot be easily duplicated in the healthcare sector before rigorous clinical validation studies are carried out. However, for the consumer m-Health wellness and fitness market, this new sector will open massive business opportunities sooner rather than later.

4.7 SUMMARY

The unprecedented advances in mobile communication systems in the last decade have reshaped how m-Health was envisaged a decade ago. Today, the mobile technology roadmap includes global stakeholders, from telecommunication operators to the mobile and medical devices industry, which are injecting multi-billion dollar investments in different areas of mobile health. These developments are numerous and varied, and we have attempted to list some of them in this chapter. The ongoing evolution of mobile communication technologies will continue to be one of the driving forces for future development of m-Health systems. Furthermore, mobile health applications will benefit from the increased availability, miniaturization, performance, enhanced data rates, and the expected increasing convergence of forthcoming wireless communication and network technologies. This will further accelerate the development of smarter m-Health services for the provision of better and more clinically viable healthcare delivery services.

In this chapter, we have presented in detail the evolution of m-Health in parallel with similar developments in the mobile communications industry that have influenced this process, from its early beginnings to the current date. We looked at this process from early GSM-based mobile health systems from the 1990s to future 5G-based systems. It is widely perceived that cellular networks and their smartphone communications are the core communication technology enablers for m-Health service delivery. We have also presented other viable networking options that can perhaps provide better accessible delivery with cost-effective solutions. The popular view of m-Health as a smartphone (cellular network operated) healthcare delivery concept is perhaps responsible for the existing ambiguity and skepticism of m-Health. This is supported by the lack of consensus and universal evidence of its clinical effectiveness so far. We will explain and clarify this notion and the relevant issues with more in-depth analysis in Chapter 5. However, the proliferation and adoption of better alternative communication and networking technologies for m-Health based on IP Internet access and connectivity are essential to mitigate, or even reverse, this trend. These can potentially lead to the development of new m-Health architectures and ecosystems to provide better clinical evidence and more universal access services to larger populations with lower cost burdens. As part of this future vision, we have presented an overview of the paradigm of an m-Health Internet of things with machine-to-machine communications, and their potential impact.

From the economic and communications policy perspectives, there is a need to propose alternative new m-Health communications policies that take account of current telecom and market driven m-Health ecosystems and services. It is also worth mentioning that most of the current m-Health communication systems operate on either licensed (e.g., WMAN, cellular) networks or in the unlicensed spectrum of available radio frequency bands (e.g., WPANs and WLANs). There are a few factors that dictate the choice between having a licensed or an unlicensed wireless m-Health network. This choice influences the economics of any particular m-Health communications infrastructure, location, and cost of the specific healthcare services to be provided. Furthermore, it is obvious that the dominating and global use of these networks operating on licensed radio frequency spectrums have so far precluded any viable economic alternatives; these would use unlicensed networks to provide cheaper and more universally accessible m-Health services. Spectrum licenses are typically owned by the cellular network operators, which empower them with the ability to have virtually private networks for their own licensed radio frequency spectrum to provide their network services over the user's smartphones. This revenue-generating model is used in most current cellular (smartphone) m-Health ecosystems, as we will explain further in Chapter 7.

These business and economic models can be successfully applied by network operators to provide many vertical markets and services, such as entertainment, banking, and transport. Perhaps the exception is for the wearables and wellness devices monitoring market (not patient and healthcare services but consumer health markets); there is no clear evidence that such models can work as successfully.

The alternative option is to develop new m-Health architectures based on unlicensed frequency technologies, combined with an all-IP access communication. This can potentially offer better financial models and reduce the operating costs on both the end users (patients, doctors, carers) and healthcare providers. This might also allow healthcare providers the flexibility to deploy their own m-Health communications infrastructures with less monopolization and dependency on cellular networks and their smartphone data dependency costs.

It is perhaps timely to address all these options as part of a policy to review the current m-Health service delivery models. The aim would be to provide better economical alternatives and clinically viable m-Health systems from what exists so far. This new notion is particularly important for providing better and more efficient healthcare services in poorer and disadvantaged areas in the world where m-Health is urgently needed. Some of the communications options presented in this chapter might provide such alternatives.

We cannot underestimate the power of smartphones and their current proliferation; their global dominance combined with their global access and coverage ensure their continued impact on the future of m-Health. Most current m-Health systems are operated on mobile telecommunications infrastructures by global communications industries whose interests are mostly in m-Health as a vertical market in their business models and services. Alternative m-Health architectures can be problematic, to say the least, for these operators. There is perhaps a stronger yet uncharted case for developing new m-Health access models collaboratively with the telecommunications industry. Future trends from the communication perspective range from developments in 5G networks, the Internet of Things, cognitive and green communication systems, and other emerging concepts. These trends may in due course influence the current market and consumer-driven approach behind m-Health and reshape its future by the end of its second decade.

REFERENCES

Adibi S (2012) Link technologies and Blackberry mobile health (m-Health) solutions: a review. *IEEE Transactions on Information Technology in Biomedicine* 16(4):586–597.

Ahmed A and Kohno R (2013) Medically reliable network using concatenated channel codes through GSM Network. In: *Proceedings of the 35th Annual International Conference of the IEEE EMBS*, Osaka, Japan, July 3–7, 2013, pp. 4755–4758.

Alam MM and Hamida EB (2014) Surveying wearable human assistive technology for life and safety critical applications: standards, challenges and opportunities. *Sensors* 14: 9153–9209.

Alinejad AN, Philip N, and Istepanian RSH (2011) Mapping of multiple parameters m-Health scenarios to mobile WiMAX QoS variables. In: *Proceedings of the 32nd Annual International Conference of the IEEE Engineering in Medicine and Biology Society*, Boston, USA, pp. 1532–1535.

Alinejad A, Philip, N, and Istepanian RSH (2012) Cross layer ultrasound video streaming over mobile WiMAX and HSUPA networks. *IEEE Transactions on Information Technology in Biomedicine* 16(1):31–39.

ANT Alliance (2014) Available at http://www.thisisant.com/ (accessed October 2014).

Ashton K (2009) *That 'Internet of Things' thing.* RFID Journal, June 22.

Atzori L, Iera A, and Morabito G (2010) The Internet of Things: a survey. *Journal of Computational Networks* 54(15):2787–2805.

Banitsas K, Istepanian RSH, and Tachkara S (2002) Applications of medical wireless LAN (MedLAN) systems. *International Journal of Health Marketing* 2(2):136–142.

Baum P and Abadie F, (Eds.) (2013) *Strategic intelligence monitor on personal health systems, phase 2. Market developments: remote patient monitoring and treatment, telecare, fitness/ wellness and m-Health*, European Commission, Joint Research Centre, Institute for Prospective Technological Studies. Available at http://ftp.jrc.es/EURdoc/JRC71141.pdf (accessed October 2014).

Bertenyi B (2014) *System standards heading into the 5G era. 3GPP global initiative: the mobile broadband standard.* Available at http://www.3gpp.org/news-events/3gpp-news/ 1614-sa_5g. (accessed November 2014).

Bluetooth (2014). Available at http://www.bluetooth.com/Pages/How-It-Works.aspx (accessed October 2014).

Bluetooth Low Energy (2013) Bluetooth SIG. Available at http://www.bluetooth.org/ (accessed August 2014).

Brydon A (2014) *HSPA+ goes from strength to strength: unwired insight.* Available at http:// www.unwiredinsight.com/2014/evolved-hspa (accessed October 2014).

Cao H, Leung V, Chow C, and Chan, H (2009) Enabling technologies for wireless body area networks: a survey and outlook. *IEEE Communications Magazine* 47: 84–93.

Cavallari R, Martelli F, Rosini R, Buratti C, and Verdone R (2014) A survey on wireless body area networks: technologies and design challenges. *IEEE Communications Surveys & Tutorials* 16(3):1–23.

Cisco (2016) *Cisco visual networking index: global mobile data traffic forecast update, 2015–2020*, White Paper, February. Available at http://www.cisco.com/c/en/us/solutions/collateral/service-provider/visual-networking-index-vni/mobile-white-paper-c11-520862.pdf (accessed, June 2016).

Cisco VNI (Visual Networking Index) (2012) *Global mobile data traffic forecast update 2011–2016*, Cisco Public Information, May 30, 2012 (accessed October 2014).

Cisco VNI (Visual Networking Index) (2015) *Global mobile data traffic forecast update, 2014–2019*, White Paper. Available at www.cisco.com/c/en/us/solutions/collateral/service-provider/visual-networking-index-vni/white_paper_c11-520862.html (accessed April 2015).

CREW (Cognitive Radio Experimentation World) (2014) European Research Project: FP7. Available at www.crew-project.eu (accessed October 2014).

DASH7 Alliance (2014) Available at http://www.dash7-alliance.org (accessed November 2014).

Debono CJ, Micallef B, Philip N, AliNejad AN, Istepanian RSH, and Amso N (2013) Cross layer design for optimised region of interest of ultrasound video data over mobile WiMAX. *IEEE Transactions on Information Technology in Biomedicine* 16(6):1007–1014.

Dimov (2014) *Human-implanted RFID chips*, InfoSec Institute. Available at http://resources.infosecinstitute.com/human-implanted-rfid-chips/ (accessed October 2014).

Eberspächer J, Vögel HV, Bettstetter C, and Hartmann, C (2009) *GSM: Architecture, Protocols and Services*, 3rd edn. John Wiley & Sons, Inc.

Elahi A and Gschwender A (2009) *ZigBee Wireless Sensor and Control Network*. New Jersey: Prentice Hall.

EnOcean Alliance (2014) Available at https://www.enocean-alliance.org (accessed October 2014).

Ergen M (2009) *Mobile Broadband: Including WiMAX and LTE*. New York: Springer.

Etemad K (2008) Overview of mobile WiMAX technology and evolution. *IEEE Communications Magazine* 46(10):31–40.

ETSI (European Telecommunications Standard Institute) (2011) *TS 102 690 V1.1.1: machine-to-machine communications (M2M); functional architecture* (accessed October 2014).

European Commission 5GPPP (2015) *5G vision: the 5G Infrastructure Public Private Partnership—the next generation of communication networks and services*, executive summary. Available at www.5g-ppp.eu/roadmaps (accessed March 2015).

Fallgren M and Timus B, (Eds.) (2013) *Scenarios, requirements and KPIs for 5G mobile and wireless systems*. Available at www.metis2020.com/wp-content/uploads/deliverables/METIS_D1.1_v1.pdf (accessed December 2014.)

Fettweis, G (2014) The tactile Internet, applications and challenges. *IEEE Vehicular Technology Magazine* 9(1):64–70.

Fiordelli M (2013) Mapping m-Health research: a decade of evolution. *Journal of Medical Internet Research* 15(5):e95.

5GNOW (2014) 5th Generation non-orthogonal waveforms for asynchronous signalling, FP7 European Union Supported 5G Project. Available at www.5gnow.eu (accessed October 2014).

Galeev M (2011) *Bluetooth 4.0: an introduction to Bluetooth low energy: Part I*, Electronic Engineering Times, July. Available at www.eetimes.com/document.asp?doc_id=1278927 (accessed October 2014).

Gao T (2007) The advanced aid and disaster network: a lightweight wireless medical system for triage, *IEEE Transactions on Biomedical Circuits and Systems* 1(3):203–216.

Garawi SA, Istepanian RSH, and Abu-Rgheff, MA (2006) 3G wireless communications for mobile robotic tele-ultrasonography systems. *IEEE Communications Magazine* 44 (4):91–96.

Gatherer A (2015) *5G and the next billion mobile users: a view from Africa*, IEEE Communication Society Technology News, December 2015. Available at http://www.comsoc.org/ctn/5g-and-next-billion-mobile-users-view-africa (accessed January 2016).

GSMA (2011) White Paper. Available at http://www.gsma.com/connectedliving/wp-content/uploads/2012/03/connectedmobilehealthdevicesareferencearchitecture.pdf (accessed October 2014).

GSMA (2012) Connected mobile health devices: a reference architecture, V1.0, January.

GSMA Intelligence (2014a) *Measuring mobile penetration: untangling subscribers, mobile phone owners and users.* C Available at https://gsmaintelligence.com/files/analysis/?file=2014-05-22-measuring-mobile-penetration.pdf (accessed October 2014).

GSMA Intelligence (2014b) *Understanding 5G: perspectives on future technological advancements in mobile, an analysis report.* C Available at gsmaintelligence.com/research/?file=141208-5g.pdf&download (accessed March 2015).

GSMA Intelligence (2015) *The mobile economy 2015*, GSMA Intelligence Report. Available at http://www.gsmamobileeconomy.com/GSMA_Global_Mobile_Economy_Report_2015.pdf (accessed January 2016).

Ha I (2015) Technologies and research trends in wireless Body Area Networks for healthcare: a systematic literature review. *International Journal of Distributed Sensor Networks* 2015: 1–15.

Halonen T, Romero J, and Melero J (2003) *GSM, GPRS and EDGE Performance: Evolution Towards 3G/UMTS.* UK: John Wiley & Sons, Ltd.

Hammond C (2011) Global standards collaboration, machine-to-machine standardization task force meeting. Wavenis open standard alliance, Grapevine, Texas, ftp.tiaonline.org/.../GSC_MSTF_20110518_010_Hammond_Wavenis_P. (accessed October 2013).

Harbert T (2012) *FCC gives Medical Body Area Networks clean bill of health*, IEEE Spectrum, June. Available at www.spectrum.ieee.org/tech-talk/biomedical/devices/fcc-gives-medical-body-area-networks-clean-bill-of-health (accessed October 2014).

Hernandez M and Mucchi L (2014) *Body Area Networks Using IEEE 802.15.6: Implementing the Ultra-Wide Band Physical Layer.* Academic Press.

Hersent O, Boswarthick D, and Elloumi O (2012) *The Internet of Things: Key Applications and Protocols.* John Wiley & Sons, Inc.

HIMSS Healthcare Information and Management Systems Society (2012) *The current state of technology in mobile healthcare: mHIMSS Road Map.* Available at www.himss.files.cms-plus.com/FileDownloads/2013-mHIMSS-Roadmap-Technology.pdf (accessed October 2014).

Hitachi Global (2006) *World's smallest and thinnest 0.15 × 0.15 mm, 7.5 µm thick RFID IC chip.* Available at http://www.hitachi.com/New/cnews/060206.html (accessed December 2014).

Huang A and Xie L (2015) SMART for mobile health: a study of scheduling algorithms in full-IP mobile networks. *IEEE Communications Magazine* 53(2):214–222.

IBM (2015) *IBM Long-Range Signaling and Control (LRSC): a highly efficient M2M infrastructure.* Available at http://www.research.ibm.com/labs/zurich/ics/lrsc/ (accessed May 2015).

IEEE 802.15.6-2012 (2012) *IEEE standard for local and metropolitan area networks: Part 15.6—wireless body area networks,* IEEE Standard Association: Piscataway, NJ, pp. 1–271. Available at http://ieeexplore.ieee.org. (accessed October 2014).

Informa Telecom and Media & The Wireless Broadband Alliance (WBA) (2012) *Global trends in public Wi-Fi next generation hotspot: moving from standardization to commercialization,* WBA- Wi-Fi Industry Report. Available at http://www.wballiance.com/wba/wp-content/uploads/downloads/2012/11/WBA_Wi-Fi_Industry_Report_Nov2012-2.pdf (accessed October 2014).

Insteon Alliance. Available at www.insteon.com (accessed October 2014).

Isaacs C (2014) *3 ways the Internet of things is revolutionizing health care,* Forbes. Available at www.forbes.com/sites/salesforce/2014/09/03/internet-things-revolutionizing-health-care/ (accessed November 2014).

Ishaq I, David Carels D, Teklemariam GK, Hoebeke J, Abeele FV, Poorter ED, Moerman I, and Demeester P (2013) IETF standardization in the field of the Internet of things (IoT): a survey. *Journal of Sensor and Actuator Networks* 2: 235–287.

Istepanian RSH, Brien M, and Smith P (1997) Modelling of photoplethysmography mobile telemedical system. *Proceedings of the 19th Annual IEEE International Conference of the Engineering in Medicine and Biology Society,* Chicago, USA, pp. 987–990.

Istepanian RSH (1998) Modelling of GSM-based mobile telemedical systems. Proceedings of the 20th IEEE Annual International Conference of Engineering in Medicine and Biology, Hong Kong, pp. 926–930.

Istepanian RSH, Woodward B, Balos P, Chen S, and Luk, B (1999) The comparative performance of mobile telemedical systems using the IS-54 and GSM cellular telephone standards. *Journal of Telemedicine and Telecare* 5(2):97–104.

Istepanian RSH (2000) Guest editorial. Special issue on mobile telemedicine and telehealth systems. *IEEE Transactions on Information Technology in Biomedicine* 4(3):194.

Istepanian RSH, Jovanov E, and Zhang Y (2004) Guest editorial. Introduction to the special section on m-health: beyond seamless mobility and global wireless health-care connectivity. *IEEE Transactions on Information Technology in Biomedicine* 8(4):405–414.

Istepanian R S H, Laxminarayan S, and Pattichis C (2006) *m-Health: Emerging Mobile Health Systems.* New York: Springer, pp. 219–236.

Istepanian RSH (2007) WiMAX for mobile healthcare applications. In: *WiMAX London,* IET.

Istepanian RSH, Philip N, and Martini M (2009a) Medical QoS provision based on reinforcement learning in ultrasound streaming over 3.5G wireless systems. *IEEE Journal on Selected Areas in Communications* 27(4):566–574.

Istepanian RSH, Sungoor A, and Earle K (2009b) Technical and compliance issues of mobile diabetes self-monitoring using glucose and blood pressure measurements. *30th Annual International Conference of the IEEE Engineering in Medicine and Biology Society,* Minnesota, USA, pp. 5130–5133.

Istepanian RSH, Hu S, Philip N, and Sungoor A (2011) The potential of m-Health Internet of things (m-IOT) for non-invasive glucose level sensing. *32nd Annual International*

Conference of the IEEE Engineering in Medicine and Biology Society, Boston, USA, pp. 5264–5266.

Istepanian RSH and Zhang YT (2012) 4G health: the long term evolution of m-Health. *IEEE Transactions on Information Technology in Biomedicine* 16(1):1–5.

Istepanian, RSH, AliNejad A, and Philip, N (2013) Medical quality of service (m-QoS) and medical quality of experience (m-QoE) for 4G health systems. In: R. Farrugia and C. Debono (Eds.), *Multimedia Networking and Coding*, IGI Global, pp. 359–376.

ISO (International Organization for Standardization) (2014) Available at http://www.iso.org/iso/home.html (accessed November 2014).

ITU (International Telecommunications Union) (2005) *ITU report on the Internet of things.* Available at www.itu.int/internetofthings (accessed October 2014).

ITU (International Telecommunications Union) (2008) *QoE challenge tackled in new ITU-T Recommendation*, ITU-T Newslog. Available at http://www.itu.int/ITU-T/newslog/QoE+Challenge+Tackled+In+New+ITUT+Rec.aspx (accessed October 2014).

ITU (International Telecommunications Union) (2012a) *Internet of things global standards initiative.* Available at http://www.itu.int/en/ITU-T/gsi/iot/Pages/default.aspx (accessed January 2014).

ITU (International Telecommunications Union) (2012b) *Impact of M2M communications and non-M2M mobile data applications on mobile networks.* Technical paper, June 15, 2012. Available at http://www.itu.int/dms_pub/itu-t/opb/tut/T-TUT-IOT-2012-M2M-PDF-E.pdf (accessed October 2014).

ITU (International Telecommunications Union) (2013) *ITU global standard for international mobile telecommunications: IMT-Advanced, Radio Communication Sector (ITU-R).* Available at http://www.itu.int/ITU-R/index.asp?category=information&rlink=imt-advanced&lang=en (accessed October 2014).

ITU (International Telecommunications Union) (2014a) *The Tactile Internet: ITU technology watch report.* Available at www.itu.int/dms_pub/itu- t/oth/23/01/T23010000230001PDFE.pdf (accessed November 2014).

ITU (International Telecommunications Union) (2014b) *Focus Group on M2M Service Layer.* Available at www.itu.int/en/ITU-T/focusgroups/m2m/Pages/default.aspx (accessed October 2014).

ITU (International Telecommunications Union) (2015) *Framework and overall objectives of the future development of IMT for 2020 and beyond*, Recommendation ITU-R M.2083. Available at www.itu.int/en/ITU-T/focusgroups/m2m/Pages/default.aspx (accessed September 2015).

Jamalipour A (2003) *The Wireless Mobile Internet: Architectures, Protocols and Services.* John Wiley & Sons, Inc.

Jamoussi B (2010) *IoT prospects of worldwide development and current global circumstances.* Available at www.itu.int/en/ITU-T/techwatch/Documents/1010-B_Jamoussi_IoT.pdf (accessed October 2014).

Jennic (2014) *JenNet Protocol Stack* Available at http://www.jennic.com (accessed October 2014).

Jones V, Halteren AV, Widya I, Dokovsky N, Koprinkov G, Bults RG, Konstantas D, and Herzog R (2006) MOBIHEALTH: mobile health services based on body area networks. In: Istepanian R, Laxminarayan S, and Pattichis CS, (Eds.), *м- Health: Emerging Mobile Health Systems*. New York: Springer, pp. 219–236.

Kim DY and Cho J (2009) WBAN meets WBAN: smart mobile space over wireless body area networks. In: *Proceedings of the 7th IEEE Vehicular Technology Conference Fall (VTC 2009-Fall)*, September 20–23, 2009, Anchorage, Alaska, USA.

Kitsos P and Zhang Y (2008) *RFID Security: Techniques, Protocols and System-On-Chip Design*. Springer.

Korhonen J (2014) *Introduction to 4G Mobile Communications*. Artech House.

Kraemer R and Katz M (2009) *Short-Range Wireless Communications: Emerging Technologies and Applications*. John Wiley & Sons, Inc.

Kumar V (2007) Implantable RFID chips: security versus ethics. In: *The Future of Identity in the Information Society: Proceedings of the Third IFIP WG 9.2, 9.6/11.6, 11.7/FIDIS International Summer School on the Future of Identity in the Information Society*, Karlstad University, Sweden, August 4–10, 2007.

Kvedar JC, Colman C, and Cella G (2015) *The Internet of Healthy Things*. Boston, MA: Partners Connected Health.

Latre B, Braem B, Moerman I, Blondia C, and Demeest P (2011) A survey on wireless body area networks. *Journal of Wireless Information Networks* 17: 1–18.

Li X, Lu R, Liang X, Shen X, Chen J, and Lin X (2011) Smart community: an Internet of things application. *IEEE Communications Magazine* 49(11):68–75.

Lloyd-Evans R (2002) *QoS in Integrated 3G Networks*. Artech House.

Markarian G, Mihaylova L, Tsitserov DV, and Zvikhachevskaya A (2012) Video distribution techniques over WiMAX networks for m-Health applications. *IEEE Transactions on Information Technology in Biomedicine* 16(1):24–30.

Mavrakis M (2014) *What are the 5G candidate technologies?* Available at http://www .telecoms.com/269842/what-are-the-5g-candidate-technologies/ (accessed October 2014).

McCue TJ (2015) *$117 billion market for Internet of things in healthcare by 2020*, Forbes Tech. Available at http://www.forbes.com/sites/tjmccue/2015/04/22/117-billion-market-for-internet-of-things-in-healthcare-by-2020/#2cb01fcd2471 (accessed May 2015).

METIS (Mobile and Wireless Communications Enablers for the Twenty-Twenty Information Society) (2014) European Research Project- FP7. Available at www.metis2020.com (accessed October 2014).

Miles S (2014) *The Internet of medical things*. MIT News. Available at http://news.mit.edu/ 2014/internet-medical-things (accessed October 2015).

MiWEBA (Millimetre-Wave Evolution for Backhaul and Access) (2014) European Research Project- FP7. Available at www.miweba.eu (accessed October 2014).

Movassaghi S, Abolhasan M, Lipman J, Smith D, and Jamalipour A (2014) Wireless body area networks: a survey. *IEEE Communications Surveys & Tutorials* 16(3):1658–1686.

Mulvaney DJ, Woodward B, Datta S, Harvey PD, Vyas AL, Thakkar B, Farooq O, and Istepanian RSH (2012) Monitoring heart disease and diabetes with mobile Internet communications. *International Journal of Telemedicine and Applications* 2012: 1–12.

Munshi MC, Xiaoyuan X, Zou X, Soetiono E, Teo CS, and Lian Y (2008) Wireless ECG plaster for body sensor network. In: *Proceedings of the 5th International Workshop on Wearable and Implantable Body Sensor Networks*, Hong Kong.

Near-Field Communications (2014) Available at http://www.nearfieldcommunication.org/ about-nfc.html (accessed October 2014).

Muzic D and Opatic D (2009) Capabilities and impacts of EDGE evolution toward seamless wireless networks, Ericsson. Available at www.ericsson.com/hr/etk/dogadjanja/mipro_2009/03_1077_F.pdf (accessed October 2014).

Neira EM (2015) *Top 10 trends in 2015*. IEEE communication society technology news special issue on the trends that tell where communication technologies are headed in 2015. Available at www.comsoc.org/ctn/ieee-comsoc-ctn-special-issue-ten-trends-tell-where-communication-technologies-are-headed-2015 (accessed January 2015).

Nuaymi L (2007) *WiMAX: Technology for Broadband Wireless Access*. Chichester, UK: John Wiley & Sons, Inc.

Onita L (2015) Internet of things network launched across Britain. *Engineering and Technology* 9(12):1–18.

Osseiran A, Braun V, Taoka H, Marsch P, Schotten H, Tullberg H, Uusitalo MA, and Schellman M (2013) The foundation of the mobile and wireless communications system for 2020 and beyond: challenges, enablers and technology solutions. *77th IEEE Vehicular Technology Conference (VTC Spring)*, June 2–5, 2013.

Osseiran A (2014) *Mobile and wireless communications system for 2020 and beyond (5G)*. ITU-R 2020 Vision Workshop, February 12, 2014, Vietnam. Available at www.metis2020 .com/wp-content/uploads/presentations/ITU-R-2020-VisionWS.pdf (accessed December 2014).

Osseiran A, Boccard F, Braun V, Kusume K, Marsch P, Maternia M, Queseth O, Schellmann M, Schotten H, Taoka H, Tullberg H, Uusitalo MA, Timus B, and Fallgren M (2014) Scenarios for 5G mobile and wireless communications: the vision of the METIS project. *IEEE Communications Magazine* 52(5):26–35.

Panayides AS, Pattichis, MS, Constantinides AG, and Pattichis CS (2013) m-Health medical video communication systems: an overview of design approaches and recent advances. In: *35th Annual International Conference of the IEEE Engineering in Medicine and Biology Society (EMBC)*, Osaka, Japan, pp. 7253–7256.

Pattichis CS, Kyriacou E, Voskarides S, and Istepanian RSH (2002) Wireless telemedicine systems: an overview. *IEEE Antennas and Propagation* 44(2):143–153.

Perera C, Zaslavsky A, Christen P, and Georgakopoulos D (2013) Context aware computing for the Internet of things: a survey, *IEEE Communications Surveys & Tutorials* 16(1):414–454.

Riegel M, Kroeselberg D, Chindapol A, and Premec D (2009) *Deploying mobile WiMAX*. Chichester, UK: John Wiley & Sons, Inc.

Roeland D and Rommer S (2014) Advanced WLAN integration with 3GPP evolved packet core. *IEEE Communications Magazine* 52(12):22–27.

Ruckus (2013) *The impact of Next Gen Wi-Fi technology on healthcare*. White Paper. Available at www.ruckuswireless.com (accessed October 2014).

Sauter M (2010) *From GSM to LTE: An Introduction to Mobile Networks and Mobile Broadband*. Chichester: John Wiley & Sons, Ltd.

Schwartz M (2005) *Mobile Wireless Communications*. Cambridge, UK: Cambridge University Press.

Sethia D, Gupta D, Mittal T, and Arora U (2014) NFC based secure mobile healthcare system. In: *Proceedings of the 6th International IEEE Conference on Communication Systems and Networks (COMSNETS)*, January 6–10, Bangalore, India, pp. 1–6.

Shelby Z and Bornmann C (2009) *6LoWPAN: The Wireless Embedded Internet*. Hoboken, NJ: John Wiley & Sons, Inc.

Shen Z, Papasakellariou A, Montojo J, Gerstenberger D, and Xu F (2012) Overview of 3GPP LTE-advanced carrier aggregation for 4G wireless communications. *IEEE Communications Magazine* 50(2):122–130.

Shnayder V, Chen, B, Lorincz, K, Fulford-Jones TR, and Welsh M (2005) Sensor Networks for Medical Care. Harvard University Technical Report, TR-08-05, pp. 1–15.

Sørensen L and Skouby KE (2009) User Scenarios 2020: A Worldwide Wireless Future. Report of the Wireless World Research Forum (WWRF). Available at www.wwrf.ch/files/wwrf/content/files/publications/outlook/Outlook4.pdf (accessed October 2014).

Stair R and Reynolds G (2014) *Fundamentals of Information Technology*, 7th edn. Course Technology.

Swedberg C (2014) *Dash7 Alliance Working on New Specification, Tags for ISO 18000-7*, RFID Journal. Available at http://www.rfidjournal.com/articles/view?7780/3 (accessed November 2014).

Sun W, Lee O, Shin Y, Kim S, Yang CKH, and Choi S (2014) Wi-Fi could be much more. *IEEE Communications Magazine* 52(11):22–29.

3GPP (Third Generation Partnership Project) (2010) Service requirements for machine type communications, TS 22.368, June 2010. Available at http://www.3gpp.org/DynaReport/22368.htm (accessed October 2014).

3GPP (Third Generation Partnership Project) (2014a) TS23.203: Technical specifications group services and system aspects: policy and charging control architecture. Available at www.3gpp.org/DynaReport/23203.htm, Release 12 (accessed March 2015).

3GPP (Third Generation Partnership Project) (2014b) *Seven standards development organizations (SDOs) are set to create a cooperative M2M standards activity*. Available at www.3gpp.org/news-events/3gpp-news/1426-Global-Initiative-for-M2M-Standardization (accessed October 2014).

Thompson J, Ge X, Wu HC, and Irmer R (2014) 5G wireless communication systems: prospects and challenges. *IEEE Communications Magazine* 52(2):62–64.

Uckelmann D, Harrison M, and Michahelles F (Eds.) (2011) An architectural approach towards the future internet of things. *Architecting the Internet of Things*. Berlin: Springer, pp. 1–24.

Ullah S, Higgins S, Braem B, and Latre BA (2012) Comprehensive survey of wireless body area networks: on PHY, MAC and network layers solutions. *Journal of Medical Systems* 36: 1065–1094.

Verma L, Fakharzadeh M, and Choi S (2013) Wi-Fi on steroids: 802.11 AC and 802.11AD. *IEEE Wireless Communications* 20(6):30–35.

Vermesan O and Friess P (2014) *Internet of Things: From Research and Innovation to Market Deployment*. Alborg, Denmark: River Publishers.

Verulkar SM and Limkar M (2012) Real time health monitoring using GPRS technology. *International Journal of Computer Science and Network* 1(3): ISSN 2277–5420.

Vilamovska AM (2010) Improving the quality and cost of healthcare delivery: the potential of radio frequency identification (RFID) technology. Ph.D. dissertation.

RAND Graduate School, California, USA (2010) Available at http://www.rand.org/content/dam/rand/pubs/rgs_dissertations/2010/RAND_RGSD239.pdf (accessed November 2014).

Violino B (2005) *The Basics of RFID technology*, RFID Journal. Available at http://www.rfidjournal.com/articles/view?1337 (accessed October 2014).

Wakefield J (2015) *The rise of the Swedish cyborgs.* BBC Technology News. Available at http://www.bbc.co.uk/news/technology-30144072 (accessed January 2015).

Wang SW, Chenb WH, Ong CS, Liu L, and Chuang YW (2006) RFID applications in hospitals: a case study on a demonstration RFID project in a Taiwan hospital. *Proceedings of the 39th Hawaii International Conference on System Sciences*, Hawaii, January 4–7, 2006.

Wannstrom J (2013) *LTE-A, 3GPP release.* Available at www.3gpp.org/technologies/keywords-acronyms/97-lte-advanced (accessed October 2014).

Woodward B, Richards CI, and Istepanian RSH (2001) Design of a telemedicine system using a mobile telephone. *IEEE Transactions on Information Technology in Biomedicine* 5 (1):13–15.

Woodward B, Rasid MFA, Gore L, and Atkins P (2006) GPRS-based mobile telemedicine system. *Journal of Mobile Multimedia* 2(1):2–22.

Woodward B and Vyas AL (2000) Software Engineering for Telemedicine. Final Report, Project No. 97TI, Loughborough University and Indian Institute of Technology Delhi, UK–India Science and Technology Research Fund 2000. Available at en.wikipedia.org/wiki/X10(industry standard).

Wu G, Talwar S, Johnsson K, Himayat N, and Johnson K (2011) M2M: from mobile to embedded Internet. *IEEE Communications Magazine* 49(4):36–43.

Yang SH (2014) *Wireless Sensor Networks: Principles, Design and Applications.* Springer.

Yong X, Yinghua L, Hongxin Z, and Yeqiu W (2007) An overview of ultra-wideband technique application for medial engineering. In: *IEEE/ICME International Conference on Complex Medical Engineering (CME)*, Beijing, China, May 23–27, 2007, pp. 408–411. Available at http://ieeexplore.ieee.org/xpl/freeabs_all.jsp?arnumber=4381766 (accessed October 2014).

Zhang Y, Ansari N, and Tsunoda H (2010) Wireless telemedicine services over integrated IEEE 802.11/WLAN and IEEE 802.16/WiMAX networks. *IEEE Wireless Communications* 17: 30–36.

Zhen B, Patel M, Lee S, Won E, and Astrin A (2008) TG6 Technical Requirements Document (TRD), Project IEEE P802.15 Working Group for Wireless Personal Area Networks (WPANs), IEEE Standard Association, Piscataway, NJ.

ZigBee Alliance (2010) ZigBee Health Care Profile Specification, Revision 15, Version 1.0, March 2010. Available at https://docs.zigbee.org/zigbee-docs/dcn/10/docs-10-5619-00-0zhc-zigbee-health-care-profile-1-0-public.pdf (accessed October 2014).

ZigBee Alliance (2014) Available at www.zigbee.org (accessed October 2014).

Z-Wave Alliance. Available at www.z-wavealliance.org/ (accessed October 2014).

5

m-HEALTH CARE MODELS AND APPLICATIONS

Perhaps the most profound change and positive impact that iPhone will make is on our health.

Tim Cook, CEO Apple, 2015

5.1 INTRODUCTION

A widely cited report, "Crossing the Quality Chasm" describes a gap between knowledge and action (Institute of Medicine, 2001). This gap is the result of a systemwide problem that requires overall redesign of the care delivery system, including current m-Health systems.

There is evidence of increasing enthusiasm by clinicians, healthcare providers, and policymakers to demonstrate tangible benefits and impact of m-Health on healthcare delivery. This enthusiasm is driven by several contributing factors that m-Health can provide a unique balancing element to the demands of improving quality of care with better healthcare access at lower cost. However, there is also a skeptical opinion that doubts these benefits and the transformative role of m-Health. These opposing views reflect the current paradoxical status of m-Health explained earlier in this book. This status can be seen more clearly in clinical evidence from numerous "m-Health" applications, clinical trials, and pilot studies conducted so

m-Health: Fundamentals and Applications, First Edition. Robert S. H. Istepanian and Bryan Woodward.
© 2017 by The Institute of Electrical and Electronics Engineers, Inc. Published 2017 by John Wiley & Sons, Inc.

far. The failure of any nationwide m-Health programs or initiatives to emerge beyond the silos and local pilot stages is therefore perplexing. This notion brings a reflective, if not an important, question as to whether the letter *m* in m-Health denotes the *medical* or the *mobile* aspects of the concept? The answer comes from the overwhelming view that m-Health is a smartphone-centric healthcare technology enabler, with potential capabilities to provide better healthcare services and outcomes using these technologies. This popular view can be better illustrated in this basic "telecom formula":

$$m\text{-Health} = \text{Mobile (smart) phone} + \text{healthcare delivery service}$$

However, this narrow but familiar interpretation is based on the provision of healthcare delivery services to patients using predominantly mobile or cellular technology, notably smartphones. This model obviously lacks the wider understanding of the scientific principles behind m-Health as a holistic concept that transcends this simple formula. The smartphone-centric interpretation of m-Health was initially and still largely driven by massive commercial markets and consumer-driven healthcare demands, reflecting potential business and market opportunities. This "service delivery model" is largely on account of the global availability of smartphones to billions of people. As a consequence of this model, there has been a decade-long increasing asymmetry between the fast-paced m-Heath innovations and the lack of rigorous clinical evidence, medical acceptability, and impact that supports these innovations. This asymmetry will be shown in the numerous m-Health intervention examples to be presented in this chapter. Whether this situation will continue is a matter of speculation, which depends on the evolution of m-Health in the next decade and whether fresh thinking of alternative approaches will prevail. However, today there is an increasing push from scientific and clinical communities that are advocating a rethink on this interpretation with a new holistic outlook and a return to the basics of m-Health and the original scientific principles behind the concept.

In this chapter, we present the clinical evidence for m-Health in its current mobile phone-centric format and evaluate the existing clinical landscape for different healthcare services (Bernhardt, 2015). Due to the limitation of space, it is not practicable to present in detail all the healthcare areas in which m-Health has been applied. Instead, we chose to focus on some examples of chronic and noncommunicable diseases and conditions that typically have high levels of prevalence and cost burdens globally, and where mobile phone interventions have been successfully evaluated. In particular, we have selected three of these chronic conditions for our detailed discussions, namely, diabetes, heart and cardiovascular diseases (CVDs), and chronic obstructive pulmonary disease (COPD). These diseases, together with cancer and stroke, constitute the most prevalent chronic diseases, with major economic and healthcare burdens globally and hence our choice. For other conditions, there has been varying levels of m-Health interventions; we also discuss some of these and the clinical evidence for these interventions in less detailed form.

5.2   MOBILE PHONE m-HEALTH SYSTEMS AND THEIR IMPACT ON FUTURE HEALTHCARE SERVICES

As discussed in Chapter 1, the evolution of m-Health since 2007 has been due to the profound impact of smartphones. This mobile technology leap has largely shaped m-Health with a perceived and still dominant view as a smartphone-centric technology tool for transforming healthcare delivery services, with estimated savings of billions of U.S. dollars (West, 2012). Other clinical views have questioned whether m-Health in its current format can transform healthcare services as widely anticipated and whether it can also provide the much-hyped clinical impact and healthcare transformation with the expected patient benefits and services (Steinhubl et al., 2014).

This debate is ongoing and will continue in the foreseeable future until a new m-Health framework is adopted. The debate is also supplemented by recent studies that have reviewed the current clinical evidence and concluded that despite hundreds of m-Health pilot studies, there has been insufficient evidence to warrant the scale-up of m-Health (Tomlinson et al., 2013). This notion also reflects the role of m-Health in the modern healthcare delivery system and its impact on both the traditional physician–patient partnership and the consumer-driven healthcare delivery processes. The current contrast between the market-driven potential of m-Health and the lack of clinical evidence for it is reflected in the hype and hope of m-Health viewed entirely as a mobile phone and/or App-centric healthcare delivery tool (Labrique et al., 2012).

However, it is obvious from the available evidence that this form of m-Health with all its limitations has provided a role in achieving tangible outcomes in different healthcare delivery areas, such as patient monitoring, patient education, disease management, and medication adherence. To illustrate some of these areas, we cite some studies from the literature of m-Health (mobile phone) interventions of different healthcare services applied in both developed and developing (low- and middle-income) countries.

For example, in a widely cited systematic review and meta-analysis study of 42 controlled trials that reviewed *mobile technology*-based m-Health interventions designed to improve healthcare delivery processes, the following outcomes emerged (Free et al., 2013a):

- m-Health intervention designed to improve healthcare delivery processes are "modestly effective," but highlight the need for more trials.
- There is a need for more trials in both high- and low-income countries that combine m-Health interventions compared with controlled interventions.
- None of the reviewed studies were of high quality trials.
- Nearly all the studies cited were undertaken in high-income countries.
- Many of these studies had methodological problems that are likely to have affected the accuracy of their findings.

Alternatively, a report by the National Health Service (NHS) in England listed several mobile phone technology (m-Health) services, including appointment reminders, mobile-enabled community nursing, SMS messaging for test results, and other examples as drivers for implementing better provision of care services supplemented with potential cost-saving outcomes (NHS England, 2012). This latter view is paralleled with many m-Health pilots and intervention studies conducted in the developing world. An extensive review analysis citing more than 676 publications on the impact of m-Health interventions in the developing world for low- and middle-income countries (LMICs) indicated the growing evidence base for the efficacy of these interventions, particularly in improving treatment adherence, appointment compliance, data gathering, and developing support networks for health workers (Hall et al., 2014). Similar findings were also supported by another meta-analysis study on the effectiveness of m-Health interventions for maternal, newborn, and child health in LMICs (Nurmatov et al., 2014).

These outcomes from different global settings provide ample proof of the inconsistent evidence and the current conundrum of m-Health interventions for the developing world. The narrow yet common interpretation of m-Health is also contributing to the increasing gap and clinical ambiguity between the benefits gained from some m-Health interventions in certain care settings and the modest outcomes in others. As a consequence, an alternative m-Health framework with clinically robust and effective outcomes is required. This new framework must encapsulate a wider understanding of the different dimensions of m-Health outside the "smartphone box" that can fundamentally change the current clinical uncertainties and contradictions associated with existing m-Health formats. At the core of this transformation is perhaps the new m-Health sphere or "m-Health 2.0" that we presented Chapter 3. This transformation is best illustrated in Fig. 5.1.

This new framework is based on the convergence from the "old m-Health" to the "new m-Health" models that aim to alter the "DNA of current m-Health systems." This framework is driven by several impact factors:

- *Societal and Demographics:* These include the ageing society and change of demographics in global populations that will have an impact on the current healthcare resources and processes.
- *Healthcare Services and Demands:* These include factors such as increasing demands on healthcare services, quality and cost of access to medical care, specific settings, an increasing shift to preventative health, and consumer health paradigms.
- *Technological and Scientific:* These include emerging innovations from digital and scientific worlds, for example, genomics, and information and digital age technologies such as new Internet, sensing, robotics, social networking, communication, and advances in personalized therapeutics.
- *Business Markets and Economics:* These include the main business and industry stakeholders driving m-Health services, ranging from the telecommunication industry to medical devices, pharmaceuticals, and others. The role of the governmental and global institutions to embrace, fund, and regulate these new m-Health services and applications in the global perspective will also be critical in this transformation process.

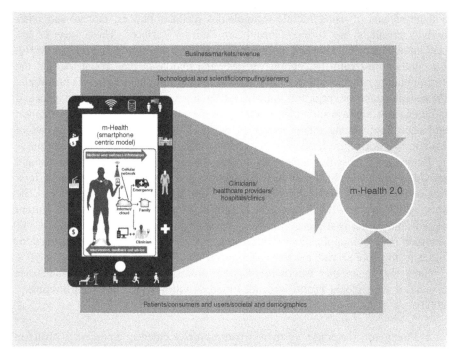

FIG. 5.1 Transformation from the current m-Health smartphone-centric model to the new m-Health 2.0 framework.

The transformation is likely to be evolutionary and might take a decade or more to mature. Most likely, it will be carried out in stepwise stages depending on the speed of recognition and adoption of such changes by all the interested stakeholders and assisted by the maturity of future enabling technologies of these new m-Health systems.

This transformation also supports the widely discussed change in existing care delivery models that are becoming increasingly inadequate in satisfying the growing health demands both in the developed and the developing world. Views on how these future models can be implemented successfully are being discussed extensively; one example advocates the following elements (PwC, 2014):

- To be focused on the patient as a consumer.
- To characterize with predictive and precise analytical methods.
- To embrace integrated and transparent approaches.
- To coordinate team-based working models.
- To be sustainable.
- To have quality-based, efficient characteristics.

This view is largely based on an increasing demand for a consumer-led healthcare culture. This is supported by popular trends of usage and awareness by both patients and physicians of using mobile technologies (cellular phones, tablets, and other mobile terminals) for different healthcare delivery services. These services are becoming increasingly applied in modern healthcare systems in the developed world. By contrast, m-Health perceptions in low- and middle-income countries, and the corresponding services offered, can differ from the list outlined earlier. For example, the general healthcare priorities in LMICs, from treatment compliance, neo-natal and child diseases, disease surveillance, emergency medical response, health promotion and education, among others (Mechael et al., 2010), demand a different set of assumptions for future healthcare delivery models embracing m-Health solutions.

Before we discuss the clinical applications of the current m-Health systems and provide a general evaluation of their applications and services, it is important to look first at the status of m-Health categorization and the associated applications. The varying views of m-Health are best illustrated by some examples of the current classifications that best relate to existing systems and models. Table 5.1 shows 12 common m-Health applications, with examples of embedded smartphone functionalities as the core service enabler (Labrique et al., 2013).

Similarly, in another extensive literature study, relevant m-Health categories associated with global health were identified as follows (Mechael and Solninsky, 2008):

- Emergency response systems (road traffic accidents, emergency obstetric care, etc.)
- Disease surveillance and control (malaria, HIV/AIDS, tuberculosis, avian flu, chronic diseases, especially diabetes)
- Human resources coordination, management, and supervision
- Synchronous and asynchronous mobile telemedicine diagnostic and decision support for clinicians at point-of-care
- Remote patient monitoring and clinical care
- Health extension services, health promotion, and community mobilization
- Health services monitoring and reporting
- Health-related m-learning for the general public
- Training and continuing professional development for healthcare workers

Other similar studies classified m-Health services into the following categories (Jadad, 2009; Vital Wave Consulting, 2009):

- *Education and Awareness:* This category includes interventions designed to provide stakeholders with the information they need to support their decisions, at the right time, at the right place, and in the right format.
- *Data Collection:* This includes the capture of facts and statistics at the community level, enabling policymakers to judge and improve the effectiveness of healthcare programs, allocate resources more efficiently, and adjust programs and services accordingly.

TABLE 5.1   Common m-Health Mobile Phone-Centric Applications

	Common m-Health Applications	Examples on the Smartphone Functionalities
1	Client education and behavioral change communication	Short message service, multimedia messaging service, interactive voice response, voice communication/audio clips, video clips, and images
2	Sensors and point-of-care diagnostics	Mobile phone camera, tethered accessory sensors, built-in accelerometers
3	Registries and vital events tracking	Short message service, voice communication, digital forms
4	Data collection and reporting	Short message service, digital forms, voice communication
5	Electronic Health Records (EHR)	Digital forms, mobile web
6	Electronic decision support systems	Mobile web, information storage
7	Provider-to-provider communication	Short message service, multimedia messaging service, mobile phone camera
8	Provider work planning and scheduling	Interactive electronic client lists, short message service alerts, mobile phone calendar
9	Provider training and education	Short message service, multimedia messaging service, interactive voice response, voice communication, audio or video clips, images
10	Human resource management	Web-based performance dashboards, global positioning service, voice communication
11	Supply chain management	Web-based supply dashboards, Global Positioning Service, digital forms
12	Financial transactions and incentives	Mobile money transfers and banking services, transfer of airtime minutes

Source: Adapted from Labrique et al. (2013).

- *Remote Monitoring:* This group refers to resources that allow one- or two-way communications to monitor the evolution of health conditions outside healthcare facilities, to maintain caregiver appointments, or to ensure adherence to medication regimens.
- *Peer-to-Peer Communication among Healthcare Workers:* This category encompasses technologies that connect health professionals with each other, improving their sense of empowerment and their ability to make decisions effectively and self-sufficiently.
- *Disease and Epidemic Outbreak Tracking:* This group of services includes the use of devices to capture and transmit data on the incidence and geographic distribution of diseases, and to guide prevention and containment efforts.
- *Diagnostic and Treatment Support:* This category includes efforts to use technology to shift diagnostic and therapeutic efforts away from healthcare facilities to homes, workplaces, schools, and the community at large, while

TABLE 5.2   m-Health Continuum and Example Applications

m-Health Continuum	Examples and Applications
Measurement	One person embedded sensor, Global Positioning System
Diagnostic	Point-of-care diagnostics, portable imaging, clinical decision support systems
Treatment/ prevention	Prevention and wellness interventions, medication adherence and tracking, chronic disease management, dissemination of health information, disaster support
Global	Access to healthcare, remote behavioral treatment, disaster surveillance, disease surveillance, medication tracking

Source: Adapted from Kumar et al. (2013).

averting expensive or unfeasible face-to-face in-person consultations with health professionals or visits to hospitals or clinics.

From a generic perspective, an m-Health continuum model, with three emerging areas of evidence, was identified (Kumar et al., 2013). This model, with example applications, is shown in Table 5.2. Alternatively, from the wider healthcare services spectrum, with potentially applicable areas of m-Health interventions, a study by The King's Fund identified the following 10 priority areas as transformative for healthcare, with varying levels of impact and implementation complexity (Naylor et al., 2013):

- Active support for self-management
- Primary prevention
- Secondary prevention
- Managing ambulatory care sensitive conditions
- Improving the management of patients with both mental and physical health needs
- Care coordination through integrated health and social care teams
- Improving primary care management of end-of-life care
- Medicines management
- Managing elective activity (referral quality)
- Managing urgent and emergency activity

Most of these priority areas can be implemented and clinically validated with specific m-Health platforms to provide better care access and cost-effectiveness, although each requires further studies. Finally, as discussed in Chapter 3, the increasing influence and use of smart health applications (Apps) can also impact on these categorizations (Sama et al., 2014; Wang et al., 2014). As an example, a review study on "smartphone Apps" and their use among physicians and medical

TABLE 5.3 m-Health Mobile Phone-Centric Themes and Example Applications

m-Health (mobile phone-centric) Areas	Example Applications
Health and wellness monitoring and disease management	Remote patient monitoring and mobile disease management, patient education, mobile EPR and EHR access
Remote diagnostic and treatment services	Point-of-care diagnostics, mobile ultrasound, and remote diagnostics
Patient education and behavioral change	SMS for patient education, appointment, and medication reminders
Home and elderly care	Ambient Assisted Living, fall detection, m-rehabilitation
Public and global health applications	Patient education and health promotion, maternity and child health, logistics management, healthcare workforce training
Medical training and learning and healthcare workflow processes and management	Remote medical education and training for healthcare workers and medical personnel
Primary and emergency care	A&E and trauma services, primary care clinics

students categorized the published articles into patient care and monitoring; health Apps for the lay person; communication, education, and research; and physician or student reference Apps (Ozdalga et al., 2012).

In conclusion, from these and other similar studies we can deduce some common healthcare themes applicable for the current m-Health (i.e., mobile or smartphone) systems. These are summarized in Table 5.3, together with some relevant examples of their applications.

5.2.1 m-Health and Patient-centered Continuum of Care

In recent years, the care concept of patient-centered care (PCC) has been increasingly adopted in many world-leading healthcare systems, particularly in the United States, Europe, and some Asian countries. PCC is generally understood to be a healthcare system that considers a patient as a whole person with biological, psychological, and social needs, and is said to improve the quality of patient care, reduce the cost of care, and increase satisfaction among nurses, physicians, and patients. There are many patients and healthcare benefits attributed to PCC (Pelzang, 2010):

- Improved coordination and integration of care process and access.
- Better patient engagement and support of self-management with emotional, psychological, and practical support.
- Timely patient access to diagnostic and prevention services.
- Delivery of more holistic care that enhances communication skills between relatives, patients, and healthcare providers, which shifts emphasis from body care to total care.

- Empowerment of clinical staff and healthcare professionals with better planning procedures.
- Facilitation of team approach, which facilitates reflection, learning, and sharing of skills and abilities among health professionals.

It is also well known that effective PCC spans across different types of healthcare systems from primary, acute, chronic, long-term, and home care systems. Future m-Health technologies are poised to provide better healthcare delivery services across these spheres, with efficient access and optimum balance between the three basic dimensions of the healthcare triangle of access, cost, and quality. It is important to consider PCC within existing m-Health applications and models and their adaptability with these services. However, this inclusion adds a degree of complexity, particularly with the following PCC challenges:

- The difficulty of identifying measurable indicators for m-Health-based PPC systems that are comprehensive and applicable globally.
- The organizational challenges and human resources required for appropriate PCC implementation plans and procedures.
- The cost-effectiveness and the health economics of implementing PPC health systems.
- The need for an appropriate PCC model to address the patient continuum requirements for developing world environments as compared to those of the developed world.

From the m-Health perspective, it was predicted a decade ago that the deployment of mobile technologies for healthcare will face these challenges in three main categories: technological, economic, and societal (Istepanian et al., 2006):

Technological Issues

- User acceptance and machine–human usability, compactness, power and battery life or batteryless devices and sensors, biocompatibility, maintainability, and reliability.
- Seamless and secure integration of increased amounts of data.
- Smart medical sensor design, integrating sensing, processing, communications, computing, and networking together into a reduced volume for wearable devices.
- Protocols for wireless medical sensor networks.
- Support for Quality of Service (QoS) and Quality of Experience (QoE).

Economic Issues

- Regulatory issues.
- Clinical costs and the other financial issues of preventive care as opposed to the usual care.

- Standardization of communications protocols and device interfaces that will significantly decrease overall cost.
- New business models and opportunities for cost-effective m-Health solutions.

Societal Issues

- Healthcare coverage, and patient participation and reimbursement.
- Legal and liability issues.
- Promotion of healthy lifestyles for diabetes, obesity, and chronically ill patients.
- Privacy and security issues and data access.

A decade on, some of these are still valid assumptions, in particular for the bridging process between the design and development of today's m-Health systems and any future m-Health framework. Ultimately, for any future categorization within the new framework, the following issues also need to be considered carefully:

- *Innovation and Technological:* This combines all the key technological developments of future m-Health systems detailed in earlier chapters and not based solely on mobile phone technology.
- *Future Healthcare Services and Clinical Care Models:* This includes all the clinical areas and healthcare services applicable with future m-Health systems.
- *Healthcare Supporting Disciplines (Scientific and Societal):* This includes the relevant scientific (e.g., biological, social, psychological) domains that will supplement the functions of future m-Health systems and services.
- *Business, Geography and Environment:* This includes all the relevant business models, stakeholders, and the global geographical environment applicable to future m-Health systems, for example, smart cities and poor and conflict regions.

The current view of m-Health is obviously diverse, with different outlooks and implementation models that are based either on existing models adopted for the developing world or those used for the developed world. The lack of any unifying options that might bridge these two outlooks contributes to the existing divergence of clinical evidence and acceptability.

In conclusion, we have highlighted some of the assumptions and basic concepts required for a new and comprehensive m-Health framework. Its development is a complex process and requires further multidisciplinary work and extensive future effort.

In the next section, we will illustrate the clinical evidence for current mobile chronic disease management systems. These were selected for our analysis, since chronic disease management constitutes both a major global health problem and an important sphere of m-Health used here as an exemplar.

5.3   m-HEALTH FOR CHRONIC DISEASE MANAGEMENT AND MONITORING APPLICATIONS

Chronic diseases include type 2 diabetes (T2D), CVD, COPD, Alzheimer's and other dementia diseases, ALS (Lou Gehrig's Disease), arthritis, asthma, cancer, cystic fibrosis, eating disorders, obesity, osteoporosis, Reflex Sympathetic Dystrophy syndrome, and excessive tobacco use (CDC, 2013). Mobile monitoring of chronic diseases started a decade ago as one of the first major clinical applications of interest for m-Health intervention. It has thrived since then as one of the most popular and commercially appealing areas. This is because the burden of chronic disease is widespread and growing globally. Also, monitoring patients' "vital signs" by remote monitoring as a means of periodically checking their condition is far more preferable, as well as being less costly and time-consuming, than asking each patient to attend a direct examination by a clinician (Mulvaney et al., 2012). This face-to-face scenario may be impractical if the patient lives a great distance from a clinic or hospital, especially in a developing country where patients do not have the means to travel. Furthermore, patients with chronic conditions make daily self-management decisions that require alternative models of patient care (Bodenheimer et al., 2002a). It is also well known that the increasing shift from acute to chronic illness has made healthcare providers and policymakers worldwide rethink how best to invest resources to manage these chronic conditions (Wilkinson and Whitehead, 2009).

Accordingly, this area has become one of the most widely recognizable areas of m-Health interventions, with some market studies predicting major healthcare and financial benefits from them (PwC, 2013). In short, there are several factors that are driving the adoption of m-Health applications:

- Demands from the ever-increasing population of patients with chronic diseases.
- Commercialization and market forces, and the push for more personalization of care.
- Economic healthcare burdens, a need for major healthcare reforms, and the adoption of new delivery models with cost savings.
- Advances in IT and mobile technologies, which are the best enablers of these new service delivery models and their acceptability and accessibility by users.

In clinical terms, chronic diseases constitute a major burden for healthcare systems globally, with increasing levels of national expenditure in both developed and developing countries. Furthermore, there is an alarming universal increase in chronic diseases listed above but particularly T2D, CVD, and COPD. This trend is poised to increase due to ageing and demographic changes, combined with poor lifestyle behavior and unhealthy eating habits. Furthermore, the global prevalence of these diseases is cited by the WHO as the leading risk of mortality worldwide across all income groups in high-, middle-, and low-income countries. For example, the WHO cited high blood pressure as responsible for 13% of deaths globally compared to tobacco use (9%), high blood glucose (6%), physical inactivity (6%), and obesity (5%) (WHO, 2009).

In the United States, individuals with chronic conditions account for 75% of healthcare expenditure, with an estimated average healthcare cost per person of $8915 in 2012, with $2.7 trillion in total spent on healthcare (CDC 2013; Meraya et al., 2015). This massive economic burden, compounded with a global shortage of healthcare workers and an increasing life expectancy, has made it a high priority of healthcare systems worldwide to develop innovative strategies to improve the care of chronic conditions and to prevent secondary complications (Wagner et al., 1996, 2005). One such strategy being widely applied is the self-management of chronic conditions using mobile health technologies. In general, self-management incorporates multiple concepts: self-care, self-monitoring, adherence, health behavioral change, patient education, and collaborative care. Self-management can apply to health promotion interventions, prevention strategies, acute care, and most often chronic illness care (Jones et al., 2012). There are numerous studies of different approaches of self-management interventions and outcomes for nearly 25 disease conditions being reported within the last decade, with diabetes and chronic conditions being the largest targeted (Jones et al., 2012).

5.3.1 m-Health and Chronic Care Models

While chronic disease management programs vary in design and implementation, almost all of them promote one or more elements of the Wagner Chronic Care Model (CCM) (Wagner et al., 1996, 2005; Bodenheimer et al., 2002b). The CCM is perhaps the best-known care framework for people with long-term conditions. The model was proposed in the 1990s as a multicomponent intervention to improve chronic care and to provide a multifaceted framework for redefining the quality of care and improved access. Over the last two decades, this model has been shown repeatedly to improve clinical outcomes, and has been implemented in most chronic care policies. Many developed countries now draw on this model to some extent (Wagner et al., 2005; Coleman et al., 2009). The six core elements or principles of this model are as follows:

- *Health System and Leadership:* This element serves as the foundation by providing the necessary structure and goals of the organizational environment that systematically supports and encourages chronic illness care through leadership and incentives results. It basically creates a culture, organization, and mechanisms that promote safe, high-quality care.
- *Community:* This element mobilizes community resources to meet the needs of people with long-term conditions, and provides community linkages for cost-effective access to them.
- *Self-Management Support:* This element emphasizes patient empowerment and self-management skills.

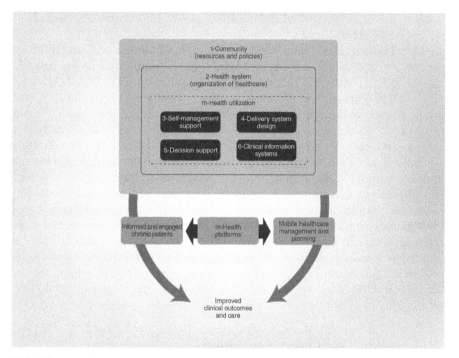

FIG. 5.2   Use of m-Health in the Wagner chronic care model. (Adapted from Wagner et al. (1996).)

- *Coordinated Delivery System Design:* This element represents the restructuring of clinical practices to facilitate team care and innovations in the delivery system process.
- *Clinical Decision Support:* This element includes assurances and incorporation of evidence-based practice guidelines into registries, flow sheets, and patient assessment as an effective method for changing provider practice mechanisms.
- *Clinical Information Systems:* This element provides timely access to data about patients and specific patient populations (e.g., access to adequate EPRs or EHRs) to enable the better delivery of more efficient and proactive care.

Most of the current mobile phone-centric m-Health chronic management systems use this model as a mere platform and not as a core element of the model. Figure 5.2 shows the use of m-Health platforms in the core elements of the chronic care model.

Also, one of the well-known and successful service delivery frameworks based on the CCM is the Kaiser Permanente Triangle model (Wallace, 2005). This integrated model has been widely adopted, especially in the last two decades in the United States and other global healthcare settings. Figure 5.3 shows that the model is stratified into three broad groups according to the level of patient support and care required, with intensive management targeted at those at highest risk. The three main levels of care are supported self-care for the majority of the chronic care population; disease and

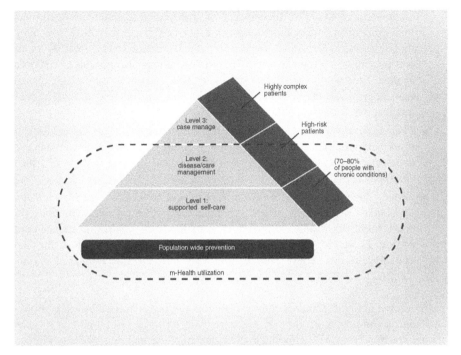

FIG. 5.3 The Kaiser Permanente Triangle illustrating different levels of chronic care applied to m-Health.

care management for high-risk patients who have multiple longest-term conditions; and case management for patients with highly complex conditions. The two levels of care at the top of the triangle require more professional intervention to be delivered effectively.

The underpinning factor in adopting this model for m-Health chronic care is the role of mobile health technologies and services in promoting better health in the population by empowering patients in lower levels of care. These technologies can provide the appropriate advice and care platforms needed for patients' treatment plans and healthy choices. Furthermore, m-Health technologies can also provide broader benefits in the context of long-term lifestyle management to prevent the condition of chronic patients from deteriorating further and consequently requiring a more intensive level of support. To emphasize the role of the patient-centered approach, and the importance of mobile health in their service delivery model and care process, Kaiser Permanente made its entire electronic healthcare system available to its nine million members via an Android App in 2012 (Thakkar, 2013).

Moreover, implementing m-Health interventions in a chronic care and service delivery model can provide the following benefits:

- *Patient Empowerment and Improving Clinical Outcomes with Better Access and Family/Community Support for Patients:* These interventions allow patient empowerment by using the powers of smartphone technologies to provide management and information feedback to patients. It can also provide seamless mobile access to physicians about their patients' disease status, treatment plans, and progress on a regular basis. These interventions also provide "behavioral change" instructions necessary for patients to adapt to their daily disease management and medication plans and to allow patients to comply with their medications for sustainable periods.

- *Provide Better Clinical Effectiveness and Efficiency Evidence:* These interventions can assist the healthcare providers in optimizing individual care plans for their patients and to redistribute resources to those patients with more critical care plan needs.

- *Reduce Healthcare Costs and the Burden of Chronic Care in Hospital and Specialist Care Centers:* These can potentially offer relative reductions and cost-benefits for healthcare providers, from hospitals, doctors, insurance companies, IT service providers, and other stakeholders involved in the care delivery process. These interventions can shift the hospital-based chronic care model toward a home-based self-care model.

As shown in Fig. 5.2, these combined benefits can be better used by integration of m-Health technologies in the core elements with an m-Health-based chronic care model (m-Health–CCM). This integration can improve and enhance chronic care facilitation. Also, the m-Health implementation of the core elements can encapsulate these benefits to overcome many of the existing barriers to routine interventions and delivery services. It is also important to note the following aspects in any future development of an m-Health–CCM based on these principles:

- The integration of new m-Health2.0 innovation platforms and not only smartphone-centric m-Health services.
- The adaptability of emerging m-Health delivery processes and their adaptation to the CCM model.
- The role of new patient-driven health data and social analytics elements in the decision support element of the existing model.

It is only by considering these issues that a comprehensive integration of future m-Health systems can lead to the development of a new and comprehensive mobile chronic care model. The full expectations of m-Health benefits can then be effectively used with sustainable improvements in chronic patient healthcare and well-being.

In recent years, numerous clinical and pilot studies have been reported on the effectiveness of mobile phone-based m-Health interventions in different chronic disease conditions. However, most were conducted in the developed world, with fewer in LMICs (Free et al., 2013b; WHO, 2011). As discussed earlier, the unprecedented proliferation of smartphone Apps targeting different chronic conditions has triggered a popular consumer-driven m-Healthcare products market, although there have been no large-scale studies or randomized clinical trials to back up the clinical effectiveness and efficacy of these products. This market is not showing any signs of slowing down.

5.3.2 Mobile Chronic Disease Management System: Design Issues and Challenges

There are a number of interpretations of disease management. These consist of a group of coherent interventions designed to prevent or manage one or more chronic conditions using a systematic, multidisciplinary approach with potential employment of multiple treatment modalities. The aim is to identify patients at risk of one or more chronic conditions, to promote self-management by those patients, and to address their condition with the best possible clinical outcome, effectiveness, and efficiency; this may involve reimbursement considerations (Schrijvers, 2009). Disease management can also be referred to as the use of an explicit systematic population-based approach to identify persons at risk, intervene with specific programs of care, and measure the clinical outcomes (Epstein and Sherwood, 1996). Moreover, disease management can also be categorized into three major parts (Dellby, 1996):

- A knowledge base that quantifies the economic structure of a disease and includes guidelines covering the care to be provided, by whom, and in what setting for each part of the process.
- A care delivery system without traditional boundaries between medical specialties and institutions.
- A continuous improvement process that develops and refines the knowledge base, guidelines, and delivery system.

For m-Health, the following questions need to be addressed in the design and development process of any mobile chronic disease management system:

- *What Are the Specific Chronic Diseases or Conditions That Need to Be Managed?* These include the diversity of chronic conditions and targeted patient populations, care model requirements, and environments. For example, consideration must be taken of patients who are at low risk and high risk, and those who have multiple chronic diseases.
- *How Can a System Be Designed to Cope with Patients with Multiple Chronic Diseases?* This is an important factor, especially for those who are vulnerable to secondary complications that can be prevented by proper adherence to self-care and medication routines. These factors also need to account for the system's ability to differentiate between adverse conditions and to translate any personalization of the care management plan into its modules without a major overhaul of its software and hardware.
- *How Flexible Can the System Work in Different Clinical Settings and Clinical Care Models?* These include the adaptability of the system to the interoperability and standardizations requirements in different care setting.

- *How Can the System Work with Current and Future Technologies That Can Be Upgradable?* The system's flexibility to reconfigure itself to fast developments in smartphone technologies, together with Internet, computing, and sensors advancements, is needed without risking serious errors or reliability problems under different care conditions. This is closely related to the interoperability factor but also to the upgradability of the system.

- *What Are the Intelligent Tools and Tasks Required for the System?* These include embedding smarter decision support and data analytical models, such as big health data analytics, and smart inference models to assist clinicians in their decision-making and personalize the care individually for each patient.

- *What Are the Cost Factors of Realizing These m-Health Management Systems and the Economic Benefit to Be Gained from Their Deployment Compared to the Usual Care Costs?*

Obviously, the translation of all these issues to practical systems is a complex task and there are no comprehensive practical answers to all the challenges. However, the fast technological and computing developments could in the near future potentially provide practical implementations to some of these challenges. Today, there are varieties of commercial mobile disease management systems available on the market. Most claim to have different capabilities and able to work with multiple chronic disease management functionalities, with adaptable modalities. Although the proliferation of the smart health Apps market is accelerating the potential of consumer-driven healthcare management systems, their large-scale adoption and clinical acceptability by patients and physicians is still debatable and yet to be seen.

Figure 5.4 shows the basic architecture of a smartphone-centric chronic disease management system. This architecture has been adopted by most mobile disease management systems in the last decade. It generally consists of the main building blocks of the m-Health monitoring system described in Chapter 2. The basic system functionalities are summarized as follows:

- To acquire a patient's medical data from mobile devices and/or wearable sensors, as required for the specific chronic condition and treatment plan. This device usually communicates data wirelessly to the patient's smartphone.

- The mobile phone then transmits the data securely to one or more remote servers, usually in a healthcare clinic or specialist center. This mobile transfer is done seamlessly, typically via wireless Internet connectivity of the mobile network linked with the servers or cloud.

- The remote server then processes the data acquired and "advises" clinicians on the necessary actions and interventional feedback, if required, via remote mobile portals and dashboards that can be accessed by them via any mobile platform. These remote cloud-based servers and analytical data base systems typically store and process the patient's electronic data records and host the relevant decision support and data analytic elements.

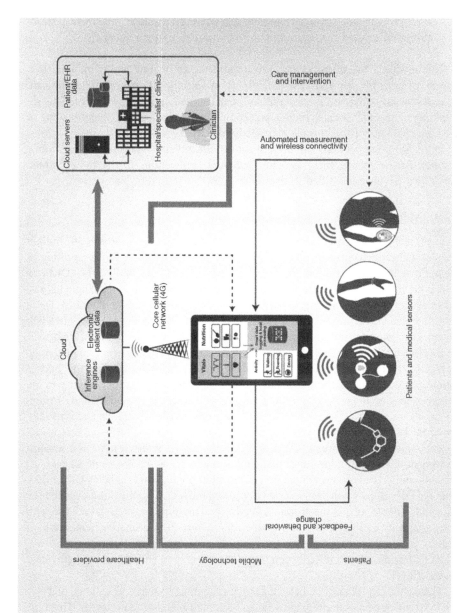

FIG. 5.4 General m-Health smartphone-centric chronic disease management system.

- These can also provide automated feedback to the patient, for example, for medication and adherence reminders or appointments.
- Smartphone Apps can provide patient education or behavioral change tasks through embedded software models for any individualized treatment instructions and revised medication plans, as informed by their specialist.

This model is based simply on utilization of the computing and communication powers of smartphones and their availability in shifting the specialist chronic care process toward more patient-centered self-management and empowerment. This m-Health system also leverages mobile and computing technologies to support some of the core elements of the chronic care model discussed earlier.

In the following sections, we present some of the clinical issues relating to mobile chronic disease management systems and their applications to the three chronic conditions we have identified as exemplars.

5.3.3   Mobile Diabetes Management Systems

Diabetes mellitus is a major global chronic disease that affects an estimated 387 million people, with an expected increase of 205 million more cases by 2035, resulting in approximately US$612 billion in global healthcare expenditure (IDF, 2014). Diabetes is also the major cause of heart failure, end-stage renal failure, and the single biggest cause of preventable blindness, as well as other causes of premature mortality. These alarming statistics represent major opportunities, but also challenges, for m-Health systems designed for not only management but also for the prediction of diabetes.

Brief information on diabetes is now presented for lay readers for completeness. Typically, type 1 diabetes mellitus (T1DM or T1D) develops when the body's immune system attacks the pancreatic cells that produce the insulin, leading to the pancreas ceasing to produce insulin. This type of diabetes is treated by daily insulin injections via a special pump or by needle injections. It is less common and is usually prevalent in adolescent and younger people.

Type 2 diabetes mellitus (T2DM or T2D), sometimes referred to as non-insulin-dependent diabetes or adult-onset diabetes, accounts for at least 90% of all cases of diabetes and is the type more common in the adult population (Diabetes UK, 2013). This type develops when the pancreas does not produce enough insulin or when the patient is unable to use the insulin effectively. This type is often associated with obesity and lack of exercise, with poor eating habits. It is primarily treated with a healthy diet and increased physical activity. In cases where these are not effective, either oral medications and/or insulin injections are recommended to control the blood glucose level.

There are other types of diabetes that are less common, such as gestational diabetes (GDM), which occurs in pregnant women with high blood glucose levels. This type usually develops in 1 in 25 pregnancies worldwide and is associated with complications to both mother and baby if not treated properly. It usually disappears after pregnancy, but there is increasing clinical evidence that women with GDM and their children are at an increased risk of developing T2D in later life (IDF, 2012).

From a clinical perspective, in a healthy person the concentration of blood glucose is normally maintained at 70–110 mg/dl by the body's natural biological control mechanisms. The level of failure of this mechanism determines the onset and type of diabetes. When diabetes onset occurs, the primary treatment is to maintain an acceptable level of blood glucose by adjusting food intake. It is well known that several factors can influence blood glucose levels; the most significant are diet, exercise, and medication. These factors are all controllable by the patient. The "prediabetic" condition means that the person's blood sugar level is higher than normal but not yet high enough to be classified as T2D; it is well known that without proper intervention or lifestyle change, prediabetes is likely to transform to T2D in 10 years or less. This condition, although important, is less well studied clinically from the m-Health perspective.

New approaches for diabetes care are being increasingly adopted with new priorities:

- Provision of better support for self-management strategies, especially for T2D.
- Use of new technological approaches for self-management, including m-Health technologies.
- Enabling new integrated diabetes care models that embrace these new technological developments.

From the self-management approach, mobile diabetes management has been one of the most prolific areas of clinical interest. This interest is reflected in terms of numerous clinical pilot studies, supported by many commercial diabetes smart Apps and management systems (Eng and Lee, 2013). The current market proliferation of mobile diabetes management systems is reflective of both the prevalence of the increase in diabetic patient populations worldwide and the commercial appeal of these m-Health systems.

These interventions and pilot studies have indicated some promising but indefinite and unclear outcomes, especially in improving levels of glycated hemoglobin (usually referred to as HbA1c), with reported results of fewer complications in diabetic patients using these systems. HbA1c is the gold standard measure for diabetes control, following two pivotal, large-scale studies in T1D and T2D, which demonstrated that intensive glucose control correlated with a decreased risk of diabetes-related complications (DCCT, 1993; UKPDS, 1998). It is interesting to note that although there is a major global market interest in diabetes m-Health management systems, so far there has been no large-scale, long-term study similar to these two earlier studies, either planned or implemented, to clinically validate their effectiveness.

Self-management and monitoring of diabetes is important in preventing some of the complications associated with diabetes, such as *hyperglycemia* or *hypoglycemia,* which occur if blood glucose is not controlled. In general, hyperglycemia means the blood glucose level is too high and is often defined as having a fasting blood glucose concentration greater than 140 mg/dl, or greater than 180 mg/dl if measured postprandial within 2 h following a meal. Typically, it does not result in immediate danger, but it is strongly associated with long-term complications. Hypoglycemia means the blood glucose level is too low and occurs when it falls below 60 mg/dl. In this complication, if immediate action is not taken and glucose levels continue to fall, it can result in unconsciousness and possibly death (Chow, 2011). Thus, measuring and knowing a patient's glucose levels and other diabetes parameters such as blood pressure, weight, food intake, and activity levels are important, particularly for diabetic patients susceptible to these complications.

Figure 5.5 shows the architecture for a typical mobile diabetes self-management system, illustrating the main functional tasks required by patients. It is similar to the more generic system shown in Fig. 5.4. The basic functions of a smartphone-based diabetes self-management system adhere to the following objectives:

- Empowering diabetic patients to self-monitor blood glucose (SMBG) or other parameters when and as needed (e.g., blood pressure, weight, and diet).
- Engaging patients in the self-management of their diabetic condition, including self-efficacy of their treatment plans.
- Achieving preset care targets and providing interactive feedback between patients and physicians on their individual disease progress and compliance.
- Leveraging patients' care procedures when and if needed by their physicians, with potential cost savings.

The systems consist of three basic modules: mobile patient end; healthcare provider or clinician end; and remote data servers, or clouds, which host the patient EHR, together with data analytics, inference engine files (for artificial intelligence), and system controllers. These servers are usually hosted in secure locations or located in hospitals or providers' private cloud system. The mobile patient end and the physician's end are usually connected via mobile network and Internet channels to transfer the blood glucose readings and other medical data. A brief summary of these functions is described next for completeness. Further details on the technical development of these systems are given elsewhere (Istepanian et al., 2009).

As shown in Fig. 5.5, the generic process consists of a structured feedback management loop that consists of the following personalized functional elements:

- The patient end, which consists of a smartphone equipped with a special diabetes App connected wirelessly via Bluetooth to a blood glucose meter and used for tracking and storing the patient's daily glycemic data and status. For hypertensive diabetic patients, there is also a wireless blood pressure device and weight scales. The patient is required to complete a basic training and educational program on using these systems and to be trained on the benefits of the daily blood glucose level measurements at specific intervals.
- The acquired blood glucose data are usually transmitted wirelessly via the individual patient's smartphone and stored in a remote portal hosted in a health clinic or hospital. This portal consists of the patient's EHR, supported with intelligent analytic programs that can process the data and produce simple

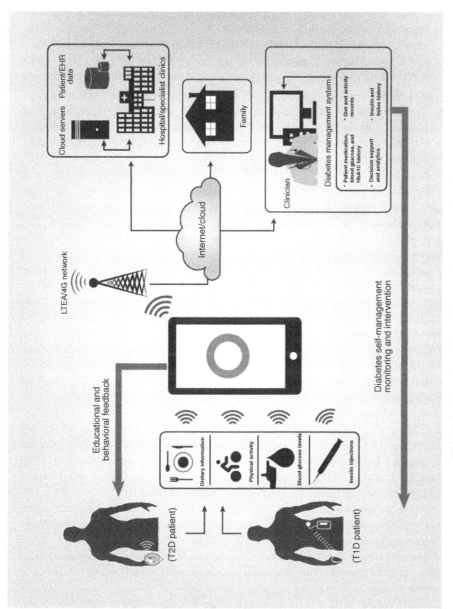

FIG. 5.5 General mobile diabetes self-management system.

illustrative graphics. These can be accessed and viewed by the patient and physician via two separate web portals, each designed and developed for separate access. The graphics provide information and decision support messages, such lifestyle change, educational notes, and medication required to improve the patient's daily management routine. The physician portal is used to view the patient's data, as well as detailed clinical history, medical analytics, and diabetes status. Critical patients can be prioritized individually via their portal for tighter monitoring and to facilitate, if necessary, communications via text messages or e-mail for any timely feedback and intervention.

• The adaptation of any medication and treatment plans required for individual patients. This is based on their self-management history, daily blood glucose profiles, treatment progress, and designated therapy protocol plans assessed regularly at 3–6 months follow-up intervals.

The schedule and frequency of taking daily blood glucose readings and other medical data is dependent on the individual care plan made for each patient by their specialist and based on their individual diabetes type, treatment, and progress. The timings of these readings are usually programmed in the patient's smartphone App as reminders to take readings according to preset schedules.

Smart devices, such as continuous glucose monitors (CGM) and continuous subcutaneous insulin infusion (CSII) pumps used by T1D patients, can also communicate wirelessly to their smartphones for monitoring their insulin injections automatically and according to the measured blood glucose levels. Patients are usually reminded to test their blood glucose several times per day, usually before meals and 2–4 h after meals. Typically for TID, the recommendation is for testing between four and eight times a day; for T2D, the recommendation is for testing two or more times a day, depending on the type and amount of insulin needed (Mayo Clinic, 2014). The following blood glucose targets and ranges are recommended (ADA, 2014a):

Average glycemia (A1c or eAG): 7% or 154 mg/dl; before a meal (preprandial plasma glucose), 70–130 mg/dl; 1–2 h after beginning a meal (postprandial plasma glucose), less than 180 mg/dl. These blood glucose targets are also dependent on other individualized factors such as duration of diabetes, age or life expectancy, comorbid conditions, known CVD or advanced microvascular complications, hypoglycemia unawareness, and individual patient considerations (ADA, 2014a).

Figure 5.6 shows an example of a new wireless Bluetooth-enabled glucose meter that is available on the U.S. market. This example module (Verio Sync, LifeScan Inc.) transmits and synchronizes the glucose data readings to an iPhone, iPad, or iPod using seamless synchronization with the patient's smartphone for automatic logging, graphic visualization, and other management functions using the device's smartphone App.

Another market example is shown in Fig. 5.7. MiniMed Connect securely transmits the patient's glucose data from an insulin pump and continuous glucose monitor to an App on an iPhone or iPod; a web display for carers shows real-time sensor glucose and insulin information. The system also delivers preset text

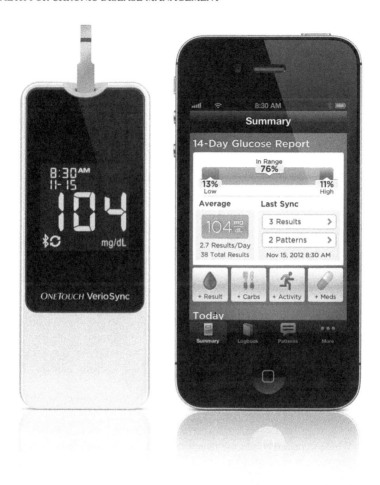

FIG. 5.6 Wireless glucose meter with smartphone connectivity (Verio Sync). (Photo courtesy of LifeScan Inc.)

notifications to carers when the glucose values go too high or too low, or when an alarm on the pump is not cleared.

It is likely that a new generation of wireless diabetes monitoring products will be increasingly available in the healthcare market in the near future. A list of these and similar smart products can be found in many of the specialist m-Health diabetes sites shown in the Appendix.

The patient empowerment process is embodied by these systems in many aspects, such as allowing patients to view their daily blood glucose levels and variation patterns on their mobile phone, daily, weekly, or biweekly as needed. Apps with

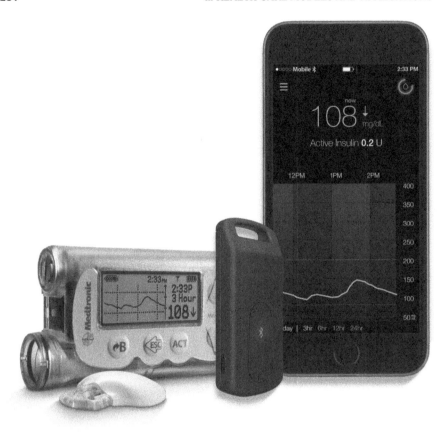

FIG. 5.7   Wireless continuous glucose monitoring with smartphone connectivity (MiniMed®
Connect). (Photo courtesy of Medtronic plc.)

specific software are programmed to alert patients with their lifestyle and treatment
choices, as well as helping them to monitor for symptoms of any potential hypoglycemic or hyperglycemic episodes. Also, in some applications it is possible to
dynamically track the average glycemia (A1c) for real-time estimation of hemoglobin
A1c levels using the self-monitoring blood glucose (SMBG) data acquired
(Kovatchev et al., 2014). It must be emphasized that there are many efficacy and
regulatory challenges associated with Apps developed for mobile diabetes self-
management (Istepanian, 2015).

Clinical Effectiveness of Mobile Diabetes Self-Management and Glycemic Control
There is general consensus within the diabetic medical community on the effectiveness of SMBG as an important modern therapy tool for the management of diabetes
mellitus. Although SMBG for T2D patients remains debatable, continuous glucose
monitoring is only approved as adjunctive to SMBG. Its role in self-management of

T1D patients is at least noncontroversial, and the literature generally supports SMBG for better glucose control (Garg and Hirsch, 2015).

It is recommended that SMBG need to be available to people receiving sulphonylurea and prandial glucose regulators because of the risk of hypoglycemia in these groups (Diabetes UK, 2013). From the clinical perspective, the results of a multicenter study including 24,500 patients from 191 centers in Germany and Australia showed that frequent SMBG is associated with better metabolic control in patients with T1D and T2D (Schutt et al., 2006). However, for an accurate SMBG process to be an integral part of diabetes management care, blood glucose meters used by patients must be accurate. Most of the current and new generations of these meters are now more accurate than the older devices. Furthermore, they meet the International Organization for Standardization (ISO) benchmark of less than 20% variability and soon all manufacturers will have to meet a benchmark of less than 15% variability (Garg and Hirsch, 2014). Further details on the relevant clinical aspects, effectiveness, and recommendations of SMBG may be found in other references (Trend-UK, 2014).

Mobile diabetes monitoring and management dates from the introduction of 3G mobile phone technologies (Istepanian et al., 2009). The introduction of smartphones and the health Apps changed the landscape of diabetes monitoring and management. Today, around 1100 diabetes-related diabetes smart Apps and a variety of mobile diabetes management modules are available commercially and on smartphone platforms. Some of these Apps are freely downloadable over the Internet and offer personalized management and education for diabetic patients. However, this proliferation has generated a heated debate on the benefits of these Apps in the absence of solid clinical studies on their efficacy and clinical effectiveness.

Current evidence on the use of m-Health for mobile phone interventions in diabetes is still evolving and has shown mixed and indefinite results. There is still ongoing debate and controversial discussions on the effectiveness of using mobile phones for SMBG. However, there is increasing clinical evidence that is tilting toward the obvious patient benefits in controlling blood glucose levels to avoid complications, especially for T2D patients with oral medication or in the context of their lifestyle behavior (Pal et al., 2014; Goyal and Cafazoo, 2013). Further examples from the developing world include a pilot study on the effectiveness of mobile diabetes management in the Kingdom of Saudi Arabia, that indicated improved HbA1C levels among Saudi diabetic patients and their disease management plans and education (Alotaibi et al., 2016).

In recent years there has been a particular increase in the numbers of published clinical pilots and randomized control trials, with extensive research projects to evaluate different mobile platforms for the management of T1D, T2D, and GDM. For example, Table 5.4 summarizes recent European research projects on diabetes management (European Commission, 2014c). Similar research and clinical trials are also being conducted in the United States and elsewhere (ADA, 2014b). Recent reviews and meta-analysis studies have reported a reduction in HbA1c levels in both T1D and T2D (Holtz and Lauckner, 2012; Liang et al., 2011). More specifically, a meta-analysis study of 22 trials conducted between 2005 and 2009 with a total of 1657

TABLE 5.4   Examples of European Diabetes Management Research Projects and Trials

EU Diabetes Project Title	Themes and Areas of Focus
AP@home (www.apathome.eu) 2010–2014	Development of communication platform between the commercially available continuous glucose monitor and insulin pump for T1D patients
Commodity12 (www.commodity12.eu) 2011–2014	To develop a platform for continuous monitoring of diabetes with emphasis on the interaction between diabetes and cardiovascular diseases
EMPOWER (www.empower-fp7.eu) 2012–2015	Self-management of diabetes patients through a modular, standards-based patient empowerment framework
METABO (www.metabo-eu.org) 2008–2012	Development of diabetes management and treatment, with integrated patient lifestyle
MISSION-T2D (www.iac.rm.cnr.it) 2013–2016	Development of patient-specific model for the simulation and prediction of metabolic and inflammatory processes in the onset and progress of the type 2 diabetes
MOSAIC (www.mosaicproject.eu) 2013–2016	Development of mathematical models and algorithms that can enhance the current tools and standards for diagnosis
REACTION (www.reaction-project.eu) 2012–2014	The development of an integrated approach to improve long-term management of diabetes with continuous blood glucose monitoring, clinical monitoring, and intervention strategies

Source: Adapted from European Commission (2014c).

participants showed that mobile phone interventions reduced HbA1c levels by an average of 0.5% (6 mmol/mol; 95% confidence interval) over a median of 6 months follow-up duration; a further subgroup analysis of 11 studies among T2D patients showed that reductions in HbA1c were more significant compared to studies among T1D patients of 0.8% (9 mmol/mol) versus 0.3% (3 mmol/mol); for $P = 0.02$) (Liang et al., 2011).

Another review of 16 randomized controlled trials of 3578 participants demonstrated that mobile phone-delivered interventions may be more effective than those delivered over the Internet (Pal et al., 2014). However, most of these trials lacked sufficient sample sizes or intervention lengths to determine whether the results might be clinically or statistically significant on larger populations. Also, the results suggest that further research is required to examine other key factors, such as provider perceptions, integration into a healthcare practice, and cost (Holtz and Lauckner, 2012).

A recent study of 72 patients to examine the effectiveness of a freely available smartphone App, combined with text message feedback from a certified diabetes educator, reported significant improvement in glycemic control in adult T1D patients (Kirwan et al., 2013). However, a meta-analysis study that investigated the impact of traditional phone call interventions on glycemic control from five randomized control trials (RCT) involving 953 patients concluded that these interventions had no more effective impact than standard clinical care in improving glycemic control (Suksomboon et al., 2014). Another review, which analyzed a total of 212 published articles and studies since 2010 on mobile and smartphone technologies for diabetes care and management, concluded that most of the m-Health interventions cited were poor, and that the outcomes of most of the interventions did not provide robust clinical evidence (Garabedian et al., 2015).

For mobile short message service (SMS) interventions, otherwise known as "text messaging," recent studies have shown that SMS can provide a simple yet effective approach for improving diabetic education and patient communications. Several RCT and pilot studies on the use of SMS for diabetes self-management and interventions have been conducted in different countries (Arora et al., 2013). In another meta-analysis of 10 trial studies involving 960 participants on the impact of mobile SMS health education and management on glycemic control of adult T2D patients, a significant reduction was reported in HbA1c levels (~50%, using the Hedges' g index) in the experimental group compared to the control group (Saffari et al., 2014). This study also reported that the effect on size was 86% among studies using both SMS and Internet for health education compared to only 44% in the studies that used text messaging only (Saffari et al., 2014). In most of these SMS studies, the technology, if used effectively, can provide improved self-efficacy, patient education, reminders, psychological support, medical appointments, and alerts (Kannisto et al., 2014; Krishna and Boren, 2008).

The clinical outcomes from some of these projects support the general trend that there is increasing, albeit cautious, evidence of the effectiveness of mobile SMS interventions in improving diabetes care outcomes. However, recommendations are also cited for more systematic studies and RCTs (Kannisto et al., 2014).

We can assume that m-Health interventions might generally improve glycemic control, particularly for T2D patients. However, most of the available evidence from comparing these interventions for both T1D and T2D patients shows poor data variability and reporting, with important methodological weaknesses; the effectiveness of m-Health interventions are therefore usually inconsistent and weak (Baron et al., 2012; Garabedian et al., 2015). Supporting this view is a large European study (Renewing Health) of three clinical clusters that investigated the effectiveness of mobile interventions for T2D patients with medium-term health coaching and life-long monitoring; it reported statistically insignificant results in lowering HbA1c levels (Kidholm et al., 2014).

It is also important to highlight the impact of self-efficacy that has been shown to be more important in diabetes management than social support, with the focus on gender differences in the self-management process (Burner et al., 2013). Recent studies on the development of a "bionic pancreas" for T1D patients are interesting to note. This new system uses a smartphone in conjunction with a continuous glucose monitor and two pumps designed to deliver precise doses of hormones to the body to control blood glucose levels with more effective outcomes than with current methods (Russell et al., 2014). These developments are at an early stage and the medical outcomes of larger clinical trials are pending at the time of writing.

Barriers to m-Health Interventions for Diabetes Management   There are several barriers to self-care management that can be generally categorized from the patient and care provider perspectives. For T2D patients, these barriers include socio-economic, psychosocial, physical, and environmental factors that are perceived to influence their ability to perform adequate self-care activities (Pun et al., 2008). For the healthcare provider, these barriers are primarily on physical, environmental, and cultural issues, as well as policy issues that are associated with impeding proper care. The recommendations to overcome these barriers, particularly for T2D self-management care, include the following:

- The provision of sufficient knowledge and robust diabetes-specific education to enhance the patient's understanding of this disease.
- The implementation of clinical requirements by T2D patients to perform all of the recommended activities that can reduce risks and prevent complications.
- The development of diabetes programs tailored to teach and support patients in acquiring and applying the required skills to facilitate positive psychosocial functioning.

These recommendations concur with existing diabetic patient education care models, particularly in primary care settings. Furthermore, there is ample evidence from several controlled clinical trials that the self-educational approach can provide better clinical outcomes and reduce care costs (Bodenheimer et al., 2002b; Norris et al., 2002).

Smart mobile interventions can potentially play a pivotal role in implementing better care for diabetes patients, provided that the appropriate behavioral change models, and long-term Apps compliance by patients, are implemented appropriately. The implementation of these solutions requires multidisciplinary effort and further work in developing new m-Health solutions outside the current smartphone-centric interventions.

There are still strong market- and consumer-driven trends on the use of mobile phone Apps for diabetes. This trend is likely to remain popular, especially among the younger and adolescence population, which is evident by the increase of the smart diabetes Apps on the market.

In conclusion, there are reservations concerning the current clinical uncertainties associated with the m-Health interventions for self-management, which are summarized as follows:

- There is increasing debate on the clinical effectiveness of mobile SMBG interventions, supplemented by a lack of m-Health studies. This reflects the current status of m-Health clinical interventions for chronic conditions in general. The available evidence is based on silos of pilots and RCTs that have generated mixed and inconclusive outcomes. Large-scale, well-designed studies are needed to clarify the current fuzzy evidence. These studies presently consider typical primary outcomes, such as lowering HbA1c, body mass index

(BMI), weight, and blood pressure. They should also include secondary outcomes, such as the impact of lifestyle changes, long-term usability and adherence, patient demographics, social and behavioral change, care barriers, language and habits, and cost-effectiveness of these interventions.

- There is ongoing clinical debate on the combined problems of diabetes and obesity, or "diabesity." However, there are so far no detailed clinical studies on the impact of mobile phone interventions, whose role in obese diabetic patients is not fully exploited and therefore needs further clinical research focus as an important but overlooked topic.

- Identification of suitable smart m-Health educational models that translate the standards of diabetes care delivery protocols suitable for different settings and ethnicity environments.

- The role of incentives and behavioral change are not yet well understood and fully realized.

- The challenges of cyber security and privacy concerns and the ownership of patients' "big health" data generated from social networking and Apps.

- The development of alternative business models and ecosystems for new diabetes m-Health care, based on recent technological advances and not solely based on smartphone-centric care models.

5.3.4 Mobile Health for Cardiovascular Diseases (CVD)

Cardiovascular diseases, according to the WHO, are caused by disorders of the heart and blood vessels. These include coronary heart disease (CHD, heart attack), cerebrovascular disease (stroke), raised blood pressure (hypertension), peripheral artery disease, rheumatic heart disease, congenital heart disease, and heart failure (Mackay and Menask, 2014). It is estimated that 17 million people globally die annually of cardiovascular diseases, including more than 600,000 in the United States alone, or one in every four deaths (Murphy et al., 2013).

The history of cardiac and wireless ECG monitoring goes back several decades, with the advent of modern telemedicine and telecardiology in the 1960s. The use of mobile phone technologies for cardiovascular and ECG monitoring is considered one of the earliest applications of mobile health and the first to be implemented (Istepanian et al., 1998). Since then there have been major advances fuelled by rapid technological developments in cardiac sensing (such as smart wearable ECG biosensors), supported by advances in low-power signal processing and short-range communications.

An extensive review of ambulatory arrhythmia monitoring devices has considered many aspects, such as cost, technical considerations, and obstacles to patient compliance (Zimetbaum and Goldman, 2010). Some of these have evolved as today's smartphone-based proprietary wearable ECG and cardiac monitoring systems available on the market. These, for example, can detect and record different arrhythmias in real time, provide diagnosis, and alert patients and their carers to such episodes (Rothman et al., 2007). Numerous smart Apps associated with smartphone-based cardiac monitoring devices can be downloaded from Apple Shop or Google Play, and

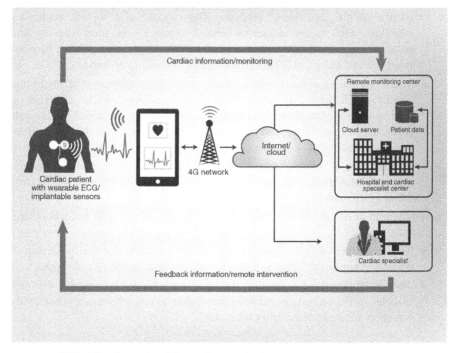

FIG. 5.8   General mobile cardiac monitoring and management system.

patients can even choose their "Top 10 Cardiac Health Apps." There is also a "superior stethoscope" that monitors the heart and sends data to the cloud for remote analysis by physicians on their smartphones (Park, 2015).

Apart from the questionable accuracy and viability of these products, as discussed earlier, commercially available cardiac monitoring devices are also not yet proven by pivotal clinical trials and rigorous studies to provide evidence of their effectiveness and efficacy. These devices include wearable and disposable ECG patches and one-touch smartphone sensing monitors and Apps. The increasing trend of consumer-driven monitoring devices contributes to blurring of the boundaries between m-Health as a scientific concept and the mobile health monitoring applications presented in Chapter 2.

The general architecture of a cardiac mobile monitoring and management system is shown in Fig. 5.8. Advances in smartphone, wearable technologies, and relevant Apps have led to numerous applications for mobile cardiac and ECG ambulatory monitoring. Technological details associated with the design and development of these systems are described elsewhere (Pantelopoulos and Bourbakis, 2010; Istepanian et al., 2006). The architecture of these systems typically falls within the general functionalities of the mobile chronic disease management and decision support system described earlier and shown in Fig. 5.4.

There is also a major global market in wireless cardiac monitoring and wearable telemetry devices for heart patients with implantable devices, with reports estimating

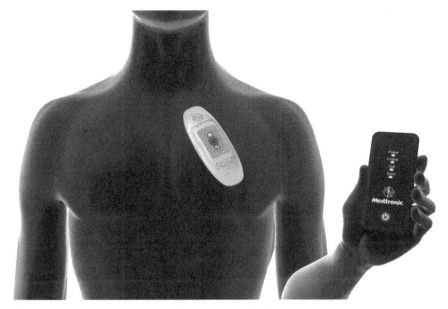

FIG. 5.9   Mobile cardiac monitoring system (SEEQ™). (Photo courtesy of Medtronic, Inc.)

the cardiac monitoring and cardiac rhythm management (CRM) market to be worth $23.3 billion by 2017 (Markets and Markets, 2013).

Figure 5.9 shows an example of a mobile cardiac monitoring system (SEEQ™) from Medtronic. The cardiac data from this system are acquired via a wearable device that can be worn for up to 30 days; the data are automatically transmitted via a cellular connection to a specialist monitoring center. The center is automatically alerted to the detection of an abnormal event, such as an arrhythmia. The system includes a trigger button that allows patients to mark symptoms of arrhythmias, which are then sent to their physicians (Medtronic Inc., 2015).

Clinical Effectiveness of m-Health Interventions for Cardiac Monitoring and Management   Numerous "telehealth" and "tele-monitoring" trials have been conducted to validate the impact of these solutions on cardiac care. The clinical outcomes are conflicting, with varying opinions among clinicians as to the clinical effectiveness of these technologies. Although there is widespread acceptance of the value of remote disease management for heart failure patients, the rates of their rehospitalization remain high when using traditional telehealth and telemonitoring systems. Systematic review studies of RCTs for "home telehealth monitoring" have indicated little evidence of the impact of telehealth solutions in improving patient outcomes by reducing rehospitalization and mortality rates (Desai, 2012; Chaudhry et al., 2010).

Similarly, the outcomes of the "Whole System Demonstrator" telehealth randomized trials in the United Kingdom have provided relatively modest clinical benefits in terms of reduced emergency admission rates and lower mortality (Steventon et al.,

2012). These interventions also did not support real economic benefits or cost-effectiveness in comparison to standard support and treatment (Henderson et al., 2013). Other studies that reviewed the outcomes of existing randomized controlled trials to evaluate telehealth interventions in CHD patients indicated a reduction in the effective risk factor and secondary prevention (Neubeck et al., 2009).

These conflicting outcomes of earlier telehealth interventions are similarly mirrored in the outcome of mobile technology interventions. One of these was a RCT called HEART, to assess the clinical effectiveness of mobile phone interventions on 171 patients with ischemic heart disease (IHD). The conclusion was that these interventions were not effective in increasing exercise capacity over and above the usual care, but with probable cost-effectiveness in their augmentation to existing cardiac rehabilitation services (Maddison et al., 2014). Another review, to study the evidence to support the feasibility and acceptability of using mobile technology for cardiac rehabilitation in IHD patients, reported better outcomes, with recommendation for further work (Beatty et al., 2013). Several studies on the effectiveness of using text messaging on cardiac patient rehabilitation have also been reported to improve adherence to lifestyle behavior (Dale et al., 2014).

In ambulatory cardiac monitoring, a multicenter study on the effectiveness of mobile cardiac outpatient telemetry (MCOT) devices versus standard loop event monitoring indicated that MCOT can provide a significantly higher yield than standard cardiac loop recorders in patients with symptoms suggestive of a significant cardiac arrhythmia (Rothman et al., 2007).

For applications of mobile health in hypertension monitoring and management, a UK study suggested the effective role of mobile phone intervention to help hypertensive patients to achieve and maintain blood pressure below their recommended targets (Earle et al., 2010).

Other interventional studies in hypertension have indicated the potential impact of mobile health applications on improving management that supports patient adherence schedules and prescribed treatments (Logan, 2013).

Another study, in Canada, addressed an assessment of the attitudes of heart failure patients and the perceptions of their healthcare providers toward the use of mobile phone remote monitoring. It concluded that patients and clinicians wanted to use these technologies, but some patients had reservations attributed to the possible difficulty of using them due to lack of visual acuity or manual dexterity. This study also concluded that the clinicians' reservations were mainly perceived in the increase of using these technologies on their clinical workload, along with possible medicolegal issues (Seto et al., 2010). However, a recent UK 6-month RCT (called HITS) of telemonitoring "uncontrolled" hypertension patients indicated that home telemonitoring was more effective at reducing blood pressure than standard care but also more expensive (Stoddart et al., 2013).

In a Mayo clinic meta-analysis and review study, researchers found that, on average, digital health interventions, including telemedicine, web-based modalities, e-mail reminders, SMS texting, mobile applications, and data monitoring interventions, reduced the relative risk for cardiovascular disease outcomes by 40%. However, the study found that not all the reviewed digital health interventions performed as effectively, with web-based SMS and telemedicine interventions acting as the most effective, while e-mails and data monitoring were the least effective. This study also found that although the interventions reduced weight loss, BMI, blood pressure, and low-density lipoprotein cholesterol, these reductions did not always obviate a cardiovascular disease risk (Widmer et al., 2015).

There are now various cardiac implantable devices, including pacemakers, cardioverter-defibrillators, and cardiac resynchronization devices, which are increasingly being used for advanced cardiac diagnostic functions, with some used to monitor patients remotely. There is increasing agreement on the clinical importance of remote monitoring after implantation of these devices. This forms part of both the device maintenance and patient care process, where it can offer the opportunity to optimize the clinic workflow at the care center and improve device monitoring and patient management (Varma et al., 2014). Some of the earlier pilot studies have shown high levels of acceptance by both patients and clinicians on the effectiveness of the mobile remote monitoring process (Marzegalli et al., 2008). However, a review study on the effectiveness of mobile phone-based management of heart failure patients with implants indicated mixed clinical outcomes and limited cost-effectiveness (Bhimaraj, 2013).

Another review concluded that patient satisfaction and compliance were increased as a result of more monitoring, with decreased healthcare costs (Afolabi and Kusumoto, 2012). Another RCT (called CHAMPION) to validate the clinical effectiveness of wireless pulmonary artery hemodynamic monitoring in chronic heart failure indicated that the outcomes concurred; it extended the previous findings in showing a significant reduction in hospitalization for patients with class III heart failure managed by wireless implantable monitoring (Abraham et al., 2011).

We can assume from the available evidence that using mobile wireless monitoring interventions for cardiac patients in different clinical and care settings is well poised to increase and to potentially provide better clinical and patient management. However, further work is required to address some of the following challenges highlighted from these studies:

- Development of smarter and more personalized mobile platforms for multitier functionalities for heart failure patients, especially for low-risk categories.
- Tailoring more m-Health interventions and solutions for cardiac management that are not possible with current telehealth solutions, especially those clinically unproven and economically viable.
- The lack of standardization and interoperability of mobile devices and relevant barriers to their routine use in cardiac care in global settings.
- Enhanced and smarter methods for mobile medication adherence for heart failure patients by using intelligent "big data" analytics that transcend current approaches.
- Adoption of intelligent self-diagnostic and low-power wireless communication technologies for implantable cardiac devices that address the shortcomings of current devices.

- Better utilization of new wearable cardiac devices with smarter Apps for home monitoring and rehabilitation of cardiac patients.
- Further investigations of the future use of nanorobotics for cardiac surgery and mobile monitoring applications.

5.3.5 Mobile Health for Chronic Obstructive Pulmonary Disease

COPD is currently ranked the fourth leading cause of death worldwide and it will rise to third place by 2030 (GOLD, 2013). In addition to generating high healthcare costs, COPD imposes a significant burden in terms of disability and impaired quality of life; unlike many other leading causes of death, its global mortality rates are projected to increase due to increasing levels of smoking (Halbert et al., 2006).

In medical terms, according to the WHO, COPD is an umbrella term used to describe chronic lung diseases that cause limitations in lung airflow. Tobacco use and second hand smoke are the primary risk factors. The Global Initiative for Chronic Obstructive Lung Disease (GOLD) defines COPD as a preventable and treatable disease, characterized by persistent airflow limitation that is usually progressive and associated with an enhanced chronic inflammatory response in the airways and the lung due to noxious particles or gases (GOLD, 2013). It is also well recognized that smoking is the most important factor for COPD onset and the key cause of morbidity and mortality from this disease. For the treatment and management of COPD, the GOLD strategy recommends that treatment plans should include smoking cessation strategies as well as the use of medication to manage the symptoms of COPD and its related complications and comorbidity conditions. Typical COPD treatment plans should include nonmedicinal interventions, such as patient counseling, pulmonary rehabilitation, and smoking cessation, which is considered the single most effective and cost-effective way to reduce exposure to COPD risk factors. In addition to smoking cessation, one of the other key challenges in COPD management is breathlessness and chronic cough. Most of the treatment measures can be implemented using mobile health platforms. In particular, for COPD rehabilitation, patients are usually trained to carry out a daily exercise program. If patients adhere to it, they should improve their health significantly. The difficulty arises once they are discharged from the supervised exercise stage and are expected to continue the program at home within their normal lifestyle (O'Shea et al., 2007). The potential role of m-Health is to demonstrate sufficient effectiveness in home care management with tailored solutions. Figure 5.10 shows a mobile health (phone-centric) COPD patient management and tele-rehabilitation system.

Despite the high prevalence of COPD, there is little reported work on m-Health interventions and monitoring. As shown in Fig. 5.10, the monitoring system needs to be specifically tailored for COPD patient care. It includes the three main building blocks of a typical m-Health smartphone monitoring system:

- Patient sensing that acquires and logs the medical data, including vital signs data, COPD parameters with a spirometer or pulse oximeter, movement data such as activity or thoracic respiration, and sputum and cough data.

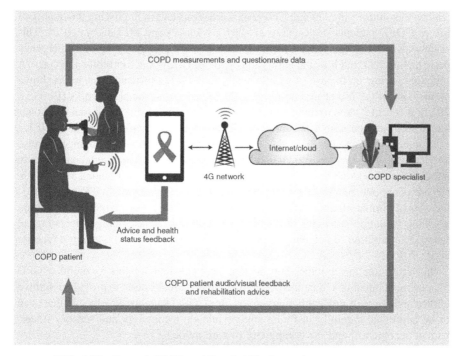

FIG. 5.10   General COPD mobile rehabilitation and management system.

- Wireless communication connectivity (mostly cellular networks) to transmit the acquired data to a remote care center and servers, or cloud.
- Data analytics to processing the received patient data and extract the clinically relevant information, which suggests the necessary interventions according to the specific COPD management and rehabilitation protocols.

Advances in wearable technologies are leading the development of new special "wearable jackets" tailored for COPD mobile management and care (Bellos et al., 2012; Patel et al., 2012). Vital signs, such as ECG, temperature, and breathing rate, are measured by embedded sensors in ergonomic garments. Medical data from these sensors are sent wirelessly from the garment to a remote monitoring portal for further analysis or for patient rehabilitation purposes. Although these "wearables" are currently in experimental and research prototyping phases, they could be adopted for COPD and other chronic conditions as part of the increasing market trends. More recently, application of wearable technologies were reported for monitoring respiratory function, heart rhythm, and heart rate variability of COPD patients using mobile and cloud technologies (Kender, 2014).

Clinical Effectiveness of m-Health for COPD Patient Care Management and Rehabilitation   Since the 1990s, numerous telemedicine and telehealth technologies have been used for COPD management. A Cochrane review indicated the potential of telehealth solutions in reducing COPD exacerbations and in improving the quality of life of COPD patients (Polisena et al., 2010; McLean and McKinstry, 2012). This review also found that telehealth interventions can significantly reduce emergency department visits and hospital admissions without changes in morbidity and costs. A literature review on the use of mobile health for COPD self-management has shown an increase in the use of mobile phones, the Internet, and social networks (Borycki, 2012). A recent system for managing COPD exacerbations using smart mobile phones has been described with a feasibility evaluation study (Der Heijden et al., 2013).

Another study on the effective use of mobile phones for home endurance exercise and training programs has shown good compliance and positive clinical outcomes in patients with moderate-to-severe COPD (Liu et al., 2008). A RCT, called MIOTC, on the use of mobile health for patients with Acute Exacerbation of COPD (AECOPD) has shown similar clinical outcomes (Zhang et al., 2013). Furthermore, research on smoking cessation strategies in COPD patients has indicated that pharmacological therapy, in addition to behavioral counseling, can be an effective strategy (Warnier et al., 2013).

In conclusion, these studies indicate that there are silos of pilots for mobile COPD management and that to date there is no pivotal large-scale study to provide definitive clinical evidence of m-Health interventions for COPD patients. The factors that need to be addressed carefully in future studies include technology and sensory issues, patient acceptance, and the complexity of care models.

5.3.6 Mobile Health Interventions in Other Chronic Diseases

In addition to the three chronic diseases described earlier, there is evidence of m-Health interventions in other chronic conditions. In particular, studies have advocated varying levels of the impact of mobile text messaging on medication adherence and clinical management for asthma, diabetes, hypertension, and cancer (Wei et al., 2011). Other studies that assessed text messaging for self-management of long-term conditions have demonstrated only limited cases providing effective benefits (De Jongh et al., 2012). Another systematic review of RCTs of mobile technology interventions for improving disease management and their use on health behavioral change has indicated mixed evidence (Free et al., 2013a, 2013b). In particular, an analysis of the outcomes of text messaging studies indicated that the pooled effect on appointment attendance using "reminder versus no reminder" was increased (Free et al., 2013a). Most of these and other reviews suggested acceptable evidence of benefits of SMS interventions in areas such as vaccine appointment attendance, and cardiopulmonary resuscitation training, in addition to some suggestive evidence of short-term benefits for asthma control, physical activity, and psychological support.

In order to explore these issues in more detail, we present for completeness a sample of recent studies that relate to mobile phone interventions in some important chronic conditions:

Cancer: Although most of the current mobile health studies are limited to their use for self-management activities carried out by cervical cancer patients, there are promising results that indicate the effectiveness of these technologies in other cancer care activities, particularly in cancer screening, including cervical cancer screening. The outcomes have indicated excellent acceptability and feasibility of the intervention by patients, with clear evidence of potential cost savings (Lee et al., 2014). Another systematic review of web-based interventions for patient empowerment and physical activity in chronic diseases, with relevance for cancer survivors, concluded that these interactive interventions have a beneficial effect on the physical activity of patients with chronic diseases (Kuijpers et al., 2013). Another RCT (Advanced Symptom Management) has investigated the perceptions of nurses toward mobile phone systems and their role in managing chemotherapy-related toxicity in breast, lung, and colorectal cancer patients; this indicated some positive outcomes (Maguire et al., 2008). However, a more recent literature review on the role of mobile technologies has indicated the unlikely evidence that these interventions can contribute to the creation of new supportive care models for cancer patients due to under-utilization factors (Nasi et al., 2015). Similarly, another study that reviewed more than 295 smartphone Apps, which deal with the prevention, detection, and management of cancer applications, reported lack of evidence of their utility, effectiveness, and safety (Bender et al., 2013). There is clearly a need to do further work to design new models for m-Healthcare and to more appropriately validate the effectiveness of these models.

Asthma: Asthma is considered as one of the most common long-term chronic conditions, with considerable health and economic impact on patients, communities, and health systems worldwide. Most of international clinical guidelines today recommend the inclusion of self-management programs in the routine management of patients with asthma. Accordingly, many reported mobile health interventions on asthma management have been reported. One systematic review, which analyzed the impact of digital interventions for asthma self-management, reported some beneficial effects but recommended further work, with emphasis on an economic analysis on the cost-effectiveness of these interventions (Morrison et al., 2014).

Another Cochrane study, which analyzed two RCTs with a total of 408 participants to assess the effectiveness of self-management interventions for asthma patients delivered via smartphone Apps, compared to those delivered via traditional methods (e.g., paper-based asthma diaries), reported insufficient evidence on their impact and recommended further research (Belisario et al., 2013). However, a feasibility study on the use of mobile phones to assess asthma symptoms and adherence levels in adolescents, using Ecological Momentary Assessment (EMA) of health behavior, reported increased adherence levels by users (Mulvaney et al., 2013).

Obesity: Obesity is emerging as a chronic condition with major global healthcare impact. In recent years, extensive work has been reported on obesity management using mobile health interventions. A recent systematic review on interventions employing mobile technology among overweight and obese patients concluded that there is consistent and strong evidence that weight loss can occur in the short term, with moderate evidence for the medium term (Bacigalupo et al., 2013). Similarly, another pilot study of 20 participants using smartphones to assist behavioral obesity treatment (self-monitoring, feedback, and behavioral skills training) reported on the potential of the technology for delivering key components of behavioral weight loss (BWL) treatment plans. This approach can be particularly advantageous for optimizing adherence to self-monitoring in BWL treatment (Thomas and Wing, 2013). A further study evaluated diet (or nutrition) and anthropometric tracking m-Health Apps, but the results indicated that these Apps received low overall scores for inclusion in behavioral theory-based strategies (Azar et al., 2013).

It is important to note that obesity treatment and management plans differ from weight loss treatments. These basically depend on several factors that include diet, the amount of energy intake, and the amount of energy expended by exercise. There are a few RCTs on obesity treatment in primary care practices, with some indications that weight-loss counseling can help clinically meaningful weight loss outcomes (Wadden et al., 2011). However, there is so far no clinical evidence from any large-scale m-Health interventional study on obesity management in primary care settings. Considering that the correlation between diabetes and obesity is well known, such m-Health studies are timely and more clinical work is needed.

Tobacco Use and Smoking Cessation: This condition also relates to the COPD patient management issues that we discussed earlier. There are numerous studies and RCTs that have indicated potential benefits of mobile phone and text messaging interventions on smoking cessation, especially in younger people (Haug et al., 2013). A meta-analysis of five studies on the effectiveness of mobile phone-based smoking cessation interventions indicated that smoking cessation delivered predominantly via automated text messaging increased the quitting rates (47–99%) compared with minimal intervention (Whittaker et al., 2012). Other review studies also indicated some evidence of the impact of mobile technologies on smoking cessation (Free et al., 2013a). A further study examined the contents of 400 popular Apps for smoking cessation (252 smoking-cessation Apps for the iPhone and 148 for the Android); it concluded that smart Apps are positively associated with scores on the Adherence Index (AI) but the study also reported that these Apps could be improved by better integration with the clinical practice guidelines and other evidence-based practices (Abroms et al., 2013).

These studies reflect the unprecedented market proliferation of m-Health Apps for smoking cessation applications, together with some tangible benefits of text messaging for this chronic condition. However, the correlation of these studies and their reported benefits for COPD complications have not yet been fully studied. Furthermore, the efficacy issues of these health Apps have also not been studied in detail and need to be considered carefully, as discussed in Chapter 3.

5.4 MOBILE HEALTH FOR OTHER HEALTHCARE SERVICES

As discussed in Section 5.2, mobile health interventions have been used extensively in many healthcare services other than for chronic diseases, as shown in Table 5.3. In this section, we present some of these examples and discuss the available evidence and impact of mobile phone technology interventions.

5.4.1 Preventative Medicine, Patient Education, and Behavioral Change

In the last decade, interest has been focused on clinical work in personalized and preventative medicine, in which m-Health can contribute significantly. The shift toward preventative care, supplemented with increasing demands for patient-centric healthcare systems, has been one of the most significant changes in medical care for many years. This transformation has been accelerated by advances in mobile health and associated technologies of sensing, computing, and communication systems.

Moreover, it is well known that poor lifestyle choices, such as lack of activity, diet, smoking, and stress, are some of key contributors in the development and progression of preventable chronic diseases. The recent proliferation of lifestyle smartphone Apps to support diet management, physical activity, mood, and others has led to increasing number of studies, particularly on the effective role of preventative medicine Apps. However, there is still debatable evidence on the effectiveness of these interventions. We cite some examples of these clinical studies and aim to analyze their outcomes and current status; for example, a RCT on coaching strategies for obesity and standard weight loss treatments. The outcome demonstrated an enhancement in short-term weight loss in combination with an existing system of care (Spring et al., 2013). Discussions on the role of m-Health for the prevention of child obesity, which is becoming a significant global health issue, advocate that more work is needed due to the lack of clinical evidence and clear economic benefits of using commercially available m-Health applications (Tate et al., 2013).

For diabetes management and behavioral changes, a recent Norwegian RCT that examined the impact of mobile health intervention for self-management and lifestyle change for T2D patients indicated no change in HbA1c levels between the two groups after a 1-year intervention. This study also indicated increased levels in skills and acquisition techniques from those who received health counseling in addition to self-management via smart Apps, compared to those who were self-counseled (Holmen et al., 2014). The results seem not to concur with an earlier U.S. cluster randomized trial looking at mobile phone personalized behavioral change intervention for blood glucose control; this demonstrated the effectiveness of combined behavioral mobile coaching, blood glucose monitoring, and self-management with lifestyle behavior in the reduction of glycated hemoglobin levels over a 1-year intervention period (Quinn et al., 2011).

Furthermore, a review of the impact of mobile phone text messaging on HIV patients and other chronic disease sufferers indicated global evidence to support the use of text messaging to improve patient education, adherence to medication, and attendance at scheduled appointments (Mbuagbaw et al., 2015). In general, these studies support other available evidence of the positive impact of text messaging interventions on other preventative care areas, such as smoking cessation, alcohol recovery, and others (Free et al., 2013a).

Overall, there is need for further work to better quantify the role of m-Health for prevention, especially for global health priorities such as obesity and diabetes. More innovative solutions for developing new intelligent "m-coaching methods' associated with "cognitive m-Health," if developed, can contribute to better use of m-Health interventions for more effective clinical impact and long-term care outcomes.

5.4.2   Remote Mobile Diagnostic and Treatment Services

The use of mobile phone technologies in store-and-forward diagnostics applied in different medical disciplines, such as mobile tele-dermatology, represents successful examples of smartphone m-Health applications (Kaliyadan, 2013). Earlier studies on tele-dermatology indicated that the agreement rate between clinic-based dermatologists and store-and-forward tele-dermatologists averaged 80% but could approach almost 95% (Whited, 2006). Other interesting applications have included the use of mobile phones with embedded cameras that can provide off-site surgeons with images for surgical care (Hsieh et al., 2005).

In women's health, there have been indications of increased use of smartphone Apps to aid weight loss, and monitor for diabetes and heart failure. However, the outcomes and the clinical evidence of these interventions were mixed, with a need for more supporting evidence (Derbyshire and Dance, 2013).

On diagnostics applications, there have been increasing numbers of accurate and cost-effective mobile point-of-care (POC) diagnostic tests that aim to control infectious diseases, such as sexually transmitted infections (STI) and the human immunodeficiency virus (HIV). These allow patients to receive immediate diagnoses and treatment advice (Gaydos and Hardick, 2014). There is also evidence of using SMS in sexual health areas that include communications between sexual health clinics and patients, partner notification and contact tracing, and contraception with reminders (Lim et al., 2008).

There is also an increasing number of commercially produced home testing and diagnostic kits approved clinically. Rigorous clinical monitoring and cost-effective evidence studies of these are still lacking. Other remote mobile diagnostic applications include pilot validation studies of handheld ultrasound scanners for routine patient referrals for echocardiography and their potential for outpatient consultations (Prinz et al., 2010). With forthcoming advances in embedded smartphone imaging and sensing capabilities, this application is poised to become the target of more clinical applications and relevant studies in different medical scenarios in the near future.

5.4.3   Home and Elderly Care

There are numerous studies on the use and clinical effectiveness of e-Health, telehealth, and telemedicine technologies for home care (Lindberg et al., 2013).

A report produced by the GSMA cited the following areas in m-Health (mobile phone) as the most appropriate for senior care: medication adherence, vital sign monitoring, activity monitoring and alert systems, wellness and rehabilitation, caregiver solutions, and remote consultation (GSMA, 2011).

Although the potential economic benefits of these m-Health interventions were highlighted in this report, more detailed clinical evidence and cost impact of the various products and services were lacking. Another study on the role of mobile health-based medication adherence, using electronic blisters in elderly patients with increased cardiovascular risk, diabetes, high cholesterol, and hypertension, demonstrated a potential increase in adherence levels that subsequently led to improved control of blood pressure and cholesterol concentrations (Brath et al., 2013).

There have also been numerous studies on the role of ambient-assisted living (AAL) for health, quality of life, and care services for the elderly (Siegel et al., 2014). In particular, there is increasing interest in m-Health applications for the prevention of falls as this is the leading cause of morbidity, mortality, and lasting functional disability among the elderly population. Various wearable devices for fall prevention have therefore been reported (Marschollek et al., 2008). Other studies on the use of wearable fall detectors have indicated that their perceived benefits may differ between older people and their carers. Furthermore, the experience of falling has to be taken into account when designing these devices, as these may influence the perceptions and how older people use them (Williams et al., 2013).

There are fewer studies that highlight the usability of m-Health for elderly carers rather than the elderly patients they care for. A recent study that assessed the use of m-Health smart Apps in alleviating a caregiver's burden at veterans' home care environments demonstrated no significant impact in the role of mobile Apps and technological intervention (Frisbee, 2014).

Further work on the impact of m-Health is clearly needed only because of the increasing population of carers. In Europe, there are many EC-funded initiatives on the "ageing well" (European Commission, 2014a). For example, some 126 research projects and pilot studies in the EU Ambient Assisted Living Joint Programme (AAL-JP) were funded between 2008 and 2013 (European Commission, 2014b). The overall clinical outcomes and benefits of these combined studies were mostly evaluated within individual projects. There is a need to consolidate these results and to translate these to more pivotal evidence on their impact in different AAL areas. A recent review on the latest technologies for AAL applications for elderly patients outlined challenges that need to be addressed for future developments, such as sensor technologies, security and privacy, human factors, and legal and ethical issues (Rashidi, 2013).

Many other aspects of medical care can use the power of home-based m-Health services, rather than in specialist centers and hospitals, such as ear, nose, and throat, orthopedics, and urology. However, there is so far no effective clinical evidence for these m-Health applications. Also, in mental health, there is no clear evidence of the impact of mobile phone and smart Apps interventions due to the lack of large clinical studies and robust evidence for the efficacy and effectiveness of these technologies

(Donker et al., 2013; Lakshminarayana et al., 2014). A recent review on the assessment of smart wearable body sensors for patient self-assessment and monitoring indicated that despite the significant innovation and progress within the monitoring device industry, the widespread integration of this technology into medical practice remains limited (Appelboom et al., 2014).

In conclusion, considering the importance of elderly care, particularly with the increasing health demands and demographics associated with the elderly population globally, there is a surprising lack of pivotal clinical evidence of mobile health interventions. Although there is a growing market for innovative AAL and elderly care m-Health applications and systems, there is obviously further room for larger, more rigorous studies to universally validate the impact of mobile health interventions and their appropriate reimbursement models and cost-effectiveness.

5.4.4 Global Public Health

There is increasing evidence of acceptable clinical outcomes and evidence of m-Health (mobile phone) interventions in global public health areas. This evidence is so far limited to pilots, and exploratory or observational studies, but there is in parallel an increasing body of work reported from the developing world and in LMICs with limited healthcare resources, on tangible benefits from m-Health applications (Vital Wave Consulting, 2009). More details on this topic will be presented in Chapter 6, but some examples of available evidence are presented here for completeness.

For example, a recent study on the impact of m-Health interventions on neonatal survival in LMICs demonstrated varied outcomes, with some showing sufficient evidence, such as in adherence to HIV/AIDS antiretroviral therapy, uptake and demand of maternal health service, and compliance with malaria treatment guidelines, but less evidence in others (Higgs et al., 2014). Another study demonstrated the impact of m-Health evidence-based modeling (called LiST) on neonatal survival rates in resource-limited settings. It reported the effectiveness in improving skilled birth attendance and facility delivery (SBA/FD) services, combined with increased facilitation of delivery as targets for better mortality impact rates relative to other intervention scenarios (Jo et al., 2014). Text messaging is becoming important and effective patient education in mobile healthcare for HIV and other chronic diseases in LMIC settings (Mbuagbaw et al., 2015).

In summary, there are promising results to show that m-Health (mobile phone) interventions can assist or benefit different public health services, in applications such as patient medication compliance, disseminating health information, remote diagnosis, and medical support. Most of this work is based on silos and pilot studies, with varying clinical outcomes. There is need for further more definitive studies for these applications and interventions, particularly in global public health. However, m-Health interventions are being increasingly used in LMIC countries, particularly for noncommunicable diseases and in epidemic areas. More efforts are required to enhance this proliferation and to tackle urgent public health demands, especially in poor and conflict areas around the globe.

5.4.5 Primary and Emergency Care

There are several important drivers for the adoption of digital health hubs in primary care and emergency medicine settings. These include, for example, the following (Baird and Nowak, 2014):

- The rapid increase in patients experiencing technology as "digital natives" with mobile health Apps and other web platforms.
- The increasing demand for primary care clinics, which include family practices to do more in reducing healthcare costs and improving healthcare outcomes

It is obvious that m-Health can play a major, if not critical, role in this care area. For example, primary care clinics can also act as a focal hub for encouraging m-Health patients to adopt more of these technologies for their health and wellness management. A recent study demonstrated that patients are more inclined to use m-Health as a complement to in-person doctor visits rather than as a substitute (Rai et al., 2013). Another study on the prevalence of mobile health use among primary care clinic patients demonstrated that m-Health technologies were less prevalent among older adults but otherwise are common among primary care patients, including those with limited health literacy and those with chronic conditions (Bauer et al., 2014). Yet another study focused on the role of mobile health in primary care practices for the treatment and management of addiction and other behavioral health conditions (Quanbeck et al., 2014). A RCT that studied the comparative effectiveness of text messaging compared to telephone reminders in reducing missed appointments in primary care indicated that text message reminders were equivalent to the usual telephone reminders but were more cost-effective (Perron et al., 2013). In emergency care settings, a pilot study examined the impact of mobile text messaging in emergency departments on patients with diabetes. This demonstrated no statistically significant improvement in HbA1c, but there were trends toward improvements in other secondary outcomes, including quality of life and improved medication adherence (Arora et al., 2014).

From these studies it is clear that there is increasing evidence of the benefits of using m-Health technologies in different primary and emergency care services. However, some of this evidence is with varying outcomes and is limited to pilot studies, with further studies still required.

5.4.6 Mobile Medical Education, Training, and Workflow Management

The majority of clinicians and healthcare providers are now "digital natives" and this trend is likely to increase during the next decade. The increasing use of micro-blogging, wikis, cloud computing, and virtual and other open web-based medical educational tools is leading to new era of medical education and training.

As described earlier, the use of smart Apps for patient education is becoming one of the major domains of consumer-led m-Health services. In addition to some of the promising outcomes for patient health promotion and chronic disease self-management, there is also an increasing level of interest and beneficial evidence for using mobile phone technologies in medical education, healthcare communications, and training scenarios. This can provide clinicians with flexible access to medical textbooks, medical guidelines, therapy standards, medications, references, and so on. Examples include the use of smartphone Apps for training orthopedic surgeons (Franko, 2011), or the use of smartphones among physicians, medical students, and junior doctors (Payne et al., 2012; Ozdalga et al., 2012). Other applications include using mobiles for general pharmacy practices and training purposes (Aungst, 2013).

In general, there are many potential benefits from the use of mobile platforms and Apps for medical education and curriculum development. These can include decreased costs, ease of use, and data mobility (Gaglani and Topol, 2014). There are also many pitfalls and challenges with this approach that need to be addressed carefully; these can potentially hinder the beneficial impact of such educational models (Redelmeier and Detsky, 2013):

- *Disrupted Clinical Communications:* These include interference due to the operational functionalities of mobile phones, such as audio distortion, faulty monologues, and other factors.
- *Social Disengagement:* This includes challenges that relate to the societal and human–machine interaction and usability of smartphones.
- *Direct Patient Harm:* This includes the effect of smartphones for nosocomial infections and the potential transmission of these infections in teaching hospitals. In addition, there is the potential breach of confidentiality and privacy of sensitive medical data.

Other uses of smart mobile health applications relate to physicians' workflow in hospitals and patient care scenarios, such as rapid response, medication error prevention, and data management and accessibility (Prgomet et al., 2009). Moreover, smart mobile phones are also becoming important for encouraging better nurse–patient communication (Blake, 2008).

In general, the use of smart mobile phones for medical learning and education is increasing, particularly where there are limited care resources, medical teaching expertise, and facilities. The challenge of whether this will lead to better educational outcomes and improved patient benefits on a larger scale is yet to be seen.

5.5 SUMMARY

In this chapter, we have discussed existing applications and clinical evidence of m-Health interventions in its mobile phone-centric model. We have also discussed in detail some of these m-Health interventions, particularly for three major chronic conditions, in addition to the role of smartphone-based healthcare delivery systems in other important healthcare applications. From these studies, the current clinical evidence of m-Health seems sketchy at best and inconclusive. Despite the large number of m-Health pilots and studies, there is insufficient systematic evidence to advocate a wider implementation and scaling-up process. Based on this status, we suggest that there is need for an alternative framework for m-Health beyond the current mobile phone-centric approach. This new framework needs to be designed to transcend the current m-Health format to a more scientific and holistic approach to overcome the existing barriers and disadvantages of the current model.

The accumulation of numerous m-Health studies and reports published in recent years does not preclude the transition to an evidence-based m-Health era. Future innovations can bridge the existing gap between the "businesses of m-Health' and the "science of m-Health" toward a more rigorous evaluation and the acceptance of m-Health as a mainstream healthcare service delivery channel. To reflect further on these findings, a market report entitled "Emerging m-Health: Path for Growths" by PricewaterhouseCoopers attributed the current status and the future of m-Health to the following factors (PwC, 2013, 2014):

- High expectations of m-Health.
- Healthcare is slow to change and to adopt m-Health innovations.
- The diversity of interests at play makes the evolving landscape even more complex.
- Emerging markets are the trail blazers in m-Health.
- Solutions, not technology, are the key to success.

These and similar analyses and other findings of m-Health market-driven studies are mostly based on the singular view of m-Health as a mobile phone-centric platform for healthcare service delivery. It is obvious from the clinical evidence presented in this chapter that this singular view is at the core of the current paradoxical status of m-Health in its second decade of evolution.

Our general conclusion from all these studies is a mixed picture on the clinical evidence and impact of m-Health. While there is promising clinical evidence in some disciplines, there is no clear and sometimes muddled evidence in others, including some important areas of healthcare delivery. With the exception of the evidence of mobile phone and text messaging interventions in some global health settings, particularly in poorer countries in the developing world, there is clear asymmetry in the clinical effectiveness and efficacy in other interventions we have cited. These outcomes indicate a need to promote larger pivotal studies with clear clinical outcomes and care targets that refrain from the current mobile phone-centric interventions. Furthermore, this asymmetry also contributes to the increasing gap between the current market potential of m-Health as a consumer-based industry and its relative lack of clinical evidence and effectiveness.

There is much hype that smartphones are becoming the hub of future medicine, driven by massive consumer- or patient-led mobile healthcare services (Topol, 2012). From the clinical evidence available so far, such enthusiasm needs to be carefully weighted, perhaps with an alternative, more cautious, approach. Furthermore, the complex traits associated with this singular view of m-Health largely justify these unclear outcomes that characterize the value of what the current interventions can achieve. These traits can also impact on the current methods used to develop measures of the quality of m-Health services, as perceived largely from the smartphone- and App-centric view (Malvey and Slovensky, 2014). Any future m-Health framework therefore needs to be developed against this background.

Such a new framework is vital if m-Health is to become a truly transformative concept as a clinically acceptable process that empowers a patient-centered approach for global healthcare delivery in the twenty-first century. This argument can be completely unpersuasive if the existing benefits and evidence of m-Health in improving healthcare delivery are clinically robust enough and universally acceptable. Existing systems will not be able to provide the much talked about transformative changes without more thoughtful use of the key building elements of m-Health described in earlier chapters.

As described at the beginning of this chapter, the recommendations for the future of healthcare listed by the Institute of Medicine may be summarized by the following, almost evangelical, statements (Institute of Medicine, 2001):

- Care is based on continuous healing relationships.
- Care is customized according to patient needs and values.
- The patient is the source of control.
- Knowledge is shared and information flows freely.
- Decision-making is evidence based.
- Safety is a system property.
- Transparency is necessary.
- Needs are anticipated.
- Waste is continuously decreased.
- Cooperation among clinicians is a priority.

These recommendations could constitute a starting point for a new redesigned m-Health framework. Most of them are either ignored or not applied in current m-Health systems in what is largely a market- and consumer-driven industry. This issue is also relevant to m-Health from the global perspective, which we will address in Chapter 6.

The recent "precision medicine" initiative announced in the United States was defined by the National Institutes of Health (2015) as "an emerging approach for disease prevention and treatment that takes into account people's individual variations in genes, environment, and lifestyle." This emerging approach can potentially provide opportunities for the new m-Health2.0 vision presented in this book to contribute significantly toward achieving these goals.

The true success of m-Health is primarily built on clinical-based evidence and not on the attainment of monetary targets. Finally, in the absence of clear clinical evidence from existing large-scale m-Health studies, we need to pose a vital question in the mist of this conundrum: Why are there still billions of U.S. dollars being invested globally on m-Health today? The answer to this question can be clarified from the following points:

- The current level of global investments in "business m-Health" is largely driven by the vested interests of global industries, notably telecommunications operators, mobile services, and medical devices and pharmaceutical industries, with their continued advocacy of the concept within their business models, and without the need for universal clinical evidence to back these on a large scale. This process has resulted in an increasing gap between medical communities and the consumer-led m-Health market.
- Healthcare policies and decision-making processes are based on much-hyped statements and projections of the economic benefits and cost savings of m-Health technologies, mostly based on the mobile phone-centric model. These are presented without rigorous studies to support or negate these policies and economic assumptions.
- There is a lack of alternative long-term strategic thinking for a new vision of m-Health outside the current "box" of mobile phone and cellular network centricity. This vision and framework is required to address the flaws, limitations, and barriers of the current models.
- The anticipated clinical and economic benefit of "Big m-Health Data and Analytics" is still at its infancy. Better understanding of m-Health beyond the mobile phone-centric models can alleviate many of the barriers expected in this context.

We cannot underestimate the role of smartphone technologies as an integral part of the digital ambiguity culture and its continuing role in transforming the perception of healthcare in this digital age. However, from what we have argued in this chapter, there is a need for realignments and a new approach from all the stakeholders interested in the future of m-Health.

REFERENCES

Abraham WT, Adamson PB, Bourge RC, Aaron F, Costanzo MF, Stevenson LW, Strickland W, Neelagaru S, Raval N, Krueger S, Weiner S, Shavelle D, Jeffries B, and Yadav JS (2011) Wireless pulmonary artery haemodynamic monitoring in chronic heart failure: a randomised controlled trial. *Lancet* 377:658–666.

Abroms LC, Lee WJ, Bontemps-Jones J, Ramani R, and Mellerson J (2013) A content analysis of popular smartphone apps for smoking cessation. *American Journal of Preventive Medicine* 45:732–736.

Afolabi AB and Kusumoto FM (2012) Remote monitoring of patients with implanted cardiac devices: a review. *European Cardiology* 8(2):88–93.

ADA (American Diabetes Association) (2014a) *Checking your blood glucose.* Available at http://www.diabetes.org/living-with-diabetes/treatment-and-care/blood-glucose-control/checking-your-blood-glucose.html (accessed November 2014).

ADA (American Diabetes Association) (2014b) *Diabetes clinical trials.* Available at http://www.diabetes.org/living-with-diabetes/treatment-and-care/clinical-trials.html. (accessed November 2014).

Alotaibi MM, Istepanian R, and Philip N (2016) A mobile diabetes management and educational system for type-2 diabetics in Saudi Arabia (SAED), *mHealth*, 2:33–36; doi: 10.21037/mhealth.2016.08.01.

Appelboom G, Camacho E, Abraham ME, Bruce SS, Dumont EL, Zacharia BE, D'Amico R, Slomian J, Reginster JV, Bruyère O, and Sander Connolly, Jr. E (2014) Smart wearable body sensors for patient self-assessment and monitoring. *Archives of Public Health* 72(1):28.

Arora S, Peters AL, Agy C, and Menchine M (2013) A mobile health intervention for inner city patients with poorly controlled diabetes: proof-of-concept of the TExT-MED program. *Diabetes Technology & Therapeutics* 14(6):492–496.

Arora S, Peters AL, Burner E, Lam CN, and Menchine M (2014) Trial to examine text message-based mHealth in emergency department patients with diabetes (TExT-MED): a randomized controlled trial. *Annals of Emergency Medicine* 63(6):745–754.

Aungst TD (2013) Medical applications for pharmacists using mobile devices. *Annals of Pharmacotherapy* 47(7–8):1088–1095.

Azar KM, Lesser LI, Laing BY, Stephens J, Aurora MS, and Burke LE (2013) Mobile applications for weight management: theory-based content analysis. *American Journal of Preventive Medicine* 45:583–589.

Bacigalupo R, Cudd P, Littlewood C, Bissell P, Hawley MS, and Woods BH (2013) Interventions employing mobile technology for overweight and obesity: an early systematic review of randomized controlled trials. *Obesity Review* 14:279–291.

Baird A, and Nowak S (2014) Why primary care practices should become digital health information hubs for their patients. *BMC Family Practice* 15:190.

Baron J, McBain H, and Newman S (2012) The impact of mobile monitoring technologies on glycosylated hemoglobin in diabetes: a systematic review. *Journal of Diabetes Science and Technology* 6(5):1185–1196.

Bauer AM, Rue T, Keppel GA, Cole AM, Baldwin LM, and Katon W (2014) Use of mobile health (mHealth) tools by primary care patients in the WWAMI Region Practice and Research Network (WPRN). *Journal of the American Board of Family Medicine* 27(6):780–788.

Beatty AL, Fukuoka Y, and Whooley M (2013) Using mobile technology for cardiac rehabilitation: a review and framework for development and evaluation. *Journal of the American Heart Association* 2:e000568.

Belisario M, Huckvale JS, Greenfield K, Car J, and Gunn LH (2013) Smartphone and tablet self-management apps for asthma. *Cochrane Database of Systematic Reviews* 11:CD010013.

Bellos C, Papadopoulos A, Rosso R, and Fotiadis DI (2012) Categorization of COPD health level through the use of the CHRONIOUS wearable platform. *4th Annual International Conference of the IEEE EMBS*, San Diego, CA, August 28–September 1, pp. 61–64.

Bender JL, Yue RY, To MJ, Deacken L, and Jadad AR (2013) A lot of action, but not in the right direction: systematic review and content analysis of smartphone applications for the prevention, detection, and management of cancer. *Journal of Medical Internet Research* 5(12):e287.

Bernhardt JM (2015) *The future of health is mobile and social.* Available at http://smhs.gwu.edu/mhealth/sites/mhealth/files/GWU%20Future%20of%20Health%20is%20Mobile%20and%20Social.pdf (accessed January 2016).

Bhimaraj A (2013) Remote monitoring of heart failure patients. *Methodist DeBakey Cardiovascular Journal* 9(1):26–31.

Blake H (2008) Innovation in practice: mobile phone technology in patient care. *British Journal of Community Nursing* 13(4):160–165.

Bodenheimer K, Lorig K, Holman H, and Grumbach K (2002a) Patient self-management of chronic disease in primary care. *Journal of the American Medical Association* 288(19): 2469–2475.

Bodenheimer K, Wagner EH, and Grumbach K (2002b) Improving primary care for patients with chronic illness: the chronic care model—Part 2. *Journal of the American Medical Association* 288(19):2469–2475(1909–1914).

Borycki E (2012) M-health: can chronic obstructive pulmonary disease patients use mobile phones and associated software to self-manage their disease? *Studies in Health Technology and Informatics* 172:79–84.

Brath H, Morak J, Kästenbauer T, Modre-Osprian R, Strohner-Kästenbauer H, Schwarz M, Kort W, and Schreier G (2013) Mobile health (mHealth) based medication adherence measurement: a pilot trial using electronic blisters in diabetes patients. *British Journal of Clinical Pharmacology* 76(Suppl.1):47–55.

Burner E, Menchine M, Taylor E, and Arora S (2013) Gender differences in diabetes self-management: a mixed-methods analysis of a mobile health intervention for inner-city Latino patients. *Journal of Diabetes Science and Technology* 7(1):111–118.

CDC (Centres for Disease Control and Prevention) (2013) *Chronic disease prevention and health promotion*. Available at www.cdc.gov/chronicdisease (accessed January 2014).

Chaudhry S, Mattera J, Curtis J, Spertus J, Herrin J, Lin Z, Phillips CO, Hodshon BV, Cooper LS, and Krumholz HM (2010) Telemonitoring in patients with heart failure. *New England Journal of Medicine* 363:2301–2309.

Chow J (2011) *Hypoglycaemia for Dummies*. New York: John Wiley & Sons, Inc.

Coleman K, Austin BT, Brach C, and Wagner EH (2009) Evidence on the chronic care model in the new millennium. *Health Affairs (Millwood)* 28(1):75–85.

Dale LP, Whittaker R, Jiang Y, Stewart R, Rolleston A, and Maddison R (2014) Improving coronary heart disease self-management using mobile technologies (Text4Heart): a randomised controlled trial protocol. *Trials* 15:1–9.

DCCT (Diabetes Control and Complications Trial Research Group) (1993) The effect of intensive treatment of diabetes on the development and progression of long-term complications in insulin dependent diabetes mellitus. The Diabetes Control and Complications Trial Research Group. *New England Journal of Medicine* 329:977–986.

De Jongh T, Gurol-Urganci I, Vodopivec-Jamsek V, Car J, and Atun R (2012) Mobile phone messaging for facilitating self-management of long-term illnesses. *Cochrane Database of Systematic Reviews* 12:12.

Dellby U (1996) Drastically improving health care with focus on managing the patient with a disease: the macro and micro perspective. *International Journal of Health Care Quality Assurance* 9(2):4–8.

Der Heijden MV, Lucas PJ, Lijnse B, Heijdra YF, and Schermer TR (2013) An autonomous mobile system for the management of COPD. *Journal of Biomedical Informatics* 46:458–469.

Derbyshire E, and Dance D (2013) Smartphone medical applications for women's health: what is the evidence-base and feedback? *International Journal of Telemedicine and Applications* 2013:1–10.

Desai AS (2012) Home monitoring heart failure care does not improve patient outcomes looking beyond telephone-based disease management. *Circulation* 125:828–836.

Diabetes UK (2013) *Self-monitoring of blood glucose (SMBG) for adults with type 2 diabetes*, Position statement, April. Available at www.diabetes.org.uk/Documents/Position%20state ments/Diabetes-UK-position-statement-SMBG-Type2-0413.pdf (accessed October 2014).

Donker T, Petrie K, Proudfoot J, Clarke J, Birch MR, and Christensen H (2013) Smart phones for smarter delivery of mental health programs: a systematic review. *Journal of Medical Internet Research* 15(11):e247.

Earle KA, Istepanian RSH, Zitouni K, Sungoor A, and Tang B (2010) Mobile telemonitoring for achieving tighter targets of blood pressure control in patients with complicated diabetes: a pilot study. *Diabetes Technology & Therapeutics* 12(7):575–579.

Eng DS and Lee JM (2013) The promise and peril of mobile health applications for diabetes and endocrinology. *Pediatric Diabetes* 14(4):231–238.

Epstein RS, and Sherwood LM (1996) From outcomes research to disease management: a guide for the perplexed. *Annals of Internal Medicine* 124(9):832–837.

European Commission (2014a) *eHealth projects: research and innovation in the field of ICT for health and wellbeing—an overview*, Brochure, Directorate-General for Communications Networks, Content and Technology, Brussels. Available at ec.europa.eu/digital-agenda/en/ news/ehealth-projects-research-and-innovation-field-ict-health-and-wellbeing-overview (accessed November 2014).

European Commission (2014b) *EU-funded projects on ICT for aging well*, Overview report and catalogue, 1-6-Brussels. Available at ec.europa.eu/information_society/newsroom/cf/dae/ document.cfm? (accessed October 2014).

European Commission (2014c) *Ambient Assisted Living*, Catalogue of projects 2013, AAL Joint Programme, The Central Management Unit (CMU), Brussels. Available at http:// www.aal-europe.eu/wp-content/uploads/2013/09/AALCatalogue2013_Final.pdf (accessed October 2014).

Franko IO (2011) Smartphone apps for orthopaedic surgeons. *Clinical Orthopaedics and Related Research* 46(7):2042–2048.

Free C, Philips G, Watson L, Galli L, Felix L, Edwards P, Patel V, and Haines A (2013a) The effectiveness of mobile-health technologies to improve health care service delivery processes: a systematic review and meta-analysis. *PLoS Medicine* 10(1):e1001363.

Free C, Philips G, Galli L, Watson L, Felix L, Edwards P, Patel V, and Haines A (2013b) The effectiveness of mobile health technology based health behavioural change or disease management interventions for health care consumers: a systematic review. *PLoS Medicine* 10(1):e1001362.

Frisbee K (2014) *The impact of mobile health (mHealth) technology on family caregiver's burden levels and an assessment of variation in mHealth tool use*, Study presentation. Available at http://www.kingsfund.org.uk/ (accessed November 2014).

Gaglani SM and Topol EJ (2014) iMedEd: the role of mobile health technologies in medical education. *Academic Medicine* 89(9):1207–1209.

Garabedian LF, Ross- Degnan D, and Wharam JF (2015) Mobile phone and smartphone technologies for diabetes care and management. *Current Diabetes Reports* 15:109.

Garg SK and Hirsch IB (2014) Self-monitoring of blood glucose: an overview. *Diabetes Technology & Therapeutics* 16(Suppl.1):S1–S10.

Garg SK and Hirsch IB (2015) Self-monitoring of blood glucose. *Diabetes Technology & Therapeutics* 17(Suppl.1):S1–S11.

Gaydos C and Hardick J (2014) Point of care diagnostics for sexually transmitted infections: perspectives and advances. *Expert Review of Anti-infective Therapy* 12(6):657–672.

GOLD (Global Initiative for Chronic Obstructive Lung Disease) (2013) *Global strategy for the diagnosis, management, and prevention of COPD.* Available at www.goldcopd.org (accessed January 2014).

Goyal S and Cafazoo JA (2013) Mobile phone health apps for diabetes management: current evidence and future developments. *Quarterly Journal of Medicine* 106:1067–1069.

GSMA (2011) *Mobile health for independent living*, Landscape Report, Mobile Health Programme, AARP and Waggener Edstrom Worldwide, February, pp. 1–15 (accessed October 2014).

Halbert RJ, Natoli L, Gano A, Badamgarav E, Buist AS, and Mannino DM (2006) Global burden of COPD: systematic review and meta-analysis. *European Respiratory Journal* 28(3):523–532.

Hall CS, Fottrell E, Wilkinson, and Byass P (2014) Assessing the impact of mHealth interventions in low- and middle-income countries, what has been shown to work? *Global Health Action* 7:25606.

Haug S, Schaub MP, Venzin V, Meyer C, and John U (2013) Efficacy of a text message-based smoking cessation intervention for young people: a cluster randomized controlled trial. *Journal of Medical Internet Research* 15(8):e171.

Henderson C, Knapp M, Fernández JL, Beecham J, Hirani PS, Cartwright M, Rixon L, Beynon M, Rogers A, Bower P, Doll H, Fitzpatrick R, Steventon A, Bardsley M, Hendy J, and Newman S (2013) Cost effectiveness of telehealth for patients with long term conditions (Whole Systems Demonstrator telehealth questionnaire study): nested economic evaluation in a pragmatic, cluster randomised controlled trial. *BMJ* 346:f1035.

Higgs ES, Goldberg AB, Labrique AB, Cook SH, Schmid C, Cole CF, and Obregón RA (2014) Understanding the role of mHealth and other media interventions for behavior change to enhance child survival and development in low- and middle-income countries: an evidence review. *Journal of Health Communication* 19(Suppl.1):164–189.

Holmen H, Torbjørnsen A, Wahl AK, Jenum AK, Småstuen MC, Årsand E, and Ribu L (2014) A mobile health intervention for self-management and lifestyle change for persons with type 2 diabetes. Part 2: one-year results from the Norwegian randomized controlled trial renewing health. *JMIR mHealth and uHealth* 2(4):e57.

Holtz B and Lauckner C (2012) Diabetes management via mobile phones: a systematic review. *Telemdicine and e-health* 18(3):175–184.

Hsieh CH, Jeng SF, Chen CY, Yin JW, Yang JC, Tsai HH, and Yeh MC (2005) Tele-consultation with the mobile camera-phone in remote evaluation of replantation potential. *The Journal of Trauma* 58:1208–1212.

IDF (International Diabetes Federation) (2012) *Global Guideline for Type 2 Diabetes.* Brussels: International Diabetes Federation. Available at http://www.idf.org/sites/default/files/IDF%20T2DM%20Guideline.pdf (accessed October 2014).

IDF (International Diabetes Federation) (2014) *IDF Diabetes Atlas: Key Findings 2014.* Available at http://www.idf.org/diabetesatlas/update-2014 (accessed January 2015).

Committee on Quality Health Care in America and Institute of Medicine (2001) *Crossing the Quality Chasm: A New Health System for the 21st Century.* Washington, DC: National Academies Press.

Istepanian RSH (2015) Mobile applications for diabetes management: efficacy issues and regulatory challenges. *Lancet: Diabetes and Endocrinology* 3(12):921–922.

Istepanian RSH, Woodward B, Gorilas E, and Balos P (1998) Design of mobile telemedicine systems using GSM and IS-54 cellular telephone standards. *Journal of Telemedicine and Telecare* 4 (Suppl.1):80–82.

Istepanian RSH, Laxminarayan S, and Pattichis C (Eds.) (2006) *m-Health: Emerging Mobile Health Systems.* London: Springer.

Istepanian RSH, Zitouni K, Harry D, Moutosammy N, Sungoor A, Tang B, and Earle K (2009) Evaluation of a mobile phone telemonitoring system for the intensification of glycaemic control in patients with complicated diabetes mellitus. *Journal of Telemedicine and Telecare* 15:125–128.

Jadad AR (2009) *Preventing and managing cardiovascular diseases in the age of m-Health and global communications: lessons from low and middle income countries*, Institute of Medicine Committee on Preventing the Global Epidemic of Cardiovascular Disease, University of Toronto, Canada. Available at http://iom.edu/~/media/Files/Activity%20Files/Global/GlobalCVD/Jadad_mHealth_IOM_090717_website.pdf (accessed October 2014).

Jo Y, Labrique AB, Lefevre AE, Mehl PT, Walker N, and Friberg IK (2014) Using the lives saved tool (LiST) to model mHealth impact on neonatal survival in resource-limited settings. *PLoS One* 9(7):e102224.

Jones KR, Daly BJ, Higgins P, Madigan E, and Moore SM (2012) *The evidence base for self-management*, White Paper, Center of Excellence in Self-Management Research, Case Western Reserve University. Available at http://fpb.case.edu/SMARTCenter/docs/The_Evidence_Base_for_Self-Management.pdf (accessed October 2014).

Kaliyadan F (2013) Teledermatology update: mobile teledermatology. *World Journal of Dermatology* 2(2):11–15,2218-6190.

Kannisto KA, Koivunen MH, and Välimäki MA (2014) Use of mobile phone text message reminders in health care services: a narrative literature review. *Journal of Medical Internet Research* 16(10):e222.

Kender D (2014) *Philips gives COPD patients a lifeline with new gadget.* Available at http://www.usatoday.com/story/tech/2014/10/13/philips-gives-copd-patients-a-lifeline-with-new-gadget/17171535/ (accessed October 2015).

Kidholm K, Stafylas P, Kotzeva A, Pedersen CD, Dafoulas G, Scharf I, Jensen LK, Lindberg I, Stærdahl A, Lange M, Aletras V, Fasterholdt I, Stübin M, d'Angelantonio M, Ribu L, Grøttland A, Greuèl M, Giannaokopoulos S, Isaksson L, Orsama AL, Karhula T, Mancin S, Scavini C, Dyrvig AK, and Wansche CE (2014) *Regions of Europe working together (renewing health): European Project Final Report—public version*, June 25, pp. 1–72. Available at http://www.renewinghealth.eu/ (accessed November 2014).

Kirwan M, Vandelanotte C, Fenning A, and Duncan MJ (2013) Diabetes self-management smartphone application for adults with type 1 diabetes: randomized controlled trial. *Journal of Medical Internet Research* 5(11):e235.

Kovatchev BP, Flacke F, Sieber J, and Breton MD (2014) Accuracy and robustness of dynamical tracking of average glycemia (A1c) to provide real-time estimation of hemoglobin A1c using routine self-monitored blood glucose data. *Diabetes Technology & Therapeutics* 16(5):303–309.

Krishna S and Boren SA (2008) Diabetes self-management care via cell phone: a systematic review. *Journal of Diabetes Science and Technology* 2:509–517.

Kuijpers W, Groen WG, Aaronson NK, and Van Harten W (2013) A systematic review of web-based interventions for patient empowerment and physical activity in chronic diseases: relevance for cancer survivors. *Journal of Medical Internet Research* 15(2):e37.

Kumar S, Nilsen WJ, Abernethy A, Atienza A, Patrick K, Pavel M, Riley WT, Shar A, Spring A, Donna Spruijt-Metz D, Hedeker D, Honavar V, Kravitz R, Lefebvre C, Mohr DC, Murphy SA, Quinn C, Shusterman V, and Swendeman D (2013) Mobile health technology evaluation: the mHealth evidence workshop. *American Journal of Preventive Medicine* 45(2):228–236.

Labrique AB, Chang LW, and Mehl G (2012) H_pe for m-health: more 'y' or 'O' on the horizon? *International Journal of Medical Informatics* 82(5):467–469.

Labrique AB, Vasudevan L, Kochi E, Fabricant R, and Mehl G (2013) mHealth innovations as health system strengthening tools: 12 common applications and a visual framework. *Global Health: Science and Practice* 1(2):160–171.

Lakshminarayana R, Wang D, Burn D, Chaudhuri R, Cummins G, Galtrey C, Hellman B, Pal S, Stamford J, Steiger M, and Williams A (2014) Smartphone- and Internet-assisted self-management and adherence tools to manage Parkinson's disease (SMART-PD): study protocol for a randomised controlled trial. *Trials* 15(1):374.

Lee HY, Koopmeiners JS, Rhee TG, Raveis VH, and Ahluwalia JS (2014) Mobile phone text messaging intervention for cervical cancer screening: changes in knowledge and behavior pre–post intervention. *Journal of Medical Internet* 16(8):e196.

Liang X, Wang Q, Yang X, Cao J, Chen J, Mo X, Huang J, Wang L, and Gu D (2011) Effect of mobile phone intervention for diabetes on glycaemic control: a meta-analysis. *Diabetic Medicine* 28:455–463.

Lim MSC, Hocking JS, Hellard ME, and Aitken KC (2008) SMS STI: a review of the uses of mobile phone text messaging in sexual health. *International Journal of STD & AIDS* 19(5): 287–290.

Lindberg B, Nilsson C, Zotterman D, Söderberg S, and Skär L (2013) Using information and communication technology in home care for communication between patients, family members, and healthcare professionals: a systematic review. *International Journal of Telemedicine and Applications* 2013:1–31.

Liu WT, Wang CH, Lin HC, Lin SM, Lee KY, Lo YL, Hung SH, Chang YM, Chung KF, and Kuo HP (2008) Efficacy of a cell phone-based exercise programme for COPD patients. *European Respiratory Journal* 32(3):651–659.

Logan AG (2013) Transforming hypertension management using mobile health technology for telemonitoring and self-care support. *Canadian Journal of Cardiology* 29(5):579–585.

Mackay J and Menask G (2014) *The Atlas of Heart Disease and Stoke*. World Health Organization. Available at www.who.int/cardiovascular_diseases/resources/atlas/en/ (accessed January 2014).

Maddison R, Pfaeffli L, Whittaker R, Stewart R, Kerr A, Jiang Y, Kira G, Leung W, Dalleck L, Carter K, and Rawstorn J (2014) A mobile phone intervention increases physical activity in people with cardiovascular disease: results from the HEART randomized controlled trial. *European Journal of Preventive Cardiology* 2:56.

Maguire R, McCann L, Miller M, and Kearney N (2008) Nurse's perceptions and experiences of using of a mobile-phone-based Advanced Symptom Management System (ASyMS©) to monitor and manage chemotherapy-related toxicity. *European Journal of Oncology Nursing* 12(4):380–386.

Malvey D and Slovensky DJ (2014) *m-Health: Transforming Healthcare*. New York: Springer.

Markets and Markets (2013) *Cardiac monitoring market (cardiac rhythm management market) worth $23.3 billion by 2017*, Report by Markets and Markets, September. Available at http://www.prweb.com/releases/cardiac-monitoring-market/09/prweb11150164.htm (accessed October 2014).

Marschollek M, Klaus-Hendrik W, Gietzelt M, Gerhard N, Gerhard N, Schwabedissen M, and Hubertus HR (2008) Assessing elderly persons' fall risk using spectral analysis on accelerometric data: a clinical evaluation study. *Annual International Conference of the IEEE Engineering in Medicine and Biology Society* 2008:3682–3685.

Marzegalli M, Lunati M, Landolina M, Perego GB, Ricci RP, and Guenzati G (2008) Remote monitoring of CRT-ICD: the multicenter Italian CareLink evaluation—ease of use, acceptance, and organizational implications. *Pacing and Clinical Electrophysiology* 31:1259–1264.

Mayo Clinic (2014) *Blood sugar testing: why, when and how*. Available at www.mayoclinic .org/diseases-conditions/diabetes/in-depth/blood-sugar/art-20046628 (accessed November 2014).

Mbuagbaw L, Mursleen S, Lytvyn L, Smieja M, Dolovich L, and Thabane L (2015) Mobile phone text messaging interventions for HIV and other chronic diseases: an overview of systematic reviews and framework for evidence transfer. *BMC Health Services Research* 15 33.

McLean SC and McKinstry B (2012) Meta-analysis on COPD: comment on home telehealth for chronic obstructive pulmonary disease: a systematic review and meta-analysis. *Journal of Telemedicine and Telecare* 18(4):242.

Mechael PN and Solninsky D (2008) *Towards the Development of mHealth Strategy: A Literature Review*, WHO and Millennium Villages Project. Geneva: WHO. www.who.int/ goe/mobile_health/mHealthReview_Aug09.pdf (accessed October 2014).

Mechael P, Batavia H, Kaonga N, Searle S, Kwan A, Goldberger A, Fu L, and Ossman J (2010) *Barriers and gaps affecting mHealth in low and middle income countries: Policy White Paper*, Columbia University and mHealth Alliance. Available at http://www.globalproblems-globalsolutions-files.org/pdfs/mHealth_Barriers_White_Paper.pdf (accessed October 2014).

Medtronic Inc. (2015) *Mobile cardiac monitoring system (SEEQ™)*. Available at http://www .medtronicdiagnostics.com/us/cardiac-monitors/seeq-mct-system/index.htm (accessed December 2015).

Meraya AM, Raval AD, and Sambamoorthi U (2015) Chronic condition combinations and health care expenditures and out-of-pocket spending burden among adults: Medical Expenditure Panel Survey, 2009 and 2011. *Preventing Chronic Disease* 12:E12.

Morrison D, Wyke S, Agur K, Cameron E, Docking RI, MacKenzie AM, McConnachie A, Raghuvir V, Thomson NC, and Mair FS (2014) Digital asthma self-management interventions: a systematic review. *Journal of Medical Internet Research* 16(2):e51.

Mulvaney DJ, Woodward B, Datta S, Harvey PD, Vyas AL, Thakkar B, Farooq O, and Istepanian RSH (2012) Monitoring heart disease and diabetes with mobile Internet communications. *International Journal of Telemedicine and Applications* 2012:195970.

Mulvaney SA, Ho YX, Cala CM, Chen Q, Nian H, and Patterson BL (2013) Assessing adolescent asthma symptoms and adherence using mobile phones. *Journal of Medical Internet Research* 15(7):e141.

Murphy SL, Xu JQ, and Kochanek KD (2013) Deaths: final data for 2010. *National Vital Statistics Report* 61:4. http://www.cdc.gov/nchs/data/nvsr/nvsr61/nvsr61_04.pdf (accessed October 2014).

Nasi G, Cucciniello M, and Guerrazzi G (2015) The role of mobile technologies in health care processes: the case of cancer supportive care. *Journal of Medical Internet Research* 17(2):e26.

National Institutes of Health (2015) *Precision Medicine Initiative Cohort Program*. Available at https://www.nih.gov/precision-medicine-initiative-cohort-program (accessed January 2016).

Naylor C, Imison C, Addicott R, Buck D, Goodwin N, Harrison T, Ross S, Sonola L, Tian Y, and Curry N (2013) *Transforming our health care system: ten priorities for commissioners*, The King's Fund, pp. 1–18. Available at www.kingsfund.org.uk/sites/files/kf/field/field_publication_file/10PrioritiesFinal2.pdf (accessed October 2014).

Neubeck L, Redfern J, Fernandez R, Briffa T, Bauman A, and Freedman SB (2009) Telehealth interventions for the secondary prevention of coronary heart disease: a systematic review. *European Journal of Cardiovascular Prevention and Rehabilitation* 16(3):281–289.

NHS England (2012) *Digital First: The Delivery Choice for England's Population*, NHS Report. Available at digital.innovation.nhs.uk/dl/cv_content/32200 (accessed October 2014).

Norris SL, Lau J, Smith J, Schmid CH, and Engelgau MM (2002) Self-management education for adults with type 2 diabetes a meta-analysis of the effect on glycemic control. *Diabetes Care* 25(7):1159–1171

Nurmatov UB, Lee SH, Nwaru BI, Mukherjee H, Grant L, and Pagliari C (2014) The effectiveness of mHealth interventions for maternal, newborn and child health in low- and middle-income countries: protocol for a systematic review and meta-analysis. *Journal of Global Health* 4(1):1–8.

O'Shea SD, Taylor NF, and Paratz JD (2007) Factors affecting adherence to progressive resistance exercise for persons with COPD. *Journal of Cardiopulmonary Rehabilitation and Prevention* 27(3):166–174.

Ozdalga E, Ozdalga A, and Ahuja N (2012) The smartphone in medicine: a review of current and potential use among physicians and students. *Journal of Medical Internet Research* 14(5):e128.

Pal K, Eastwood SV, Michie S, Farmer A, Barnard ML, Peacock R, Wood B, Edwards P, and Murray E (2014) Computer-based interventions to improve self-management in adults with type 2 diabetes: a systematic review and meta-analysis. *Diabetes Care* 37(6):1759–1766.

Pantelopoulos A and Bourbakis NG (2010) A survey on wearable sensor-based systems for health monitoring and prognosis. *IEEE Transactions on Systems, Man, and Cybernetics, Part C: Applications and Reviews* 40(1):112–117.

Park A (2015) *The 25 best inventions of 2015*. Available at http://time.com/4115398/best-inventions-2015/ (accessed January 10, 2016).

Patel S, Park H, Bonato CL, and Rodgers M (2012) A review of wearable sensors and systems with application in rehabilitation. *Journal of Neuro Engineering and Rehabilitation* 9(21):1–17.

Payne KB, Wharrad H, and Watts K (2012) Smartphone and medical related App use among medical students and junior doctors in the United Kingdom (UK): a regional survey. *BMC Medical Informatics and Decision Making* 12:121.

Pelzang R (2010) Time to learn: understanding patient-centered care. *British Journal of Nursing* 19(14):912–917.

Perron JN, Dao DM, Righini NC, Humair JP, Broers B, Narring F, Haller DM, and Gaspoz JM (2013) Text-messaging versus telephone reminders to reduce missed appointments in an academic primary care clinic: a randomized controlled trial. *BMC Health Services Research* 13:125.

Polisena J, Tran K, Cimon K, Hutton B, McGill S, Palmer K, and Scott RE (2010) Home telehealth for chronic obstructive pulmonary disease: a systematic review and meta-analysis. *Journal of Telemedicine and Telecare* 16(3):120–127.

Prgomet M, Georgiou A, and Westbrook JI (2009) The impact of mobile handheld technology on hospital physicians' work practices and patient care: a systematic review. *Journal of the American Medical Informatics Association* 16(6):792–801.

Prinz C Voigt JW, and Oeynhausen B (2010) Diagnostic accuracy of a hand-held ultrasound scanner in routine patients referred for echocardiography. *Journal of the American Society of Echocardiography* 24(2):111–116.

Pun SPY, Coates V, and Benzie IFF (2008) Barriers to self-care of type 2 diabetes from both patients' and providers' perspectives: literature review. *Journal of Nursing and Healthcare of Chronic Illness* 1:4–19.

PwC (Pricewaterhouse Cooper and GSMA) (2013) *Socio-economic impact of m-Health: an assessment report of the European Union*, PwC. Available at www.gsma.com/connectedliving/wp-content/uploads/2013/06/Socio-economic_impact-of-mHealth_EU_14062013V2.pdf (accessed October 2014).

PwC (Pricewaterhouse Cooper) (2014) *Healthcare delivery of the future: how digital technology can bridge time and distance between clinicians and consumers*. Available at http://www.pwc.com/en_US/us/health-industries/top-health-industry-issues/assets/pwc-health care-delivery-of-the-future.pdf (accessed May 2015).

Quanbeck AR, Gustafson DH, Marsch LA, McTavish F, Brown RT, Mares ML, Johnson R, Joseph Glass JE, Atwood AK, and McDowell H (2014) Integrating addiction treatment into primary care using mobile health technology: protocol for an implementation research study. *Implementation Science* 9:65.

Quinn CC, Shardell MD, Terrin ML, Barr EA, Ballew SH, and Gruber-Baldini AL (2011) Cluster-randomized trial of a mobile phone personalized behavioral intervention for blood glucose control. *Diabetes Care* 34(9):1934–1942.

Rai A, Chen L, Pye J, and Baird A (2013) Understanding determinants of consumer mobile health usage intentions, assimilation, and channel preferences. *Journal of Medical Internet Research* 15(8):e149.

Rashidi P (2013) A survey on ambient-assisted living tools for older adults. *IEEE Journal of Biomedical and Health Informatics* 17(3):579–590.

Redelmeier DA and Detsky AS (2013) Pitfalls with smartphones in medicine. *Journal of General Internal Medicine* 28(10):1260–1263.

Rothman TA, Laughlin JC, Seltzer J, Walia JS, Baman RI, Siouffi SY, Sangrigoli RM, and Kowey PR (2007) The diagnosis of cardiac arrhythmias: a prospective multi-center randomized study comparing mobile cardiac outpatient telemetry versus standard loop event monitoring. *Journal of Cardiovascular Electrophysiology* 18(3):1–7.

Russell SJ, El-Khatib FH, Sinha M, Magyar KL, McKeon K, Goergen LG, Balliro C, Hillard MA, Nathan DM, and Damiano ER (2014) Outpatient glycemic control with a bionic pancreas in type 1 diabetes. *New England Journal of Medicine* 371:313–325.

Saffari M, Ghanizadeh G, and Koenig HG (2014) Health education via mobile text messaging for glycemic control in adults with type 2 diabetes: a systematic review and meta-analysis. *Primary Care Diabetes* 8(4):275–285.

Sama RP, Eapen ZJ, Weinfurt1 KP, Shah BR, Kevin A, and Schulman KA (2014) An evaluation of mobile health application tools. *JMIR mHealth uHealth* 2(2):e19.

Schrijvers G (2009) Disease management: a proposal for a new definition. *International Journal of Integrated Care* 9:12.

Schutt M, Kern W, Krause U and DPV Initiative (2006) Is the frequency of self-monitoring of blood glucose related to long-term metabolic control? Multicenter analysis including 24,500 patients from 191 centers in Germany and Austria. *Experimental and Clinical Endocrinology & Diabetes* 114:384–388.

Seto E, Leonard K, and Ross H (2010) Attitudes of heart failure patients and health care providers towards mobile phone-based remote monitoring. *Journal of Medical Internet Research* 12(4):1–15.

Siegel C, Hochgatterer A, and Dorner TE (2014) Contributions of ambient assisted living for health and quality of life in the elderly and care services: a qualitative analysis from the experts' perspective of care service professionals. *BMC Geriatrics* 14(112):1–13.

Spring B, Duncan JM, Janke AE, Kozak AT, McFadden G, DeMott A, Pictor A, Epstein LH, Siddique J, Pellegrini CA, Buscemi J, and Hedeker D (2013) Integrating technology into standard weight loss treatment: a randomized controlled trial. *JAMA Internal Medicine* 173(2):105–111.

Steinhubl SR, Muse ED, and Topol EJ (2014) Can mobile health technologies transform health care? *Journal of the American Medical Association* 310(22):2395–2396.

Steventon A, Bardsley M, Billings J, Dixon J, Doll H, Hirani S, Cartwright M, Rixon L, research Knapp M, Henderson C, Rogers A, Fitzpatrick R, Hendy J, and Newman S (2012) Effect of telehealth on use of secondary care and mortality: findings from the Whole System Demonstrator cluster randomised trial. *BMJ* 344:e3874.

Stoddart A, Hanley J, Wild S, Pagliari C, Paterson M, Lewis S, Sheikh A, Krishan A, Padfield P, and McKinstry B (2013) Telemonitoring-based service redesign for the management of uncontrolled hypertension (HITS): cost and cost-effectiveness analysis of a randomised controlled trial. *BMJ Open* 3(5):pii: e002681.

Suksomboon N, Poolsup N, and Lay Nge Y (2014) Impact of phone call intervention on glycemic control in diabetes patients: a systematic review and meta-analysis of randomized, controlled trials. *PLoS One* 9(2):e89207.

Tate EB, Spruijt-Metz D, O'Reilly G, Jordan-Marsh M, Gotsis M, Pentz MA, and Dunton GF (2013) mHealth approaches to child obesity prevention: successes, unique challenges, and next directions. *Translational Behavioral Medicine* 3(4):406–415.

Thakkar P (2013) *Mobile technology re-strengthens the health industry*, Healthcare Global. Available at http://www.healthcareglobal.com/tech/1322/Mobile-Technology-Restrengthens-The-Health-Industry (accessed October 2014).

Thomas JG and Wing RR (2013) Health-E-Call, a smartphone-assisted behavioral obesity treatment: pilot study. *JMIR mHealth and uHealth* 1(1):e3.

Tomlinson M, Rotheram-Borus MJ, Swartz L, and Tsai AC (2013) Scaling up mHealth: where is the evidence? *PLoS Medicine* 10(2):e1001382.

Topol E (2012) *The Creative Destruction of Medicine: How the Digital Revolution Will Create Better Health Care*. Basic Books.

Trend-UK (2014) *Blood Glucose Monitoring Guidelines: Consensus Document [Version 1.0].* Available at http://www.trend-uk.org/documents/TREND_BG_Consensus_May_Final_HIGHRES.pdf, (accessed January 2015).

UKPDS (UK Prospective Diabetes Study Group) (1998) Intensive blood-glucose control with sulphonylureas or insulin compared with conventional treatment and risk of complications in patients with type 2 diabetes (UKPDS 33). UK Prospective Diabetes Study (UKPDS) Group *Lancet* 352:837–853.

Varma N, Ricci RP, and Morichelli L (2014) Remote monitoring for follow-up of patients with cardiac implantable electronic devices. *Arrhythmia & Electrophysiology Review* 3(2): 123–128.

Vital Wave Consulting (2009) *mHealth for Development: The Opportunity of Mobile Technology for Healthcare in the Developing World.* Washington, DC and Berkshire, UK: United Nation Foundation and Vodafone Foundation. Available at http://unpan1.un.org/intradoc/groups/public/documents/unpan/unpan037268.pdf (accessed October 2014).

Wadden TA, Volger S, Sarwer DB, Vetter MLMD, Tsai AG, Berkowitz RI, Kumanyika S, Schmitz KH, Diewald LK, Barg R, Chittams J, and Moore RH (2011) A two-year randomized trial of obesity treatment in primary care practice. *New England Journal of Medicine* 365(21):1969–1979.

Wagner E, Austin B, and Von Korff M (1996) Organizing care for patients with chronic illness. *Milbank Quarterly* 74(4):511–44.

Wagner EH, Bennett SM, Austin BT, Greene SM, Schaefer JK, and Vonkorff M (2005) Finding common ground: patient-centeredness and evidence-based chronic illness care. *Journal of Alternative and Complementary Medicine* 11 (Suppl.1):S7–S15.

Wallace PJ (2005) Physician involvement in disease management as part of the CCM. *Health Care Financing Reveiw* 27(1):19–31.

Wang A, An N, Lu X, Chen, Li C, and Levkoff S (2014) A classification scheme for analyzing mobile apps used to prevent and manage disease in late life. *JMIR Mhealth Uhealth* 2(1):e6.

Warnier JM, Riet EV, Rutten FH, De Bruin LM, and Sachs AP (2013) Smoking cessation strategies in patients with COPD. *European Respiratory Journal* 41:727–734.

Wei J, Hollin I, and Kachnowski S (2011) A review of the use of mobile phone text messaging in clinical and healthy behaviour interventions. *Journal of Telemedicine and Telecare* 17(1): 41–48.

West D (2012) How mobile devices are transforming healthcare. *Issues in Technology Innovation* 18:1–14.

Whited JD (2006) Teledermatology research review. *International Journal of Dermatology* 45:220–229.

Whittaker R, McRobbie H, Bullen C, Borland R, Rodgers A, and Gu Y (2012) Mobile phone-based interventions for smoking cessation. *Cochrane Database of Systematic Reviews* 11: CD006611.

Widmer JR, Collins MN, Collins CS, West CP, Lerman LOMD, and Lerman A (2015) Digital health interventions for the prevention of cardiovascular disease: a systematic review and meta-analysis. *Mayo Clin. Proceedings* 90(4):469–480.

Wilkinson A and Whitehead L (2009) Evolution of the concept of self-care and implications for nurse: a literature review. *International Journal of Nursing Studies* 46:1143–1147.

Williams V, Victor CR, and McCrindle R (2013) It is always on your mind: experiences and perceptions of falling of older people and their carers and the potential of a mobile falls detection device. *Current Gerontology and Geriatrics Research* 1–7.

WHO (2009) *Global Health Risks: Mortality and Burden of Disease Attributable to Selected Major Risks*. World Health Organisation, pp. 1–70.

WHO (2011) *mHealth: New Horizons for Health Through Mobile Technologies*. World Health Organisation, pp. 1–112.

Zhang J, Song YL, and Bai CX (2013) MIOTIC study: a prospective, multicenter, randomized study to evaluate the long-term efficacy of mobile phone-based Internet of things in the management of patients with stable COPD. *International Journal of Chronic Obstructive Pulmonary Disease* 8:433–438.

Zimetbaum P and Goldman A (2010) Ambulatory arrhythmia monitoring choosing the right device. *Circulation* 122:1629–1636.

6

m-HEALTH AND GLOBAL HEALTHCARE

Mobile health will save lives, help overpopulations.
Bill Gates, Chairman of Bill and Melinda Gates Foundation,
Keynote Speech, 2010 m-Health Annual Summit, Washington, DC

6.1   INTRODUCTION

The purpose of this chapter is twofold. First, we highlight a general overview of the current landscape of global m-Health initiatives in different developing countries. In the second part of the chapter, we analyze the barriers and challenges to these initiatives, and the potential of m-Health in areas of conflict and natural disasters. We also present some recommendations for a new global m-Health framework based on future technological advancements discussed in earlier chapters.

The importance of m-Health from the global perspective is reflected by the numerous major initiatives developed in recent years by the UN, WHO, World Bank, and telecommunications and industry alliances. These aim to harness different m-Health solutions for addressing healthcare delivery services in the developing world. Predominately, most of these initiatives have emerged from exploiting the power of mobile phone technologies, tailored for global healthcare challenges. They are also driven by the massive growth in the use of mobile phones in most of the developing world, particularly in low- and middle-income countries (LMICs). This

m-Health: Fundamentals and Applications, First Edition. Robert S. H. Istepanian and Bryan Woodward.
© 2017 by The Institute of Electrical and Electronics Engineers, Inc. Published 2017 by John Wiley & Sons, Inc.

growth is continuing, and the revenue from mobile phone use is forecast to reach $49 billion by 2020 (Grand View Research, 2014).

The growth of mobile traffic indicates that approximately half the global population uses mobile communications, reaching 3.6 billion mobile subscribers worldwide in 2014 (GSMA Intelligence, 2015). However, mobile monetary projections have not materialized proportionately with a parallel scaling up of m-Health programmes in the developing world. This process remains largely fragmented and complex, with much-debated barriers remaining in achieving such objectives. These include, for example, the lack of clear, measured impact and clinical evidence, and an understanding of the diverse economies, and social and cultural norms in the countries concerned. Most important, the narrow translation of m-Health as a mobile phone-centric healthcare service is one of the key factors that has affected this process.

The global m-Health benefits are widely documented (UN Foundation–Vodafone Foundation Partnership, 2009), but challenges remain when attempting to quantify the impact of m-Health in terms of its effect on global health status (Malvey and Slovensky, 2014). The importance of m-Health may be summarized from a global m-Health survey of 112 countries as follows (WHO, 2011):

- The vast majority (83%) of the 112 states that completed the survey reported at least one m-Health initiative in their country.
- Eighty-three percent or most member states reported implementing four or more types of m-Health initiatives.
- Responding low-income countries (77%; $n = 22$) reported at least one m-Health initiative compared to 87% ($n = 29$) of high-income countries.

It is most likely that the number of m-Health initiatives has increased in some of these countries since this survey. However, there is a clear gap between quantifying the global impact of m-Health associated with the role of mobile phone technologies and its proliferation in the developing world. This gap can be interpreted by the current perspective of m-Health globally, with two contradicting features:

- The dominant interpretation of mobile health is largely a mobile phone-based technology facilitator for improving care access and delivery processes.
- The existence of too many localized pilots and initiatives report healthcare benefits based on this model. There is generally a clear lack of clinical evidence and global efficacy standards that transcend these pilots to large-scale adoption and scaled-up processes, especially in the developing world and poorly resourced countries.

The main challenges of assessing global m-Health, by identifying and measuring global health outcomes, remain unanswered (Malvey and Slovensky, 2014). Moreover, the long-term benefits and sustainability strategies of any global m-Health initiative remain ambiguous at best. These challenges can best be illustrated from these examples:

- There are many reported m-Health initiatives on maternal and child health in collaboration with global mobile network operators (MNO) and other stakeholders in developing world countries (World Health Organization and United Nation Foundation, 2015). These studies have not resulted in clinical evidence supported by sustainable long-term strategies and national plans in most of the LMICs targeted by the various initiatives. Neither is the impact of these initiatives clearly demonstrated in the UN Millennium Development Goals (MDG) on child health and improving maternal health (United Nations, 2014).
- There is a lack of quantifiable measures of global m-Health impact. This is best obtained from the 2014 West African Ebola virus outbreak, particularly in the most affected areas of Guinea, Sierra Leone, and Liberia. This outbreak was one of the largest in history, claiming tens of thousands of victims, with an estimated cost of US$33 billion, or nearly 2.5 times the entire size of the three affected economies (World Bank, 2014a). Surprisingly, this catastrophic health crisis received far less attention from global m-Health institutions, with a timid and disproportionate response. The failure of any robust m-Health initiative to predict or alert on the magnitude of this outbreak is an example of the shortcomings of existing m-Health strategies and their sustainability plans. This example reinforces a grim situation: the lack of any global m-Health initiatives or realistic action plans in conflict zones and refugee hot spots are not yet properly addressed and funded.
- The 2016 outbreak of the mosquito-borne Zika virus that exploded in Latin America affected more than 28 countries and territories and resulted in more than 1.5 million people infected in Brazil. In addition, there are estimates of a further three to four million people who are likely to be infected in the Americas (Walsh and Sifferlin, 2016). The outbreak of this latest global epidemic necessitates further need for a radical rethinking of existing global m-Health strategies.

Generally speaking, the developing world is way, way behind so-called developed countries in terms of medical facilities and qualified personnel for the delivery of much-needed healthcare services in these areas. What might be routine in the developed world is likely to be far too expensive or impossible to do in places like India, the Middle East, or Africa. The global need for m-Health in developing countries and poor regions of conflict is every bit as urgent as elsewhere, and special consideration has to be given to the relevant m-Health system design and implementation process in these areas, with the need for these to be tailored to their specific healthcare needs and demands, social norms, cultures, and topographies.

In 2011, 2.2 billion people lived on the equivalent of less than US$2 a day. Over 80% of the extremely poor lived in South Asia (399 million) and Sub-Saharan Africa (415 million), while fewer than 50 million lived in Latin America, the Caribbean, Middle East and North Africa, and Eastern Europe and Central Asia combined (World Bank, 2014b). The number of people living in extreme poverty globally remains unacceptably high. Technology can improve global health, but it is doubtful if most people living in extreme poverty can afford costly smartphones.

6.2 m-HEALTH TECHNOLOGIES FOR GLOBAL HEALTH

Historically speaking, the first foray into using mobile health technologies for the developing world was reported with a number of global initiatives (Mechael et al., 2010; Hall et al., 2014). These pilot initiatives were followed by other m-Health programs, driven mainly by their potential benefits, particularly in LMICs. The high mobile phone use in these countries also acted as the magnet for these initiatives. The post-smartphone era marked a substantial increase in these initiatives, especially in Africa, Latin America, Asia, and South-East Asia (Donner, 2008). Most of these initiatives, pilots, and local projects were, and still are, interpreted on the simple "telecom formula" of m-Health that we described earlier:

m-Health = mobile(smart)phone + healthcare delivery service

This formula became increasingly synonymous with m-Health, such as the WHO definition of m-Health cited by the Partnership for Maternal, Newborn & Child Health (PMNCH), as *the provision of health services and information via mobile technologies such as mobile phones* (WHO, 2012a). This interpretation of m-Health, as using the power of mobile phone technologies for healthcare delivery services in an overlapping context of e-Health and other ICT for health information systems, is shown in Fig. 6.1.

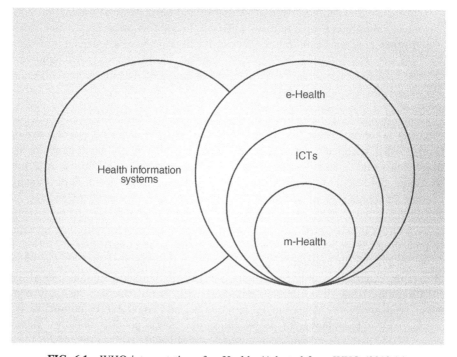

FIG. 6.1   WHO interpretation of m-Health. (Adapted from WHO (2012c).)

The use of mobile phones or smartphones to provide global m-Health services, such as text messaging, access to patient records, mobile asset management, and other applications that are often referred as "m-Health," is incomplete and represents an omission of the other pivotal elements of m-Health, such as the innovations in medical sensors, new communications mechanism, and powerful computing elements. Such use is at odds with the canonical definition of m-Health explained in Chapter 1.

Further, m-Health is understood as the evolution of telemedicine and e-Health from traditional desktop platforms to wireless and mobile networking configurations (Bashshur et al., 2011). It also brings together different mobile communications, wireless networking, and Internet technologies together with medical sensing for "connected healthcare" anytime and anywhere (Istepanian et al., 2004, 2006; Istepanian and Zhang, 2012; WHO, 2010).

The most common global m-Health pilots, projects, and initiatives are summarized in Table 6.1. Most of the themes and applications are typically implemented using the general configuration shown in Fig. 6.2.

This configuration follows an existing m-Health model that is based on three basic functional "elements," namely, patient, healthcare system, and healthcare provider (WHO, 2010, 2012b). The technology element in these solutions predominately uses mobile phones as the primary technology, acting as the communicating link for delivering specific healthcare services and connecting with the two other elements. The mobile patient is the recipient of the healthcare service provided by the healthcare provider via the mobile communication link.

The advantages and disadvantages of this model are still debatable, with mixed clinical and healthcare impact outcomes. The proponents of this approach cite growing efficacy and clinical evidence in some m-Health interventions, such as

TABLE 6.1   Common Themes of Current Global m-Health Services and Example Applications

Global m-Health Theme	Examples
Patient education, health promotion, and diagnostic communications	Using SMS text messaging for pregnancy and maternal care education, remote cancer diagnosis, social m-Health
Health records management, improved service delivery, and treatment compliance	Mobile EHR and EPR for family and birth registration, HIV and STI management, and case registration for mobile workers
Population surveillance, disease outbreaks, health monitoring and prevention	Mobile surveillance of HIV/AIDS, malaria, and other communicable disease treatments, SARS surveillance, monitoring birth and infant complications, watery diarrhoea conditions, and prevention approaches using text messaging and Apps
Mobile management of logistics, finance, and human resources	Mobile follow-up scheduling, medication management, appointment reminders
Medical learning and healthcare training	Midwife training, medical m-learning

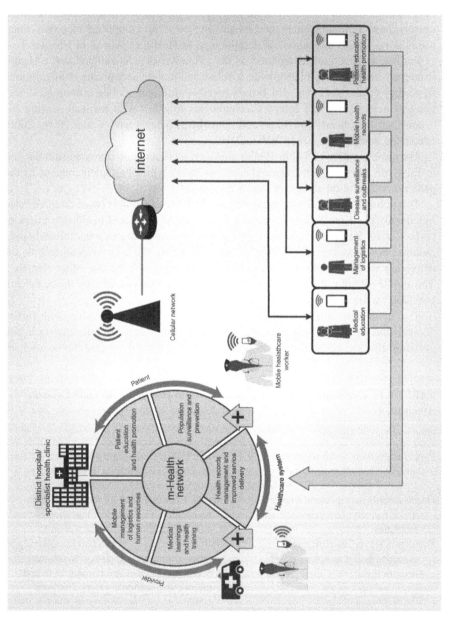

FIG 6.2 General configuration of current global m-Health services.

improving treatment adherence, appointment compliance, data gathering, and developing support networks for health workers. However, the pivotal quantity and quality of the clinical impact and evidence, particularly in LMICs, remains limited (Hall et al., 2014; Beratarrechea et al., 2014). These limited clinical outcomes and lack of rigorous studies in the developing world require further work to achieve new sustainable, affordable, and clinically effective global m-Health solutions, with obvious long-term impact and tangible economic benefits.

Most of today's health technology is focused on the needs of the wealthy. For developing countries a mobile phone is beyond the reach of most people, which is why Google for example has introduced a smartphone tailored for emerging markets, the so-called gray market, with a handset costing less than $50 (Mohammed, 2015). More frugal technologies, combined with innovations to provide the greatest health-care impact, including low-cost diagnostics, are much needed for the world's poorest people (Howitt et al., 2012; Bates, 2015). The majority of the mobile phone-centric m-Health technologies that have been adopted for applications in different global settings have yielded mixed economic and cost-effective benefits. At best, these have shown ambiguous clinical evidence and unclear outcomes.

In response to the increasing burden of noncommunicable diseases (NCD), such as cardiovascular conditions, cancers, chronic respiratory diseases, and diabetes, with their large share of mortality levels in the developing world, the WHO and ITU announced a joint m-Health strategic work plan (WHO, 2012c). This 4-year scalable m-Health plan focused primarily on eight developing countries and three NCD aspects: prevention, treatment, and enforcement. These mostly use basic m-Health solutions, such as text messaging and applications (Apps) for a range of services, including m-awareness, m-communication, m-training, m-behavioral change, m-services, m-treatment, and m-screening.

To evaluate these and other similar mobile phone-centric m-Health initiatives, it is worth looking at the outcomes of the eight MDGs set by the United Nations in 2000. These goals reflected the major global health challenges facing the developing world and low-income countries. The health-dominant MDGs were summed up as reducing child mortality (MDG4), improving maternal health (MDG5), and combating HIV/AIDS, malaria, and other diseases (MDG6). In 2014, the UN MDG Progress Report summarized the health-related outcomes (United Nations, 2014) by stating that MDG targets have been largely met in the fight against malaria and tuberculosis. Further, substantial progress has been made in most areas, but much more effort is needed to reach the set targets, particularly with regard to the following:

- Chronic undernutrition among young children, which has declined, with one in four children still affected.
- Child mortality, which has been almost halved.
- Maternal mortality.
- Antiretroviral therapy.

Other targets relating to eradicating extreme poverty and hunger (MDG1), achieving universal primary education (MDG2), promoting gender equality to

TABLE 6.2   Example of Global Distribution of m-Health Projects Related to Health MDGs

m-Health Projects and Initiatives	Addressing MDG4	Addressing MDG5	Addressing MDG6	Total
Global distribution	46	58	92	196
m-Health projects in Asia	10	20	17	47
m-Health projects in Sub-Saharan Africa	31	41	63	135
m-Health projects in Latin America	3	4	9	16

Source: Adapted from m-Health Alliance (2012a).

empower women (MDG3), ensuring environmental sustainability (MDG 7), and developing a global partnership for development (MDG8) were also cited as either met or with substantial progress achieved (Lomazzi et al., 2014).

We discuss next the m-Health initiatives associated with the different MDGs. As an example, more than 200 projects and initiatives were identified globally and related specifically to MDG 4, 5, and 6 in LMICs of Asia, Latin American, and Sub-Saharan Africa by the end of 2012 (m-Health Alliance, 2012a). These are shown in Table 6.2.

Only two of the selected nine countries from these areas mentioned national strategic m-Health documents, compared to five mentioning national e-Health strategies (m-Health Alliance, 2012b), as shown in Table 6.3. These statistics also illustrate the relative disparity in the distribution of m-Health adoption levels from one region to another, and also between individual countries within the same region. These disparities can be attributed to several contributing factors that are discussed later in the chapter.

Examining the outcomes, it is evident that meeting these targets is still "work in progress." Moreover, the role of existing global m-Health initiatives in achieving these goals is timid at best, although some MDGs have benefitted from substantial investments, global research efforts, and pilot projects. Although mobile technology has represented the crux of the driving technology for many of the targeted global healthcare delivery services so far, these are not yet reflective of the desired health outcomes. An example of the main barriers to greater use of technology for global health (Howitt et al., 2012) are summarized as follows:

- *The Necessary Technology Is Not Available*: This can be attributed either to insufficient funds for developing the technology or to nonexistence of achievable scientific breakthroughs.
- *The Technology Exists But Is Not Accessible*: These can be attributed not only to cost factors but also to other barriers such as challenges in distribution, inadequate human resources, and unreliable energy supplies.
- *Accessible Technology Is Not Adequate*: This barrier can be attributed to the sustainable use of technology and its conflict with issues such as cultural resistance and reluctance to change.

TABLE 6.3 Distribution Levels of m-Health in Nine Focal Countries Relevant to the MDGs

Region	Focal Country	m-Health Programs Identified	Level of m-Health Activity	Number of m-Health Initiatives Addressing MDGs			Adoption of e-Health and m-Health National Strategies	
				MDG4	MDG5	MDG6	e-Health National Strategy	m-Health National Strategy
Asia	India	16	High	1	6	6	Yes	No
	Bangladesh	7	Medium	5	5	0	Yes	No
	Vietnam	1	Low	0	0	0	No	No
Latin America	Guatemala	4	High	0	0	0	No	No
	Peru	2	Medium	0	0	2	Yes	Yes
	Panama	1	Low	0	0	0	Yes	No
Sub-Saharan Africa	Tanzania	21	High	5	8	7	Yes	Yes
	Nigeria	5	Medium	1	2	3	No	No
	South Sudan	2	Low	1	0	0	No	No

Source: Adapted from m-Health Alliance (2012b).

Overcoming these barriers is important in the successful realization of any technology for global health (Howitt et al., 2012). Most of the common health delivery services and applications are implemented using mobile phone technologies and their associated business models. The inherent deficits of the clinical outcomes of these models are mostly reflected in the barriers above, resulting in the poor global m-Health impact so far. It is also evident that there have been many global m-Health projects that use mobile phones for chronic disease management, to facilitate drug adherence and diagnosis of HIV and malaria, to monitor polio outbreaks, to implement mobile platforms for electronic health records (EHR) and electronic personal records (EPR), to provide advice and education to pregnant women about maternal health, and many others.

But to what extent are these projects effective clinically? Do they produce the desired healthcare result? Are they beneficial to the overall healthcare in the countries being piloted? These are the kind of pragmatic questions posed by journalists but they need to be addressed in order to understand the current trend and to provide alternative options and appropriate solutions to overcome the existing barriers. It would be contrary to human nature if claims made by the teams that have carried out these projects were not mainly positive, but systematic objective studies have shown mixed effectiveness and only modest benefits so far. These shortcomings are also unable to translate existing solutions to scaled-up plans with economically viable m-Health strategies and service delivery mechanisms; nor can they adapt successfully to the various socioeconomic and political conditions in LMICs.

6.3 GLOBAL m-HEALTH INITIATIVES FOR THE DEVELOPING WORLD: HEALTHCARE CHALLENGES AND IMPACTS

Apart from the role of innovative technologies for global health (Howitt et al., 2012), there have been numerous examples of applications in the developing world and LMICs that illustrate the potential benefits of m-Health in diverse environments and challenging healthcare scenarios. Before looking at some of these examples, it is instructive to think about the choices people have in rural and remote areas in any developing country if they are unwell and do not have immediate medical aid (Benjamin, 2013). They could do nothing, and hope they become well again, which is often the best option! They could walk to a public clinic, which might take hours, or take a bus, which they have to pay for, and although quicker, it might still mean taking time off work or arranging childcare. They might then wait for hours for a 5-min session with a busy healthcare worker or nurse who might (or might not) diagnose and treat their condition correctly before the journey home. Finally, they could seek the services of a local traditional healer, who may offer soothing words, but not have any genuine medical knowledge.

It is this backdrop of inaccessibility to conventional healthcare that has heralded the use of m-Health scenarios to increase people's access to primary care facilities, enhance preventive healthcare, and provide information or treatment. While billions of people in the developing world now have mobile phones (over 95% in South

Africa, for example), it should be remembered that many do not have them, and these may well be among the most disadvantaged people.

It is not possible to list all the previous and ongoing global m-Health initiatives, as this topic has been discussed and cited extensively. We also do not claim to have included every project that has ever been started or is still running. Instead, we aim to outline a selected variety of m-Health projects initiated in different so-called developing countries or LMICs, although the definition is a loose one. The common factor among most of these projects is their use of conventional mobile phone technology, usually with no frills. However, "low-tech" is a necessity in these countries, where the latest equipment and services are impossibly expensive, but phone calls and, more likely, text messages (or the short message service, SMS) have the power to reach the people in the greatest need of healthcare.

Most applications are invariably at the lower end of the technological scale, mainly because of the limited budgets available to run them. They include the use of low-cost sensors for simple diagnostic tests that can withstand warm temperatures indefinitely, without the use of refrigeration or electricity, as mentioned in Chapter 2 (Bates, 2015).

The use of mobile phones is widespread throughout the world, so it is natural to try to exploit their use for healthcare applications, of which many have been launched in Africa, India, Latin America, and the Middle East. Clinically, most of existing pilot initiatives are driven by a need to address the increasing prevalence of communicable and noncommunicable diseases such as HIV/AIDS, TB, diabetes, obesity, hypertension, and heart diseases (Paradis and Chiolero, 2011).

There are also recent related activities, such as the setting up a specialist innovation task force within the Innovation Working Group (IWG) and the World Economic Forum's Global Agenda Council on Digital Health. This initiative aims to prepare the guidelines and appropriate framework for cost-effective digital innovations that can accelerate the progress of MDGs; in particular, the main objective is to establish a global strategy for women's and children's health and to deliver "Digital Health Knowledge Exchange" (Ferguson, 2014).

Further, the clinical evidence and impact of most existing global m-Health applications have drawn increasing debate, particularly on issues such as the lack of clinical evidence and rigor and on their transformation from pilots-only stages to large-scale adoption (Free et al., 2013; Hall et al., 2014). In the main, most projects started as pilot studies, with a view to scaling up to major services, but whatever their status, most are extremely valuable because they can make a difference to a large number of people at a modest cost.

It is also evident that perhaps the major barriers to progress when trying to introduce m-Health services is resistance to change. This is a battle with entrenched views, attitudes, and behavior that have been acquired over generations. Suddenly introducing what amounts to a new way of life for people who are illiterate or constrained by tribal customs, especially women, is an innovation that cannot be overstated. Until recently, many people in these countries have never used a mobile phone, or may not have even seen one before, and would have no idea how it works or how it can help them. If they are illiterate, they cannot read a text message, even in their own language.

The "behavioral change" challenges represent important factors that have been underestimated in most of the existing global m-Health studies, especially for health promotion, disease prevention and management, treatment compliance, and appointment reminders (Kwan et al., 2013). Whatever the project's aims and objectives, the following questions have to be asked:

- What projects addressing behavioral change have been introduced or are ongoing?
- What cultural, social, political, or other factors need to be considered?
- What advantages can mobile phones (or smartphones) offer and how will they be used?
- Which people are targeted, and how and where does the project start?

The significance of health impact and cost savings were also issues identified in another report that examined the role of mobile technology in healthcare transformation in the developing world. It identified the need for more demonstrative studies of m-Health projects to show clear health outcomes, especially in reducing the burden of chronic conditions, infant mortality, and infectious diseases (West, 2012).

Other important factors in launching and carrying out a global m-Health project have been considered:

- The project must support a national priority, which means that government at national, regional, and district levels must be engaged from the start in all aspects, from the basic idea through to field work.
- When dealing with government, adherence to protocol and respecting cultural diversity is important, because this can lead to formal endorsement and approval.
- An awareness of other projects in the same area, and liaising with the partners involved, may avoid overburdening the government bureaucracy.

The complementary partnership and selection process of the key collaborations plays an important factor in facilitating successful outcomes for any m-Health project in the developing world. Most projects are funded by various global institutions, nongovernment organizations (NGOs), and government-sponsored institutions in partnership with academic institutions and telecommunications industries. Liaison between these partners is vital to the success of any project, as evidenced for example by the Grameen Foundation working in Ghana (Grameen Foundation, 2014; Osborn, 2013) and Cell-Life working in South Africa (Benjamin, 2013). The m-Health Alliance and m-Health Knowledge are examples of organizations that aim to promote m-Health globally and facilitate knowledge transfer between stakeholders (m-Health Knowledge, 2014).

We next present some m-Health projects from different regions representing the African, South American, and Asian continents. Some of the projects listed are presented in detail, while others are outlined briefly for completeness. More comprehensive lists of global m-Health projects can be accessed elsewhere (Malvey and

Slovensky, 2014; Donner and Mechael, 2013; McQueen et al., 2012; Free et al., 2013; Hall et al., 2014; m-Health Knowledge, 2014; WHO, 2011).

6.3.1 Global m-Health Initiatives in Africa

In general, the African continent has witnessed the most extensive m-Health initiatives compared to other global areas. According to a recent study by PwC, it was indicated that Africa is bound to witness tremendous growth in the m-Health sector (Levy, 2012). The WHO global m-Health survey estimated 75% of the African nations (29 of them) reported at least one m-Health initiative (WHO, 2011).

Most of these initiatives are either pilots, local projects, or case studies based on the mobile phone m-Health service delivery model discussed earlier. Their long-term sustainability, clinical efficacy, and cost-effectiveness are subject to ongoing debate and scrutiny (Free et al., 2013; Hall et al., 2014; Folaranmi, 2014). However, few rigorous clinical evaluations of these interventions exist, and while some are starting to emerge, there is not much evidence to support replication of most of these projects on a long-term and sustainable basis (Folaranmi, 2014). The recent Ebola outbreak and epidemic in West Africa and the failure of existing m-Health programs in the region to mitigate or predict such outbreaks is a testament to these shortcomings and the need to reconsider existing m-Health methodologies and interventions.

Next we illustrate for completeness some these m-Health initiatives in different parts of the African continent.

South Africa   There is general consensus that the South African healthcare system is facing a widening divide between the public and private healthcare sectors. The country also has an alarming record of HIV/AIDS, and any means of reducing the associated mortality rates must be welcomed (Bassett et al., 2014). Further, mobile phone ownership in the country is considered one of the highest in the world. Some of the mobile health initiatives in the country and projects include the following:

Cell-Life, a not-for-profit initiative based in Cape Town, contributes to developing open-source computer systems to initiate and promote HIV communications (Benjamin, 2013). The initiatives started from a project called *Cellphones for HIV* in 2007, partly funded by Vodacom and the RAITH Foundation, to exploit mobile communications in tackling information dissemination about HIV, and joined by the Communications Task Team of the South Africa National AIDS Council. By 2010, their services included SMS, USSD (text menus to allow information retrieval and data collection), MXit (an instant messaging service to provide HIV information), location-based services (knowing where a mobile phone is located to tell someone where their nearest clinic, hospital, or doctor can be found), *Please Call Me* (a free call to anyone with no airtime to initiate a callback service), and *Cellbooks* (for downloading large amounts of texts to a mobile phone). The result of using these services included the following:

- Information for positive living.
- Mass communication about HIV prevention using text messages.
- Informing patients of appointments at clinics.

- Text counseling by linking patients with the National AIDS helpline.
- Building organizational capacity by providing subsidized or free services to government, NGOs, and community HIV groups.
- Monitoring and evaluating the information sent and received.

Cell-Life is just one of several organizations in South Africa that have launched m-Health projects in attempts to promote their importance. These include GeoMed, Centre for Scientific and Industrial Research, Medical Research Council, Praekelt Foundation, Stellenbosch University, Right to Care, and the Reproductive Health Research Unit (Benjamin, 2013). As in other countries, projects generally use text messaging, data collection, and smartphone Apps, but there has not always been much evidence of proven medical benefit by carrying out randomized controlled trials (RCTs), or by cost-benefit analysis.

To measure the impact of mobile phones, *Cell-Life* carried out detailed clinical impact and cost-effectiveness studies, including *SMS to Test*, which involved sending two kinds of text messages to encourage people to take HIV tests. One was informational, containing HIV statistics and public health material to people; the other was motivational, which prompted people to take responsibility for their lives and tell them how HIV might affect their family. Interestingly, it was the motivational messages that inspired an increase of HIV testing.

Another m-Health project in South Africa is called *Project Masiluleki*, which is also to encourage people to submit themselves to HIV tests and to take antiretroviral medicines. The support offers daily *Please Call Me* text messages, allowing free access to HIV call centers and to the National AIDS Helpline in Johannesburg. Closely allied with this project is *LoveLife's MYMsta*, which is a social network concerned with spreading information on HIV prevention (Kwan et al., 2013).

The Pan-African m-Health Initiative in South Africa was launched in 2013 to develop a suite of interoperable m-Health services addressing maternal, newborn and child health across South Africa (GSMA South Africa m-Health Initiative, 2013).

A further project in South Africa is called *SIMpill*, which is a system to alert patients and carers, via text messages, when medication is due or has not been taken, in this case for tuberculosis. The way it works is to embed a tiny SIM card and transmitter in the cap of a medication bottle, so that a text is sent to a computer when the bottle is opened; if the bottle is not opened during the prescribed time window, a warning message is sent back to the patient or carer. This is likely to be used more widely in both developed and developing countries because of the huge waste of drugs that are prescribed but not taken.

In a report compiled by GSMA, 83 m-Health initiatives and services were listed in South Africa. These included HIV/AIDS treatments (42), women and child health (18), community health workers' data collection (18), and other user-focused projects (GSMA South Africa m-Health Initiative, 2013). The report highlighted the following challenges of m-Health in the country:

- Integration of current services into national systems with multiple m-Health services that have had problems reaching national scale.

- Regulation of m-Health policy and financing the lack of articulated, tested standards for m-Health.
- Standards and interoperability challenges with lack of business cases for commercial roll out of m-Health services.

The opportunities for m-Health in South Africa were also cited in this report as:

- Evidence of similarities in aims and objectives across different m-Health services.
- Availability of expertise and lessons learnt across the m-Health implementation value chain.
- Commitment and interest from mobile operators as well as health service providers for collaborative action.

East Africa   This region of Africa has also seen major activity on m-Health projects, particularly in Uganda, Kenya, Tanzania, and Ethiopia. Some of these are as follows.

Uganda   The trigger to launch *Text to Change* (TTC) was the realization that "HIV/AIDS was a rampant scourge that was eroding Uganda's population" (Van Beijm and Hoefman, 2013). The idea, similar to initiatives elsewhere in Africa, was to use mobile phones to remind people of appointments or to take their medication, and to offer information on how to control the disease. The starting point was a pilot phase carried out by the Uganda Demographic and Health Survey (UDHS) in 2006, which revealed that only 30% of women and 40% of men were knowledgeable about HIV/AIDS (Uganda Bureau of Statistics, 2013; Van Beijm and Hoefman, 2013). The aims of the project were to improve knowledge and awareness of HIV/AIDS and to increase the uptake of HIV voluntary counseling and testing throughout the country.

The project leaders set up text messaging quizzes as part of a package of health communication activities, with the aim of demonstrating that mobile phones could be effective in disseminating information to create personal health awareness, and encouraging testing and drug compliance. TTC has lived up to its promise and has expanded its activities beyond Uganda to Kenya, Tanzania, Namibia, Cameroon, Sierra Leone, and the Democratic Republic of Congo. It is also operating in South America. From modest beginnings, the project now has many partners and sponsors, including USAID, Airtel Bharti, UNICEF, Family Health International, IICD, Infectious Diseases Institute, Malaria Consortium, Health Child, Jpiegho, and the Dutch Ministry of Foreign Affairs.

Some interesting and unexpected problems came to light during the project. One of these was that TTC innocently registered a short code with the number 666, but potential clients saw this as "a number of the devil!" The number was then changed to 777, which caused confusion. Further problems arose because all the messages were in English and there was no acknowledgement in them of the service provider, AIC. The more positive outcome was the creation of a service called Mobile for Reproductive Health (M4RH).

The lessons learned from TTC have much in common with projects launched in other developing countries, so they are worth reproducing here (Van Beijm and Hoefman, 2013):

- Low literacy, especially in rural areas, means that recipients who are unable to read and write need to share the information imparted by text messages with friends or family members who are more literate; this clearly has an impact on personal confidentiality.

- Mobile phone service companies are in business to make profits, and although they may offer reduced tariffs, someone else has to pay. The clients, that is the people the project tries to reach out to, are usually unable or reluctant to pay, so the cost falls to sponsors, government, ministries, or other benefactors. TTC has negotiated subsidized tariffs in Uganda, Kenya, and Tanzania, and Orange provides support in countries where their network operates.

- Africa has too many m-Health pilot projects that do not easily scale up or extend to other countries, unlike TTC, whose model has long-term funding agreements by Western agencies so that needy clients do not have to pay for calls.

- Low penetration of mobile phone ownership and usage in rural areas means that people in the most need of the TTC service are often sidelined. A family may have a mobile phone but this is invariably owned by a husband, so that messages directed to his wife may not reach her or they may be questioned by the husband; such entrenched cultural practices therefore present a barrier. A further obstacle is the scarcity of electricity to charge mobile phones, so TTC has found it necessary to supply solar chargers in some cases.

- There is no central database for health information, and it is a challenge to disseminate messages that are limited to 160 characters. The aim has been to accumulate data that are relevant to cultural, religious, and demographic differences.

These represent good recommendations and lessons to be considered for any future initiatives similar to this one, particularly in other parts of Africa.

Tanzania   There is a well-established NGO infrastructure in Tanzania, with m-Health initiatives backed by a number of internationally recognized funding bodies. Some of the initiatives include malaria clinics helped by the *SMS for Life* initiative that uses a text messaging program to efficiently deliver the malaria vaccine (Barrington et al., 2010). The GSMA Mobile for Development m-Health in Tanzania includes the m-Nutrition program as a child and maternal nutritional health initiative (Merry 2014).

Kenya   Kenya is a major African country with 41 million people and US$783 GDP per capita. It also has major health challenges including the following (Cargo, 2012):

- High HIV prevalence rates (6.3%)
- Extremely high rate of maternal and infant deaths (the proportion of women who die in childbirth is 1 in 39)
- A high burden of tropical parasitic and bacterial infections.

Working with the GSMA and the Global Health Initiative, the Kenyan government's strategy is focused on addressing three objectives: moving from disease-specific treatment to preventive care, finding cost savings, and doubling the number of mothers who have access to a skilled birth attendant and to emergency obstetrical care (Cargo, 2013).

One such initiative is the *WelTel Kenya*, a RCT to evaluate the use of text messaging for adherence to antiretroviral treatment, as part of a strategy to improve health interventions. The trial used 538 HIV-infected adults, initiating antiretroviral therapy (ART) in three clinics in Kenya (Lester et al., 2010; WelTel, 2014). The outcome was that patients who received SMS support had significantly improved ART adherence and rates of viral suppression compared with the control individuals. Another RCT addressed the effect of text message reminders on Kenyan health workers' adherence to malaria treatment guidelines, with a total of 119 health workers receiving mobile intervention. The study recommended the use of simple one-way text messaging to health workers for improvement in the quality of artemether–lumefantrine for malaria management (Zurovac et al., 2011).

These m-Health pilots remain largely localized, with many challenges for larger adoption, such as lack of appropriate scaling-up strategies, change of existing clinical practices, long-term funding, cost, and implementation policies.

Ethiopia   In recent years, the Ethiopian government has invested heavily in expanding the capacity of its healthcare system by considering the many socioeconomic and healthcare challenges, such as maternal, malaria, HIV/AIDS, tuberculosis, and poor nutrition. A detailed study of the potential for m-Health in Ethiopia and a list of the existing projects are cited elsewhere (Vital WaveConsulting, 2011). m-Health projects in Ethiopia include Wegen AIDS Talkline, which is a toll-free hotline that provides information, telephone counseling and a referral service on HIV/AIDS, STIs, and TB-related topics, all anonymously. The hotline has language-specific counseling available, covering 14 local user mobiles to save lives and money. Other initiatives were also listed in the Ethiopian Health Extension Program (HEP), a 20-year development plan introduced in 2003 to respond to challenges of healthcare access of people with communicable diseases.

A small feasibility study of m-Health applications was conducted in which primary healthcare workers used smartphones to collect routine health data relevant to maternal health, with a total of 952 patient records collected over 6 months (Medhanyie et al., 2015). This study also recommended a better patient identification system, a robust standardization policy of healthcare, and better mobile network coverage as some of the suggested prerequisites for any scaled-up use of similar services in Ethiopia.

Central and West Africa

Congo   In the Democratic Republic of Congo, *La Ligne Verte* was introduced in 2005 as a free cellphone hotline concerned with family planning and contraceptive methods (Corker, 2010; Kwan et al., 2013). It was set up by the mobile providers

Vodacom and Zain, and funded by Population Services International and its affiliated partner, Association de Santé Familiale, as part of a Family Planning Project (Corker, 2010). It enables callers to speak in confidence to trained mobile educators in Kinshasa about contraceptive methods and side effects, and to get referrals to clinics or sales outlets in the callers' own neighborhoods. The hotline number and hours are printed on items such as pocket calendars that are given away during FPP's information–education communication activities in 8 of the country's 11 provinces. The surprising outcome to those running the service was that 80% of the respondents were men!

Another m-Health pilot project for the use of commercial mobile phone-based diabetes management systems was tested in two cities in the eastern part of Congo with a total of 40 type 2 diabetic patients. This study reported a reduction of HbA1c levels from 8.67% (before the start of the trial) to 6.89% levels after the end of the 2-month RCT (Takenga et al., 2014).

Ghana   As with most developing countries, antenatal and neonatal care is in need of improvement to ensure the healthcare of mothers and newborn babies, especially in rural areas where access to basic information is very limited and local myths and damaging cultural practices are rife. With the aim of reducing the high maternal and child mortality, Ghana set up the Mobile Technology for Community Healthcare (*MoTeCH*) initiative in 2008, in partnership with the Ghana Health Service, the Grameen Foundation, and the Mailman School of Public Health at Columbia University, with funding from the Bill & Melinda Gates Foundation (MoTeCH, 2014; Grameen Foundation, 2014; Osborn 2013).

One of the ongoing services is *Mobile Midwife*, launched in 2010 to enable pregnant women and their families to receive text messages or prerecorded voice messages on their mobile phones, offering weekly advice and appointment reminders as pregnancy progresses. The two aspects of the service were the "share ability" of messages with family and friends, and the personalization of messages tailored to the woman receiving it, for example, their stage of pregnancy, care history, medical history, language, address, and their preference of when and where they access advice. The service included a hotline manned by healthcare professionals who have found that the level of interest among patients was greater than they had expected, but while people were happy to receive information by voice they were reluctant to receive text messages. A further discovery was that some 20% of the callers were men, and 62% of the enquiries were more concerned with child health than with pregnancy, which led to the service being continued for a year after each child's birth to ensure adherence to postnatal healthcare and vaccination schedules.

Mobile Midwife revealed a number of interesting communication barriers that are undoubtedly common to other developing countries:

- Women often thought that their husbands prevented them from seeking healthcare because they did not appreciate the medical need for it, but in reality the husbands were more concerned about the cost, not realizing that it was free.

- There was a need to translate into different Ghanian languages and to localize the content to different cultures, which have many cultural beliefs, myths, and practices across regions.
- The voice on recorded messages was of vital importance; a male voice talking about savings and finance was acceptable to both men and women, but on pregnancy information women wanted to hear an older woman's voice, which they tended to find more trustworthy, sympathetic, and respected, but the accent found most acceptable was neither too rural nor too educated.
- Some languages do not have formally written forms.
- Mobile phones are often shared among a family, and SIM cards are changed to obtain a better signal with one network over another; as a result, women had to make a call to receive messages.
- Mobile phones were often switched off or "dead" because power cuts prevented their recharging, and people could not remember their own number, which was a problem when registering it in the service's data bank.

A further service is *Nurse Application*, which enables community health nurses to record electronically the care given to pregnant women they have found in their area. Here, further barriers became apparent:

- Older nurses, unused to mobile phones, did not know how to send or retrieve text messages, which necessitated training.
- Some nurses were poor at keying in the correct syntax, so their messages could be ambiguous or wrong.
- Some nurses shared their phone with one or more family members, which was a potential or actual breach of patient privacy, and they were reluctant to use their own phone for work. It turned out because the General Packet Radio Service (GPRS) was much cheaper than the SMS, and it was cost-effective to buy dedicated project phones. When issued with Java-enabled phones (Nokia 1680) under a Ghana Health Service agreement, nurses had to keep them at their clinic or where they were working. If their phone was lost or stolen, there was a fine imposed, shared among all the nurses at the clinic!
- Some simple phones had low storage for messages, many different types of phone were in use, and the message center settings were sometimes wrong.
- Nurses needed incentivizing to carry out all their data recording, which meant a shift in their work practices and culture.
- Nurses saw the *MoTeCH* initiative as something temporary, rather than as part of their normal work, so to overcome this a technical working group was formed to improve their perception of the Ghana Health Service, whose staff worked with the nurses to ensure closer liaison and mutual understanding (Grameen Foundation, 2014).

The Grameen Foundation and Ghana Health Service have expanded the *MoTeCH* applications in five languages across the four regions of Ghana, with support from the Saving Lives at Birth Partners, United States Agency for International Development (USAID), the Government of Norway, the Bill & Melinda Gates Foundation, Grand Challenges Canada, and the World Bank (Grameen Foundation, 2014).

Another m-Health project reported in Ghana is the *Text Me! Flash Me!* helpline, which was launched in 2008. As with initiatives elsewhere, especially in Africa, the organization is complex. This project was created out of the Ghana Strengthening HIV/AIDS Response Partnerships (SHARP), which is managed by the Academy for Educational Development (AED), through collaboration with the Ghana National AIDS Control Program, the Ghana Health Services (GHS), and nine local NGOs, with USAID funding (Kwan et al., 2013; Text Me! Flash Me!, 2014). The service uses mobile phone technology to connect the most-at-risk populations (MARP) to a helpline for HIV/AIDS information, referrals, and counseling services from qualified carers. The targeted people are men who have sex with men and female sex workers.

The *Text Me!* component prompts a person to send specific words relating to HIV to receive detailed information about the aspect of interest. The *Flash Me!* component allows someone to contact a helpline councilor, who phones back within 24 hours, so that the call is free to the inquirer.

Fewer m-Health initiatives have been reported in West Africa, notably in countries affected by the Ebola outbreak (Guinea, Sierra Leone and Liberia). These initiatives mainly focused on maternal and neonatal health (Royal Tropical Institute, 2011; m-Health Alliance, 2012a). However, in the post-outbreak period, calls for setting up different initiatives were increased substantially, with urgent demands to use m-Health technologies for fighting and mitigating the impact of the Ebola epidemic (O'Donovan and Bersin, 2015; Wicklund, 2015). This global health crisis perhaps acted as a wake-up call and for lessons to be learnt in reshaping and prioritizing the current m-Health strategies, policies, and deployment plans in Africa.

Conclusions   From this snapshot of the many m-Health projects in Africa, wider opportunities exist, with some ongoing and expanding, but most remain as local initiatives, with major barriers to large-scale adoption. Some of these barriers can be summarized as follows (Folaranmi, 2014):

- Lack of clear standardization and regulatory frameworks to guide the scale-up process in the continent.
- Absence of rigorous and pivotal clinical studies that can evaluate the efficacy and cost-effectiveness of m-Health interventions best suited to healthcare in African countries.
- Shortage of sustainability and affordability plans of m-Health programs on a long-term basis. Most of the current initiatives are either funded by private–public partnerships, or overseas institutions and NGOs.
- The lack of long-term and coordinated governmental m-Health policy and strategy in most of the African countries.

In a report by the Economist Intelligence Unit on the future of m-Health in Africa, five potential scenarios depicting the possible health landscape on the African continent in 2022 were proposed (Economist Intelligence Unit, 2012):

- Refocusing on primary and preventive care.
- Empowering communities as healthcare providers.
- Implementing universal coverage.
- Making telemedicine ubiquitous.
- Encouraging local suppliers.

The West African Ebola epidemic of 2013 is considered one the longest and most serious disease outbreaks in decades. The critical failure of the WHO to properly predict, contain, and manage the epidemic was attributed to several shortcomings within the organization (WHO, 2015). The design and implementation of proper m-Health strategies focusing on community engagement, preparedness, and implementing tailored m-Health monitoring and management intervention could have alleviated some of the problems posed by this major epidemic. The challenges of other similar global health scenarios can be alleviated by using proper m-Health implementation plans and coordinated policies that can mitigate the associated risks. Developing a new Pan-African tailored m-Health model and implementation framework for future epidemic outbreaks is one such possibility.

The burdens of conflicts, and poor healthcare systems, compounded by an increasing population with high chronic disease prevalence, inadequate health workforce, widespread rural population, and limited financial resources are some of the many factors that face most African countries. The demand for revised strategies with more coordinated efforts to develop new m-Health systems with more effective and large-scale outcomes are vital for the success of m-Health in Africa.

6.3.2 Global m-Health Initiatives in South America and Latin America

In the last two decades, the healthcare challenges and the medical landscape in South America and Latin American countries have been changing rapidly. The demands of ever-increasing population numbers, combined with varying economic conditions and increasing unevenness in wealth and inequalities compounded with crowded cities and slums, are all instigating new long-term healthcare improvement in these developing regions. Some of the countries, for example Brazil, have implemented universal healthcare systems with tangible healthcare outcomes, including a 50% drop in infant and maternal mortality rates, with better access to care in the poorest and remotest areas (Francis, 2011). However, the majority of other countries still lack such progress in their healthcare system, with poor universal access to quality care remaining largely unfulfilled.

In parallel, mobile phone users in Latin America are predicted to reach 374 million people in 2017, with the biggest markets expected in Brazil with more than 112 million mobile users, followed by Mexico with 97.6 million, and Argentina with 53 million (GSMA Intelligence, 2013). This contradicting yet typical scenario of poor healthcare access with higher use of mobile phones offers the necessary opportunities for some m-Health services. Some of the m-Health initiatives in the region are presented next.

Brazil and Mexico   According to a PwC study, the most common chronic diseases in both countries that m-Health can impact effectively are type 2 diabetes (T2D), COPD, obesity, hypertension, and CVD. These diseases, in addition to prenatal and maternal care, represent major challenges in both countries and can be effectively mitigated by appropriate adoption of preventive and cost-effective mobile healthcare solutions (PwC, 2013). The report also predicts that by 2017, m-Health could enable 28.4 million people to have healthcare access in Brazil, with an additional 15.5 million people in Mexico, without having to add a single doctor and with an estimated cost saving of US$14 billion in Brazil and US$3.8 billion in Mexico, respectively (PwC, 2013). Another report that tracked the downloads of smart health and fitness Apps in Brazil reported an average of 238,000 Android phone App downloads compared to 92,000 for iPhone Apps (Research2Guidance, 2014).

Although these are mostly market predictions, the indications are that there is increasing awareness of m-Health benefits in different countries in this region. Some of these projects included an 18-month pilot study conducted in urban areas of the Santa Marta community in Rio de Janeiro where mobile healthcare workers were equipped with backpack systems that consisted of medical tools designed to enable health workers to take essential measurements from elderly patients. These included blood pressure and glucose levels, and also provided portable ultrasound scan services by visits to more than 100 elderly patients suffering from chronic diseases and mobility problems. The outcome of this pilot study indicated significant clinical impact and reduction of prevalence of T2D, hypertension, and heart failure in the intervention patients (New Cities Foundation, 2013). A review of the m-Health research initiatives in Brazil listed more than 42 m-Health projects, of which 86% were focused on health surveys, surveillance, patient records, and monitoring. However, other areas such as compliance of treatment, educational awareness, and decision support systems were lacking (Iwaya et al., 2013).

A pilot study on the use of mobile phone text messaging for alerting HIV/AIDS patients in Mexico City, together with the patients' drug compliance and medication intake, has been reported (Feder, 2010). Other mobile health initiatives have also been reported, such as door-to-door visits by teams of volunteer doctors to the rural Chiapas areas of Mexico to identify cases of chronic conditions. The doctors were equipped with tablets with an Android App called *CommCare* that allowed them to enter patient information and provide the appropriate remote advice and care (Partners in Health, 2013).

However, large-scale and sustainable mobile health projects remain limited in these two countries, due to barriers that are common themes in other developing world regions. These include economic, regulatory, structural, and technological challenges

(PwC, 2013). In addition, the societal disparities, remoteness of some areas, and economical inequality, necessitate the urgency for adopting more radical m-Health plans and policies in Brazil and Mexico.

Chile Although Chile has a high level of mobile phone users, with an estimated 85.4% ownership of smartphones, the country has relatively few mobile health applications (Oleaga, 2014). Chile, as with other countries in the region, has a high prevalence of chronic diseases, particularly diabetes and obesity; tackling diabetes is a national priority. A study by the School of Nursing at Pontifica Universidad Cathólica de Chile, supported by the Chilean Ministry of Health, working in Puente Alto, Santiago, reported several mobile management initiatives for diabetes (Lange et al., 2010; Lange, 2013). The approach has been to use tele-self-management to aid early diagnosis and reduce the risk of serious complications. Without telecare, a patient with a chronic condition can go to a primary care center, receive urgent care, or go to a hospital. The work has been carried out as four projects:

Project 1 was aimed at the development of a telephone-mediated chronic care-support model to improve self-management. The ATAS Model (Apoyo Tecnológico para el Automanejo de Condiciones Crónicas de Salud) was conceived to improve self-management, metabolic compensation, and satisfaction of Chilean patients with type 2 diabetes. The original model comprised a training program for primary care clinicians, a telecounseling service, a self-management tele-support guide, and special software for self-care information, management, and follow-up advice. The training program introduced primary healthcare clinicians to what the project can offer. Telecounseling, the main component of ATAS, comprised preclinic visit calls, postclinic visit calls, and special event counseling calls, mostly via mobile phones or occasionally via landlines. The self-management guide was designed to help patients improve their lifestyle by exercise, which may have come as a shock to many diabetes sufferers who have never done any exercise in their lives. The systems' software also provided self-care information to patients for stabilizing their blood sugar levels (Claudia et al., 2012).

Project 2 was initiated to replicate the ATAS model in seven primary care centers. It was during this project that the value of using mobile phones was apparent because fixed phone services were often unavailable as a result of the copper wires having been stolen! The problems that emerged from using mobile phones was that the cost per minute was three times higher than for fixed lines, and that patients needed to organize themselves to be ready for a telecare call, when they were in a quiet place where they would not be interrupted.

Project 3 used the ATAS model in the "real world" of a poor commune, Puente Alto, with an emphasis on patient-centered chronic disease management, motivational interviewing, behavioral change, self-care, and decision support. This project was seen as a demonstration site where carers and clinicians could come to learn how to improve chronic care for patients with type 2 diabetes and other cardiovascular conditions in their own communities.

Project 4 provided support and self-management using mobile phones for automated calls and text messages. The ATAS intervention, in low-income primary care centers, significantly increased the probability of stabilizing the metabolic control of patients with type 2 diabetes, and improved their use of health services. To people in developed countries this may seem routine but to the people of poor communes in Chile, especially old people (as elsewhere in developing countries), it was truly a breakthrough. Text messages can be received at any time at no cost to the patient, and they can be reread and shared with the patient's family, who in turn may be able to take advantage of the advice offered. The vision is now to expand the whole program to a national scale, with technical support from the Pan American Health Organization and the Chilean Ministry of Health.

Overall, the outcomes of the ATAS intervention reported effective levels of stabilized hemoglobin, reduced emergency room visits, and improved adherence to planned visits of patients to their primary care center (Claudia et al., 2012). There is also increasing use of smartphones for healthcare information and advice in Chile (Oleaga, 2014). This large-scale use of mobile phones combined with increased health awareness represents positive indicators of the popularity and potential acceptance of mobile health solutions in the near future. However, there is much work to be done, particularly on healthcare policy and wider m-Health strategies.

Nicaragua   According to the WHO, like other Latin American countries, Nicaragua faces many healthcare challenges, such as inequality of access to healthcare services, poor child nutrition, high maternal mortality, and an increase in communicable and noncommunicable diseases (Foundation for Sustainable Development, 2014). Although there is a pressing need for m-Health solutions to alleviate these problems, few have been reported.

One of the more sophisticated projects for the developing world is *Adhere.IO,* developed by the Massachusetts Institute of Technology and introduced in Nicaragua to improve patients' adherence to taking medication for tuberculosis (TB) using behavioral diagnostics (Gomez-Marquez, 2013). The system uses colorimetric diagnosis that reveals an alphanumeric code when exposed to the so-called metabolites of the medication. For TB, the metabolite chosen was isonicotinic acid, which is secreted in urine, so when a test strip is exposed to a patient's urine, a code is revealed that is sent as a text message to a database in a distant clinic. If the expected code is not received within a certain time window, a reminder is sent to the patient; if the expected code is received, the patient is rewarded with mobile phone minutes, which are transferred electronically as soon as the "correct" result is recorded. This is one of the few m-Health examples in Nicaragua.

Here, the test strip is a sensor, so the system conforms more closely to a true m-Health system, according to our definition in the Introduction, more than any other that we have come across for use in the developing world. It is curious that anyone should need to be rewarded for taking medication that can save their lives, but incentives seem to be necessary to overcome many people's basic lack of self-discipline, forgetfulness, or ignorance. In general, there is an urgent need for improved efficiency in care services in the country and for adopting new solutions, such as m-Health applications that can yield cost-effectiveness and better healthcare outcomes (Broughton et al., 2014).

Peru   Although the economy in Peru is considered one of the fastest growing in the region, according to a report published by the Peruvian National Institute of Statistics and Information (INEI) in 2012, 37.4% of Peruvians do not have health insurance, with 34% suffering from chronic conditions, and only 52.2% receiving some form of treatment (Goldberg, 2012). Yet Peru has one of the highest levels of mobile phone usage in Latin America. These challenging healthcare conditions, combined with growing mobile phone use, enhance the case for adoption of appropriate mobile health solutions and services, especially in remote areas and poorer regions.

One of the m-Health initiatives reported in Peru includes *Cell-PREVEN,* an interactive computer system using cell phones for real-time data collection and surveillance of adverse events related to metronidazole administration among female sex workers in Peru (Curioso et al., 2005). Other projects include the use of mobile phones for community care, maternal care, and antiretroviral treatment adherence (Kwan et al., 2013). Further, Peru seems to have a national m-Health strategy, as shown in Table 6.2. However, most existing m-Health initiatives need scaling up and also targeting prevalent diseases, especially in rural areas.

In conclusion, the demographic and healthcare landscape of Latin American is changing rapidly. Over large geographical areas, people are still plagued with acute socioecomonic inequality, poverty, and poor access to proper healthcare facilities. Moreover, some of these countries are facing major healthcare challenges in both infectious and noncommunicable diseases, and disease surveillance remains a major problem. These healthcare priorities are becoming increasingly acute economic burdens against the rapid development. A massive increase in mobile phone owner-ship has led to increasing advocacy for developing effective mobile healthcare solutions, particularly in poor and remote regions and in densely populated cities. While there is considerable interest in m-Health services in Latin America, the effort so far has been with isolated pilots and silo studies, without a comprehensive long-term national m-Health strategy, except in Peru.

As in the rest of the developing world, Latin America suffers the same barriers that contribute to the slow adoption of scaled-up m-Health services. These include cost constraints, long-term strategies and planning, disparities, corruption, lack of medical expertise, and IT skills. Some Latin American countries, such as Brazil, are considered to be leading models in economic growth, with major healthcare trans-formation initiatives in key challenging areas. But there is much to be done to scale up these achievements, with large-scale m-Health initiatives that can supplement and sustain healthcare outcomes and act as workable models for the other countries in the region. The recent outbreak of the Zika virus in the continent justifies funding large and focused m-Health initiatives to tackle this epidemic and contain it.

6.3.3 m-Health Initiatives in the Indian Subcontinent

The Indian subcontinent, with its huge population, economic resources, and social, ethnic, and economical diversity, is one of the most needy places in the world for m-Health to be adopted to alleviate the major healthcare challenges. Yet there are fewer than expected m-Health projects in this vast region, and these are mostly localized. With the available resources and human capabilities, these projects can be potentially scaled up and implemented over a wider area with minimum cost, provided that the current barriers are adequately considered. In this section we outline some of these m-Health initiatives in India, Pakistan, Bangladesh, and Nepal.

India   India is one of the largest and most populous countries in the world, with an estimated 1.2 billion people living in a diverse geographical landscape. It is estimated that more than 300 million people use mobile phones in India, with more than 16 million urban Indians having Internet access on their phones almost on a daily basis (Dass, 2015). Although different m-Health projects have been implemented, the full potential of these services and their benefits is as yet largely untapped. There are a number of private and public healthcare programs in India that utilize mobile health technologies (Davey and Davey, 2013). There are also limitations due to the slow adoption of communication standards above 2G; even 3G has never really become widely used, so the large-scale adoption of 4G and 5G may well be a long time off (Gatherer, 2015).

Recent review publications list several m-Health projects and pilot studies, mostly using SMS and mobile phones for HIV/AIDS and TB patient education/awareness, patient monitoring, and connecting healthcare workers (Chigona et al., 2013; Davey and Davey, 2013). The review outcomes of these pilots, mostly funded by donors and private institutions, indicated a lack of large-scale plans, clinical rigor, and the necessary conceptual efficacy framework.

Other m-Health services in India have been initiated by private healthcare entities and mobile phone operators, such as providing healthcare and emergency advice, and mobile phone-based remote consultations (Kappal et al., 2014). Some of the other m-Health initiatives and projects include:

Freedom HIV/AIDS, which uses an unusual initiative with game-based programs on mobile phones to inform young people about HIV/AIDS, particularly sexual interactions, myths, misconceptions, discrimination, testing, and treatment. It has been developed by ZMQ, a Technology-for-Development company, in partnership with the Delhi State AIDS Control Society and the mobile phone operator Reliance Infocomm (Quraishi and Quraishi, 2013). The strategy was to design games with memorable messages and instructions that targeted youths, including school children, and people in rural communities, with the aim of promoting good behavior in matters of personal health, and changing bad or risky behavior. ZMQ has referred to the interaction with real-world scenarios in a risk-free environment as a "real-world risk reduction method using games mechanics." Since talking about sex in general and HIV/AIDS in particular are taboo subjects in India, the games were designed in a culturally sensitive format.

Other games include *AIDS Messenger*, an adventure game, *Life Choices*, a role play life skills games for girls, and *Great Escape*, a role play detective game; all are designed to carry "mobile persuasive messages," mainly concentrating on the avoidance of HIV/AIDS.

The ZMQ Software designers have attempted to evaluate their games and have used studies to assess learning. One study involved qualitative analysis, random surveys, and interviews to try to assess behavior change, and the success was measured in terms of certain indicators:

- An increase in safe sex by condom usage, abstinence, and faithfulness to a single partner.
- An increase in ability and skill to oppose peer pressure.
- An increase in demand for disposable syringes at healthcare centers.
- A reduction in myths and misconceptions about HIV/AIDS among young people.
- A reduction in stigma and discrimination against people living with HIV/AIDS.

While the incidence of HIV infections has fallen, there is no way that the games can have been the only reason for this change: playing games on a mobile phone to generate warnings and advice is one thing, but to carry these through to changes of behavior in real life is quite another, and the success can only be assessed in the most subjective way.

Another initiative in India used mobile persuasive messages for rural maternal health in the state of Orissa, and has focused on Accredited Social Health Activists (ASHAs), who are employed by the government in their home villages to promote free health services and offer counseling on many issues such as contraception, pregnancy care, and breastfeeding (Ramachandran, 2013). The difficulty faced by ASHAs is that their advice is often contrary to entrenched social attitudes, myths, and traditions. This project was not strictly what we would describe as being in the m-Health domain, because the methodology relied on the use of videos viewed by target communities along with the ASHAs, rather than transmitted by mobile technology. Otherwise, its intended outcome was the same: to improve maternal health using persuasive messages to overcome social barriers. The interactive videos included instructional messages, basic information, material from health handbooks, illustrations sketched by local artists, and with local language voice-overs.

Did it work? The author of the study could not be sure, because a mixture of theory, practice, and intuition is needed to develop a behavior change model and this takes time, more time than was available, but the important lessons learned are common to any initiative to promote healthy behavior by persuasive messages, and here we quote these six guiding principles:

- Focus the message on action items, not on broad topics of information.
- Address local myths and barriers, and provide convincing corrections and solutions, respectively.
- Create opportunities for structured, persuasive dialog between people, keeping in mind that persuasion is still largely a social phenomenon in rural communities.
- Include reminders about the positive rewards for changing behavior, paying close attention to local values.
- Capture the most persuasive local language and prosody style, even if it is counterintuitive.
- Do not assume reactions are honest; persuasion takes time.

Another study that explored the feasibility and acceptability of SMS and mobile phones for delivering healthcare interventions among 488 users in a village in rural Bangalore concluded that this mode of m-Health used for receiving health information and supporting healthcare was acceptable in rural India (DeSouza et al., 2014).

Another project (*mSakhi*) is led by IntraHealth International, a U.S. government-funded NGO funded by the Bill & Melinda Gates Foundation. It began in 2012 to test a multimedia mobile phone App designed to provide ongoing support to local Indian healthcare workers. Meaning "mobile friend" in Hindi, *mSakhi* provides 65 health messages pertaining to prenatal care, postpartum mother and newborn care, immunization, family planning, and nutrition. The smartphone App employs a combination of text messages, audio clips, and illustrations to deliver its messages (Farrell and Bora, 2012). Other m-Health programs for improving maternal, neonatal, and child health in India are cited in relevant sponsorship reports (m-Powering Frontline Health Workers, 2015). Successful m-Health service delivery applications in India are mostly based on the mobile phone-centric model discussed earlier. This is due to the availability of low-cost mobile phones and cellular networks. A survey on affordable m-Health services in India outlined several challenge barriers (Lunde, 2013):

- Security and privacy concerns.
- Poor mobile network coverage.
- Complexity of m-Health applications due to the diversity of cultures and languages.
- Cost and willingness to pay, especially in low-income population.

One example of the use of m-Health technology for chronic disease management is the UKIERI project (Panel 1). A list of the most relevant m-Health attributes led to the development of new and cost-effective solutions that are tailored for rural areas of India (Mulvaney et al., 2012a, 2012b).

Case study: A Global m-Health Project in Rural India

The concept of exploiting global mobile communications networks to link "anywhere to anywhere" so that a patient can be monitored remotely by a clinician is nowhere more relevant than in India. The project is aimed at improving healthcare provision, particularly for patients suffering from heart disease and diabetes, which (with asthma) are ever-increasing globally (Mulvaney et al., 2012a). The project has been funded since 2007 by the British Council under the United Kingdom–India Education and Research Initiative (UKIERI). It is an ongoing m-Health partnership between Loughborough University and Kingston University in the United Kingdom and the Indian Institute of Technology Delhi, Aligarh Muslim University, and the All India Institute of Medical Sciences in India, and initially led by the authors of this book.

The methodology, based on the system shown in Fig. 6.2, is to transmit biomedical data, typically the electrocardiogram (ECG), blood pressure, oxygen saturation, and blood glucose level, using "smart" sensors and a miniprocessor carried by the patient. This "body area network" (BAN) is linked via a modem to mobile networks and the Internet to a hospital computer, enabling doctors to monitor patients remotely. Such a "mobile disease management system" is overdue in view of the proliferation of applications in mobile data communications.

The work plan was divided into tasks that require liaison between the UK and Indian teams to ensure the interchange of ideas, designs, software, hardware, and data, with the aim of demonstrating m-Health systems during clinical trials in the United Kingdom and India. The main tasks of the project were as follows:

Task 1: Identification of clinical needs, to establish m-Health requirements for monitoring chronic diseases remotely.

Task 2: Development of sensors and processors, leading to "smart" sensor interface chips for biomedical signal transmission by converting existing sensors and associated circuits using microelectronic techniques; also the design and development of a BAN as well as medical signal processing, data compression, and encryption.

Task 3: Network development, to exploit the mobile communications infrastructure for transmitting biomedical data by adopting mobile systems (e.g., GSM/GPRS/3G/4G) for use with the Internet and, in India, a national fiber-optic network; also the development of communications protocols and servers.

Task 4: Technical and clinical trials with hospitals in London and Delhi to test and evaluate evolving m-Health systems, first to establish any technical limitations, then with patients living remotely from hospitals; also advanced data analysis to identify abnormal trends.

The deliverables from these tasks were also the proposed project outcomes:

- Visits by UK staff and students to Indian institutions in Delhi and Aligarh for meetings to launch an overall plan of operation, to coordinate logistics, and to decide on system design, communications protocols, experimental techniques, scope of technical and medical trials, and so on.
- Familiarization with medical services and communication networks in India.
- Visits by Indian staff and students to UK universities to liaise with appropriate groups working on related aspects of the project.
- Seminars and workshops in the United Kingdom and India to disseminate ideas and discuss progress alternately in each country at 6-month intervals.
- Visits to regional hospitals in small towns and to basic clinics in remote villages in India and to discuss health issues with local doctors, nurses, and paramedics.
- Visits by UK staff to India to assist in technical and clinical trials.
- Preparation of a technology implementation plan.

The project is just one of many funded by the British Council's UKIERI programme to link education and research teams in the United Kingdom and India. Following 6 years of good progress in the design and development of an m-Health system to monitor heart disease and diabetes, the project funding was extended by the British Council in 2013 for a further 3 years.

Bangladesh   Healthcare services in Bangladesh, as with similar LMIC economies elsewhere, are faced with major challenges of accessibility and affordability, especially in the rural regions of the country. Numerous m-Health initiatives have been reported, and the "m-Health knowledge" repository cites some examples (m-Health Knowledge, 2014), as follows:

mCare: A pregnancy and neonatal health information system that connects rural health workers and facilities with pregnant women and their newborns in Bangladesh, in partnership with Johns Hopkins University in the United States.

MAMA Bangladesh (Aponjon): This project, Mobile Alliance for Maternal Action, is called *Aponjon*, which means "close one" or "dear one" in Bangla; it empowers the delivery of vital health information to new and expectant mothers using text messaging services. The *Aponjon* program was launched nationally in December 2012 by Bangladeshi social enterprise, Dnet, in partnership with the Government of Bangladesh MOHFW, USAID, and Johns Hopkins University (MAMA Aponjon, 2013).

Most of the cited projects in the country were sponsored by the private sector, NGOs, academic institutions, and other funding institutions.

Another survey in Bangladesh identified more than 225 different e-Health and m-Health initiatives and projects since 1990 and categorized these in the following themes (Ahmed et al., 2014): health education (10); health financing (5); health awareness (30); disease and epidemic outbreak and surveillance (15); patient monitoring and support (80); health management and information systems (35); and point of care and diagnostics (50). This study concluded that there is still uncertainty and a dearth of clinical evidence and effectiveness for these projects in terms of healthcare impact. It recommended a need for the adoption of a national governance framework of m-Health and e-Health, with clear plans for the financial viability of projects and their wider integration into the country's existing health system.

Pakistan   More than half of the population of Pakistan live in rural areas, with challenging healthcare services, insufficient facilities, and lack of infrastructure. The poorly organized healthcare system, with lack of necessary clinical expertise and manpower, especially in primary care settings, necessitates (among other plans) developing modern mobile health initiatives in the country. According to a report prepared by Boston Consulting Group, Pakistan is aiming to strategically shore up its overall healthcare service provision by strengthening stewardship functions in this sector to ensure reliable service provision, equitable financing, and accountability (Boston Consulting Group, 2012).

For example, Pakistan, like India and Bangladesh, aims to achieve several of the MDGs by reducing child mortality, improving maternal health, and combating HIV/ AIDS, malaria, and other diseases with evidence-based policymaking and strategic planning in the health sector (Boston Consulting Group, 2012).

Examples of some of the m-Health initiatives in Pakistan include *The HealthLine* project, which is an automated health information and education service used by illiterate community health workers in Pakistan. In this system, workers can call at any time, free of charge, from a mobile phone or landline and talk to the system in their own language to ask questions and learn about symptoms and treatments (Boston Consulting Group, 2012). The *Lady Health Worker* pilot project is another example that aims to bring low-cost mobile communications to remote areas to empower midwives or community workers to access emergency consultation (GSMA Development Fund, 2013).

The *PRIDE* is another mHealth initiative in Pakistan, which is aimed to increase access to maternal and child health information and services (Kwan et al., 2013). It uses interactive voice response (IVR) and text messaging to offer women information about antenatal care, immunizations, and postnatal care.

Nepal   Few m-Health studies and projects have been reported in Nepal. A study addressing the shortage of experienced healthcare workers in rural areas of Nepal and their use of mobile phones to report to the district hospital has been reported (Morrison et al., 2013). Another study on the empowerment of community healthcare workers (CHW) with mobile phones working in remote areas highlighted some of the major barriers in adopting m-Health services (Chib et al., 2012). The 2015 Nepal earthquake, with its devastating impact on human lives and its effect on the country's infrastructure, represents an example of the urgent need for the international health community to adopt m-Health policies and strategic plans for m-Health services and delivery models in such natural disaster areas.

In conclusion, the opportunities to harness m-Health solutions in the Indian subcontinent are immense but also challenging. There are numerous barriers that can impede the vision of implementing large-scale m-Health initiatives in this region. These include, for example, the cost and the lack of appropriate national policy plans, and abiding by the governmental bureaucracy among others. However, there have been some successful m-Health initiatives with tangible outcomes. Overall, most of the existing m-Health initiatives in this region remain largely pilots and localized studies conducted by different NGOs, telecommunications sponsors, local enterprises, and market entities, with less government-sponsored initiatives. Nevertheless, there is increasing interest in adopting m-Health services by clinicians and healthcare providers.

6.3.4 Global m-Health Initiatives in Other Developing Regions

There are other m-Health initiatives in various developing and LMICs that were not listed in the previous sections, such as the Caribbean and West Indies, and southeast Asia (Malaysia, Philippines, Thailand, Indonesia). The details of the these and other similar initiatives are outlined in more specialist studies, reports, and compendiums cited earlier (WHO, 2011; GSMA Development Fund 2013; Kwan et al., 2013; Boston Consulting Group 2012). Some brief examples are given here for completeness.

West Indies   The *MediNet Project* (short for Medical Networks) was set up in 2007 by the University of West Indies with Microsoft sponsorship. The aim was to design and develop a remote patient monitoring system for Trinidad and Tobago for the self-management of chronic conditions, notably diabetes and cardiovascular disease, which are highly prevalent in the West Indies, with a view to scaling up for other chronic diseases such as asthma and cancer (Sultan and Mohan, 2009, 2013).

The system architecture has three components: a patient interface, various healthcare web services, and a healthcare provider interface. Readings from a glucometer (LifeScan One Touch) and a blood pressure meter (A&D) are transferred via a Bluetooth wireless link to the patient's mobile phone and these are sent along with data about food intake and exercise. The phone has a software electronic diary application called My Daily Record to save all the data. The web services enable data from the mobile phones of all patients in the network to be saved on a remote server. The system allows a doctor or carer to monitor any patient's individual health status remotely, which ensures good routine health management and enables personalized feedback to each patient.

MediNet is envisaged to be extendable across the whole Caribbean region, although the services available to patients would be centered in their own country but could be linked to other countries. The project is one of the few in LMICs that offers a true m-Health service, that is, it enables patients to be monitored remotely and empowers those patients to manage their own medical condition. An important aspect of the project is that it does not replace doctors; it simply extends the reach of patients to the healthcare system, which is then more "patient-centric" than "provider-centric."

Haiti   This is another country to have suffered a recent devastating earthquake, which struck in 2010. The need for m-Health services, especially in the most affected areas, is considered vital. Since 2010 there have been some examples of m-Health initiatives sponsored by different NGOs and other organizations working in the country. One example is the project conducted by the Operational Medicine Institute (OMI) for a qualitative trial of a modified version of the off-the-shelf EHR application "iChart"; this was applied at the Fond Parisien Disaster Rescue Camp during the 2010 earthquake (Callaway et al., 2012). This trial demonstrated the applications of simple electronic records to provide improvement over existing patient tracking and facility management. Another initiative is the *Health eVillages*, a global m-Health program launched in 2011 by Physicians Interactive and the Robert F. Kennedy Center for Justice & Human Rights; this provides various m-Health services, particularly for infant and maternal health, internal medicine, tropical medicine, HIV/AIDS, and pediatrics (Health eVillages, 2014).

Malaysia, Philippines, and Thailand   These southeast Asian countries, sometimes dubbed as the "Tiger Club Economies," face diverse challenges such as growing population, and socioeconomic changes compounded with increasing healthcare expenditures and growing economies. Some m-Health projects in these countries are listed next.

In Malaysia, the JHU-IKU M-CHILD program is funded by the National Institute of Health in collaboration with Johns Hopkins University. This project aims to strengthen mobile health capacity in assessing the risks and improving the prevention of child injuries at home through an innovative mobile health service and delivery model. This is a key national priority in Malaysia, focusing on feasible, measureable in-home interventions (Hayder and Bachani, 2014). Malaysia is also facing a growing elderly population, adding the elderly care burden on the existing stretched healthcare services. *Love on Wheels* is a mobile nurse programme that provides mobile nursing services for the elderly population that needs care and attention in capital areas such as Kuala Lumpur (Pak, 2013).

In the Philippines, the Department of Health announced a 5-year e-Health strategic framework for the period 2013–2017 and implementation plans by 2020 (Philippines Department of Health, 2013).

Several joint venture m-Health projects have also been reported in the country. The Leprosy Alert and Response Network and Surveillance (LEARNS), the country's first mobile phone-based leprosy referral system, was launched in 2014 by a consortium of the Department of Health, Philippine Council for Health Research and Development, and Novartis Healthcare Philippines (Novartis Philippines, 2014). This system enables healthcare practitioners to refer possible Philippino leprosy patients by sending a picture of the skin lesion and patient details from their mobile phones.

Another m-Health example is *Safe Motherhood*, a mobile health system led by Molave Development Foundation, a not-for-profit organization based in the Philippines. The program, aimed to reduce maternal mortality rates by using text messaging to inform and educate pregnant mothers, was funded by the International

Development Research Center (IDRC), Canada, under the Pan-Asian Collaboration for e-Health Adoption and Application (PANACeA) Network Research Initiative (Banks, 2011).

In Thailand, several m-Health pilot projects have been reported in recent years. In 2007, a study of 60 patients from Chiang Mai province led by the local public health department piloted a program in which TB patients were provided with mobile phones that allowed them to receive daily reminder calls for their medication. The project was not only effective but also inexpensive, with an estimated cost of just 100 baht ($3) per person (Compendium of mHealth Projects, 2014). Other studies cited the effectiveness of SMS methods in treatment and management of Thai patients with TB and diabetes (Iribarren et al., 2014; Navicharern et al., 2009).

China   China is the most populous country in the world, with more than 1.3 billion people. Recent statistics indicate that about 8% of the population is aged 65 years or older and the expectations are that this level will increase to 33.3% by 2050. This is paralleled with a massive growth in mobile and broadband users, with an estimated 1.17 billion mobile devices and nearly 100% of mobile phone use in 2013 (Xiaohui et al., 2013). Further, the Chinese mobile health and applications markets are growing rapidly, with revenues estimated to exceed $2.5 billion by 2017 (Xiaohui et al., 2013).

China is facing annual increases in healthcare expenditure (more than 5% of the GDP in 2011) and rising. In addition to an ageing population, there is increasing prevalence in chronic conditions, notably heart and cardiovascular disease, cancer, diabetes, COPD, and arthritis (Xiaohui et al., 2013). Further, more than 80% of specialist medical healthcare centers are concentrated in the big cities, including a high percentage of the highly qualified healthcare providers (Wang et al., 2007). These combined healthcare challenges compared to the high percentage of mobile phone users and large m-Health markets necessitate the case for developing large-scale affordable m-Health services. Although large-scale m-Health developments and initiatives are still in their early stages, there are promising examples of m-Health services and applications (Li et al., 2014). Other m-Health services include the *Wireless Heart Health* project that was launched in 2011 for the prevention and care of cardiovascular diseases in underserved communities. This program was designed to support rural communities with limited access to specialist cardiac care. It involved distributing smartphones with built-in ECG sensors, web-based EMR software linked to Internet to connect community health clinics in rural areas with specialist centers. These clinics used m-Health to perform cardiovascular screenings on more than 10,000 patients, 1700 of whom were referred to specialist clinics for further treatment (Ni et al., 2014). These limited initiatives remain disproportionate compared with the size of the population and the associated healthcare challenge, which include the following (Ni et al., 2014; IBM, 2006):

- Low utilization of mobile health applications and services.
- Regulatory and standard challenges in using mobile health applications.
- The slow adaptability and affordability of mobile health applications among the wider population, especially for the elderly and chronic patients.

- Inefficient use of healthcare resources.
- Lack of high-quality patient care and access to affordable healthcare services especially in rural areas.

There is some evidence of further m-Health initiatives from the government and private sectors, to be developed for the provision of affordable mobile health services across the country. These are expected to grow substantially during the next few years due to the sustainable low costs of mobile phones and medical sensors, combined with an increased awareness among users and clinicians, together with increased clinical evidence associated with the tangible economic benefits of m-Health.

6.3.5 Global m-Health Initiatives in Conflict, Postconflict, and Natural Disaster Areas

Health is closely related with conflict, directly or indirectly affecting the health of the population in conflict zones. According to research by the Center for Research on the Epidemiology of Disasters (CRED), there are at least 172 million people worldwide who were directly affected by global conflicts in 2012, with 87% of those affected residing in conflict zones, rather than refugees who fled from violence (Save the Children Federation, 2014). The latest statistics from the United Nations High Commissioner for Refugees (UNHCR) show that there are more than 42 million refugees and displaced people around the globe, the highest level seen since World War II (UNHCR, 2015). Some of the general priorities in conflict zones include provision of security, health, shelter, food, clean water, and prevention of violence against vulnerable populations of women and children. In many environments of ongoing instability and large-scale population displacement, emergency healthcare delivery will continue to be the focus of international agencies and NGOs (Wood and Richardson, 2013).

There is a clear lack of awareness and a necessity for adopting m-Health in these troubled areas, with disproportionate response from NGOs and other organizations supporting global m-Health initiatives. These are not commensurate with the increasing levels of global conflicts, refugee crises, and the resultant healthcare challenges. The conflict zones are usually seen as a difficult context for m-Health for various reasons, such as lack of security, availability of expert medical staff, policy, and other factors. However, m-Health can be seen as essential for providing vital emergency and healthcare services. For example, with a record number of more than 45 violent global conflicts, with millions of casualties, refugees, and displaced victims (Roome et al., 2014), there are surprisingly very few studies that have addressed the role and potential of m-Health in conflict or postconflict areas (Istepanian et al., 2014).

From the natural disaster perspective, similar healthcare challenges are increasing. For example, the devastating Nepal earthquake in 2015 highlighted the urgent need to develop resilient m-Health systems to provide effective emergency relief and health-care delivery services for postnatural disaster areas, especially in poor regions with difficult access.

Moreover, in current global conflict regions, there are numerous examples of demographic shifting, with socioeconomic changes affecting the displaced population and their health and well-being. According to the WHO, there is increasing evidence from conflict countries, such as Iraq, Syria, Yemen, and Afghanistan, of high levels of prevalence of communicable diseases (Roberts et al., 2012). Similar trends exist in conflict areas of African countries, with an increasing prevalence of child and maternal health diseases (Bustreo et al., 2005). The fragile national health systems in these areas will continue to face multiple long-term challenges in their postconflict periods. While the Middle East and North Africa MENA region, especially Gulf Co-operation Council (GCC) member states, have the highest levels of mobile phone users globally, there is dearth of any national strategies for large-scale m-Health services in these countries. For example, the prevalence of diabetes and obesity in the GCC is one of the highest globally (Alhyas et al., 2012), yet there are no coordinated and national m-Health plans for tackling and managing these diseases.

However, there are many windows of opportunity to engage in innovative m-Health initiatives that can provide cost-effective and resilient healthcare delivery solutions with impact. The challenge for m-Health planners is to translate WHO's basic building blocks of health system leadership and governance, cited as health services, health information, human resources, financing, and access to essential medicines, vaccines, and technologies (WHO, 2012b), into successful m-Health initiatives. The short-term deployment of global m-Health solutions and programs similar to the numerous examples cited earlier from the developing world can provide the initial evidence for the effectiveness of m-Health. However, the broader planning and development of their fragile healthcare systems requires more long-term strategies and sustainable mechanisms in addressing the massive healthcare challenges in these regions.

These global health strategies are subject to several policy framework interventions and recommendations for the healthcare needs in conflict and postconflict regions (Spiegel et al., 2010; Haar and Rubenstein, 2012; Witter 2012; Roome et al., 2014). Table 6.4 shows an applicability grid between the common global m-Health themes identified from Table 6.1 and the challenges in conflict regions.

The financing, implementation, and mapping of the services needed remains a complex task, and requires extensive collaborative effort. This can also formulate new "anatomy of m-Health" system, tailored for conflict and postconflict regions. This anatomy requires the following ingredients:

- Understanding the healthcare environments and realities and their resultant socioeconomic conditions in these regions.
- Mapping the most appropriate m-Health services and healthcare priorities to complement the existing healthcare structure and organizations.
- Financial planning of services and their motivational strategies for the healthcare policymakers.

The effectiveness and the contribution of m-Health technologies for the huge healthcare problems in a conflict zone can be achieved successfully if the barriers and

TABLE 6.4 Applicability Grid Between Common Global m-Health Themes and Associated Healthcare Challenges in Conflict Zones

Global m-Health Themes	Global Healthcare Challenges in Conflict and Postconflict Areas						
	Delivery of Healthcare Services	Addressing Chronic and Noncommunicable disease	Improved Healthcare Services in Urban Areas	Surveillance, Measurement and Monitoring	Health Financing and Cost Models	Leadership and Governance	Health Workforce and Distribution
Patient education and health promotion and diagnostic communication	X	X	X	X			
Health records management and improved service delivery and treatment compliance	X	X	X				X
Population surveillance and disease outbreaks and health monitoring	X			X			
Mobile management of logistics, finance, and human resources					X	X	
Medical learning and healthcare training			X			X	X

available resources are coordinated collaboratively, even in a challenging conflict and postconflict country like Iraq (Panel 2).

Case Study: Global m-Health Project in Conflict/Postconflict Region

Current status of healthcare services in Iraq

The Middle East is considered one of the main conflict hot spots in the world. Yet the growth expectation in the healthcare market in this region is estimated to reach US$100 billion in the next 15 years (Pillai, 2012). This is complemented by the highest penetration of mobile phone usage worldwide. Yet there are surprisingly few m-Health initiatives in this region, and particularly in the conflict zones.

Iraq was chosen for this study because of this country's continued conflict and postconflict cycles since the 1980s, particularly since 2003 with its internal crisis. These continued wars and conflicts have fundamentally affected the health of the country's population and transformed what was once a country with a flourishing national healthcare system between the 1950s and 1970s to a fragile and fragmented system that barely functions.

The Ministry of Health (MoH) in Iraq remains the main provider of healthcare, both for curative and preventive aspects. The current health infrastructure is based mainly on primary healthcare centers and secondary care hospitals. There are 229 general and specialized hospitals, including 61 teaching hospitals, and also 92 private hospitals. There are also 2231 primary healthcare centers (PHC), and the vast majority of health services are provided centrally, with no reliable referral system (Ali et al., 2011). In addition, there is an estimated average of 6.2 doctors per 10,000 of the population, and 50% of Iraqi doctors are based in Baghdad.

However, since the 2014–2015 internal conflicts and refugee crisis, some of these hospitals and primary health centers have either been destroyed or made inaccessible. The current PHC structure is not based on cost-effective public health interventions that would achieve maximum health gains for the money spent, as its services only partially meet the health needs of the population. When combined with a low perceived quality of care, people tend to seek care at the secondary and tertiary levels, bypassing the primary level.

One major health burden in the country is child and maternal health. The WHO estimated that in 1990, the maternal mortality rate in Iraq was around 93 deaths per 100,000 live births, while in 2008, it was estimated to be around 75 deaths per 100,000 live births, with recent estimates predicting further reduction (Jabir et al., 2013). This level might have leapt again in recent years, especially with the continued conflicts and refugee crises in the country.

Iraq also suffers from a double burden of both acute and chronic disease, including acute conditions such as diarrhea infection, acute respiratory infection, malaria, tuberculosis, and leishmaniasis, as well as an increased prevalence of chronic conditions such as cardiovascular disease, diabetes, malnutrition, and cancer, which are major causes of mortality (Al Hilfi et al., 2013).

Furthermore, Iraq's Ministry of Health, Iraq (MoH, 2006), in collaboration with the WHO, reported in their joint chronic noncommunicable diseases risk factor survey similar high prevalence of diabetes, hypertension, and cardiovascular diseases, with 40.4% of the adult population (25–65 years old) having raised blood pressure, 10.4% with hyperglycemia, and 37.5% with hypercholesterolemia (WHO, 2009). In addition, this survey estimated that 66% of the population were overweight and 33% were clinically obese, with a higher prevalence among women (38.2%) than men (26.2%). Unhealthy lifestyles are the underlying reason for this medical nightmare, with 21.9% of the population being smokers, 90.1% having low fruit and vegetable consumption, and 56.7% reporting low levels of physical activity. These results explain that serious disease burdens reflect the increased levels in morbidity and mortality attributed to NCDs (MoH Iraq and WHO, 2006).

The resulting severe drop in Iraq's gross domestic product, and consequently its public expenditure on health, has led to deterioration in the quality of healthcare services, including shortages of drugs and supplies. Moreover, the ongoing conflict and the poor security situation has physically damaged the health infrastructure, resulting in the loss of many health professionals and impaired access to basic health services.

The Potential of m-Health in Iraq

Given the above problems, healthcare in Iraq needs urgent innovative measures to improve the delivery of services. Simply identifying the potential and raising awareness of the benefits of m-Health to improve healthcare services through publication of literature reviews of benefits in developing countries is not enough. There is a need for proof-of-concept studies to explore and demonstrate the practicalities of delivering urgent healthcare and educational services to influence Iraqi decision-makers in healthcare. Further, promoting patient awareness and exploring their views about the utility of m-Health services through pilot studies is important for the future development of such services.

The potential of m-Health in Iraq has been successfully studied in a collaborative "e-Health Iraq" project between United Kingdom and Iraqi partners. The project was one of the first m-Health projects in a conflict region and was funded by the UK Development Partnerships in Higher Education (DelPHE-Iraq British Council). The 2-year project (2010–2012), led by the authors of this book, was an educational and research partnership between Kingston University, Cardiff University Medical School, and Loughborough University in the United Kingdom, and the Universities of Baghdad and Basra in Iraq. Its aim was to introduce innovative and strategic e-Health plans for medical education and research in Iraq, with the following objectives:

- To introduce innovative and strategic m-Health plans for medical education policy and research for Iraq.
- To identify and implement two proof-of-concept e-Health projects, with trials to classify the most needed healthcare services in Iraq.

- To make recommendations and propose future plans for Iraqi healthcare policy adoption, together with wider deployment and integration of e-Health in existing medical education and healthcare.

By 2012, two units of excellence, Baghdad University Al-Kindy Medical School and Basrah Medical School, were established with the general aim of building further capacity for e-Health and m-Health services, and two m-Health text messaging pilot studies were identified, planned, and carried out in the two main cities in Iraq.

The technical details of the mobile health pilots are described elsewhere (Haddad et al., 2014; Istepanian et al., 2014; Mulvaney et al., 2012a), but a brief summary is given next for completeness. The m-Health systems used were developed jointly by the UK and Iraqi teams to ensure the interchange of ideas, designs, software, hardware, and data, with the aim of demonstrating the potential of m-Health systems during the clinical pilots in Iraq.

In the first pilot, the feasibility and acceptability of SMSs to support Iraqi adults with newly diagnosed type 2 diabetes in the southern city of Basrah was studied. Fifty patients from a teaching hospital clinic in Basrah within the first year of diagnosis were selected to receive weekly SMSs relating to diabetes self-management over a period of 29 weeks. Numbers of messages received, acceptability, cost, effect on glycated hemoglobin (HbA1c), and diabetes knowledge were documented. Forty-two patients completed the study, receiving an average 22 of 28 messages. Mean knowledge score rose from 8.6 (SD 1.5) at baseline to 9.9 (SD 1.4) 6 months after receipt of SMSs ($P = 0.002$). Baseline and 6-month knowledge scores correlated ($r = 0.297$, $P = 0.049$). Mean baseline HbA1c was 79 mmol/mol (SD 14 mmol/mol) (9.3% [SD 1.3%]) and decreased to 70 mmol/mol (SD 13 mmol/mol) (8.6% [SD 1.2%]) ($P = 0.001$) 6 months after the SMS intervention. Baseline and 6-month values were correlated ($r = 0.898$, $P = 0.001$). Age, sex, and educational level showed no association with changes in HbA1c or knowledge score. Changes in knowledge score were correlated with postintervention HbA1c ($r = -0.341$, $P = 0.027$). All patients were satisfied with text messages and wished the service to be continued after the study. The cost of SMSs was approximately US¢6 per message (Haddad et al., 2014).

In the second pilot study, a mobile diabetes management system was developed based on the system shown in Fig. 5.5 and described elsewhere (Istepanian et al., 2009). The effectiveness of the system, entitled DIAR, was evaluated in a feasibility study, with 12 Iraqi patients ($n = 12$) with type 2 diabetes. The patients were randomized into two groups, intervention ($n = 6$) and control ($n = 6$) groups. The selected patients were recruited from the outpatient clinic in Al-Mawane Hospital in Basrah. The intervention period was for 3 months with a follow-up period of 6 months for each group. All patients were within their first year of diagnosis of type 2 diabetes, regardless of microvascular complications. The clinical outcomes had shown that the mean baseline HbA1c was 8.95% [0.73] and decreased to 8.05% [1.31] [$P = 0.115$] after the mobile intervention for the intervention group, compared to HbA1c reduction from 8.95 [2.17] to only 8.7 [1.7] for the control group (Istepanian et al., 2014).

These two feasibility studies represent the first evidence of successful m-Health interventions in Iraq and also in a conflict/postconflict region. The studies were particularly important in the medical education and healthcare service sectors, promoting research with a focus on clinical evidence that had been lacking. The long-term strategy is to modernize the current healthcare system by exploiting communications and technology with innovative solutions.

Opportunities and Challenges of m-Health in Iraq

The long-term strategy is to modernize the current healthcare services in Iraq by exploiting the potential of m-Health with mobile communication technology with innovative solutions. There are several current opportunities for the successful implementation and deployment of m-Health opportunities:

- The interest and support of leading medical practitioners to embrace m-Health services.
- The support and understanding of the health policymakers on the need in establishing e-Health and m-Health services in the country.
- The motivation and willingness of Iraqi patients to embrace new m-Health technologies.
- The availability of two e-Health centers of excellence in Iraq.
- The availability of medical data from which future m-Health studies can be planned.
- The trend toward public health modernization using new communications technology.

There are major obstacles to the wider deployment of these services in Iraq. Some of these share commonality with the other healthcare challenges in LMICs described earlier in this chapter. These include the following:

- The lack of awareness of the m-Health concept and relevant technologies in the medical establishment and institutions in the country.
- The incomplete availability of comprehensive e-Health and m-Health policies and guidelines.
- The lack of the required level of management to develop relevant strategies such as the organization and coordination required for a health partnership between the public and private sectors.
- The absence or lack of clear financial resources and support to provide the required quality and quantity of m-Health services, especially after 2014.
- The lack of specialists among the m-Health educational experts and relevant m-Health medical training programme.

- The lack of a proper electronic patient medical record infrastructure and referral system.
- An incomplete national vision of strategic e-government, especially for the healthcare sector, in addition to the high level of official corruption and bureaucracy in the country.

The outcome of the DelPHE-Iraq project has highlighted the opportunities and challenges for m-Health services in Iraq in particular and in developing countries in general. Finally, we summarize the key recommendations necessary for successful deployment of m-Health systems and services in Iraq. These present template recommendations and lessons to be learnt for future initiatives in other countries with similar conflict and health conditions:

- There is a need to identify national healthcare priorities and services that could benefit from the use of e-Health technologies.
- Strategic and sustainable national e-Health and m-Health plans are required for Iraq.
- The establishment of a national e-Health strategy and working group to lead and implement national e-Health and m-Health services in Iraq.
- The capacity for building on existing e-Health centers established as part of this project needs expanding along with the establishment of new e-Health centers and initiatives elsewhere in the country.
- The development of the digital health sector from private and public partnership to ensure successful and sustainable m-Health services for the long term.
- The establishment of a governmental regulatory body to supervise these services and to oversee the relevant security, privacy, and ethical issues.

The application of m-Health in conflict and refugee areas has been largely overlooked. In line of the recent global conflicts and displacement areas in different parts of the globe, the use of m-Health services in these areas is becoming important if not urgent. Many challenges remain, with some being difficult and others less complicated and achievable. The introduction of appropriate m-Health services tailored to local circumstances is an urgent and necessary global healthcare need, especially for the tens of millions living in war-torn countries and refugee camps.

m-Health in Natural Disaster Areas   There has been an alarming frequency of natural disasters such as earthquakes, tsunamis, and hurricanes, especially in the last two decades. These include, for example, Hurricane Katrina (2005), the earthquake in Haiti (2010), earthquakes and tsunamis in Japan (2011), and the more recent earthquake in Nepal (2015). To reflect on the importance of m-Health in these circumstances, several mobile inventions and applications have been reported (West and Valentini, 2013). Most of these initiatives address how different mobile disaster aids and relief systems are applied in response to these natural disasters. However, to date there are no systematic studies of the economic impact of m-Health and the outcomes of its use in disaster area and relief scenarios.

In this context, a recent report by the ITU on disaster relief systems, network resilience and recovery, provided general information on the appropriate tele-communication service options with the relevant mitigating mechanisms, from potential damage caused before, during, and after the disaster to the existing communication infrastructures (ITU-T, 2014). However, the translation of these technologies and mechanisms for practical and cost-effective m-Health services warrants further studies. The clinical evaluations and benefits of using the technologies under these conditions are also lacking, and warrant further research, considering the importance of the topic.

In conclusion, there are increasing levels of refugee crises, growing numbers of global conflicts, and a greater frequency of natural disasters worldwide. These events are generating major socioeconomic and health crises in the affected areas. The urgent need for proper healthcare planning and for better access and delivery services becomes paramount in these circumstances, and m-Health can contribute immensely by providing the necessary services. However, there is a surprising dearth of m-Health applications. This lack of interest can be attributed to several factors, such as a shortage of funding, security, the allocation of resources, and m-Health leadership issues.

This important, underreported area of global m-Health that is highlighted in this section needs to be addressed and revised by global health institutions, NGOs, academia, and other humanitarian institutions interested in m-Health. We cited an example of the effectiveness of m-Health in a major conflict area in the Middle East as a case study, and for the necessity of introducing similar m-Health initiatives in other conflict/postconflict areas worldwide.

Overall, the combination of the global m-Health studies and initiatives presented in the preceding sections claim some common beneficial healthcare outcomes that include the following:

- Increased access to healthcare and health-related information, particularly for hard-to-reach populations.
- Increased efficiency and a lower cost of service delivery.
- Improved ability to diagnose, treat, and track diseases.
- Timely, actionable public health information.
- Access to ongoing medical education and training for health workers.

Many challenges remain in quantifying the long-term impact and clinical benefits of m-Health gained from existing initiatives globally, particularly in conflict or postconflict and natural disaster areas. Furthermore, the clinical effectiveness and impact measures of the different global m-Health initiatives cited under the umbrella of the United Nation's Millennium Development Goals to be achieved by 2015 are yet to be accessed adequately.

6.4 GLOBAL m-HEALTH FOR THE DEVELOPING WORLD: BARRIERS AND RECOMMENDATIONS

The fact that there are a multitude of factors and barriers ("potholes") that can obstruct the path for wider and sustainable global m-Health programs targeting the developing world is one of the many issues that are igniting the current debate on the effectiveness of m-Health. Most of the current global m-Health models are based on the axiom of mobile phone connectivity and features. It is possible to review these models and develop new ones that can be outside the "cellular connectivity blind spots." Most of the current m-Health models advocate service delivery mechanisms using mobile phone connectivity as the predominant technology conduit for m-Health services (Levy, 2012; PwC, 2013; Philbreck, 2013; Boston Consulting Group, 2012). In this context, voluminous studies of global m-Health initiatives that we have highlighted throughout this chapter provide mixed findings about the current global m-Health status. These include a lack of clinical rigor and large-scale evidence, continued resistance to change from existing practices, lack of funding, limited scope of scaling up, and others. Moreover, a recent survey of 1700 m-Health projects showed that none of these services provided essential, actionable, off-line guidance for direct use by citizens on a larger scale or addressed the range of acute healthcare situations commonly encountered in low-resource settings, and very few provided any such content at all (Royston et al., 2015).

The constraints of the use of mobile phones in the developing world warrant alternatives with more cost-effective m-Health solutions and configurations that can, for example, be developed specifically for delivering health services in conflict and disaster areas where the circumstances of collapsing telecommunications networks is most likely. The future challenge for m-Health is how best to utilize these themes in successful, efficient, and cost-effective service delivery mechanisms and not to rely only on mobile phone connectivity as the main communication element.

There is also a need for developing appropriate m-Health strategies that can tackle the increasing global burden of cancer, with innovative and sustainable solutions tailored to the developing world and low-income countries. The focus should be on preventative solutions and innovative service delivery models (Kulendran et al., 2013).

Different recommendations for the implementation of m-Health in the developing world and poor resource settings have been reported, which include the following proposals (Mechael, 2009):

- Assessing the current state of e-Health, telemedicine, and m-Health to identify m-Health opportunities, gaps, and barriers.
- Documentation of existing e-Health and telemedicine initiatives and systems and to ensure their interoperability and proper ecosystem implementation options that support mobile technologies.
- Identification of priority diseases and health conditions.
- Identification of the role of voice and visual data or other media and channels.
- Development of short- and long-term strategic plans, implementation plan, and relevant budget.

- Development of a set of guidelines, policies, and accountability systems.
- Establishment of targets and measures of success, together with monitoring and evaluating procedures to adapt plans of findings.

These are mostly generic recommendations. The development of any specific national m-Health strategy for LMICs or in poorly resourced countries depends on the consideration of a complex set of conditions and determinants. These include, for example, economic realities, healthcare system structures, the diseases being targeted, patients' demographics, social norms, and cultural mobility. For example, the implementation of specific m-Health strategies and applications according to the WHO's guidelines for the prevention of noncommunicable diseases in primary healthcare in low-resource settings (WHO, 2012b) differs from the strategies of m-Health interventions targeting infectious diseases in the developing world (WHO, 2014).

The recent technological developments that we have presented in earlier chapters can be the impetus for a new vision to crystallize this change to better and more effective global m-Health solutions. For example, the development of cost-effective ($10 or less) smart m-Health solutions can go a long way in alleviating some of these barriers. This is feasible and not impossible, provided that the current stakeholders embrace realistic economic realities, with fewer for-profit m-Health models. These frugal solutions can represent a significant leap in the future evolution of m-Health, particularly for the developing world and its impact on global health.

Figure 6.3 shows a high-level representation of a global m-Health framework. In this framework, the business component constitutes the main stakeholders: mobile technology (typically the mobile phone industry and cellular network operators), healthcare system (patient, healthcare providers, medical equipment, supply chain,

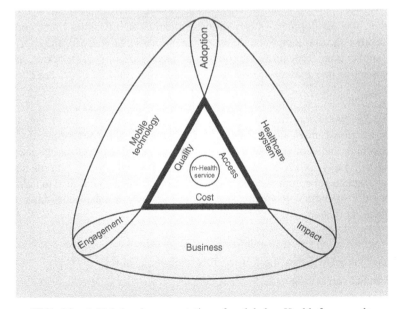

FIG. 6.3 A high-level representation of a global m-Health framework.

and service applications), and business elements (funders, donors, private investors, insurance companies, and telecoms). The attributes of this framework from the m-Health perspective include levels of adoption, engagement, and impact. The main components associated with this generic healthcare system are quality, access, and cost, which constitute the drivers of the specific m-Health service. The multilayer functioning of this framework is based on the balance of each of its components, determinants, and constraints. In order to understand the limitation of the existing global m-Health frameworks, some of the barriers and other factors associated with existing systems are described next.

6.4.1 Barriers to the Current Global m-Health Framework

In general, there are several factors that impact on the performance of a global m-Health framework and reflect the complex elements involved. These include the following:

- Development of the evaluation, planning, and strategy, which includes elements such as establishing clear strategic and clinical objectives, setting business plan and project management priorities, identifying the particular m-Health intervention, and specifying the connection between the interventions and outcomes, together with appropriate documentation of the methods and results.
- Identification of the most appropriate validation quality, access, and cost elements of m-Health incorporated in the targeted setting and specific healthcare evaluation plan.
- Establishing a valid continuum strategy and transformative process from pilots to large-scale adoptions.

Further, the development of a rigorous evaluative and planning framework also includes other factors, such as readiness of the mobile technology for the specific m-Health application, local financial conditions, localized political priorities, and the development of new markets. Some of these and other relevant issues are discussed next.

Barriers for m-Health Planning and Improving Evaluation There are several barriers to global m-Health systems; these include technology, clinical, social, organizational, and economical barriers, which are explained extensively elsewhere (PwC, 2013; Levy, 2012; Mechael et al., 2010). We will briefly present here the important aspects for completeness. Some of these barriers are identified to unlock the full potential of current m-Health systems, particularly for the developing world, which include (Cargo, 2013):

- Fragmentation of service delivery
- Lack of scale across the full reach of mobile networks
- Limited replication

- Misalignment of the value proposition between mobile and health stakeholders, with inability to create long-term sustainability of services

Others include lack of user engagement; reimbursement costs; healthcare and organizational structures; privacy and security; workflow structures and training (Chib, 2013).

Another analysis of m-Health (mobile phone model) projects in Africa categorized the following successful and unsuccessful factors and their associated attributes, together with the reasons for failure (Aranda-Jan et al., 2014):

- *Successful Factors*: High availability of mobile phones, overall positive perceptions and potential of positive outcomes at small scale, adaptation at local scale, and provision of incentives.
- *Unsuccessful Factors*: Unclear long-term outcomes and benefits, scarce large-scale studies, risk of bias reporting and weak evidence, unknown cost-effectiveness, lack of integration to healthcare systems, poor evidence of standards, and privacy and legal issues.
- *Reasons of Failure*: Lack of adequate planning and poor project design, limited funding of long-term projects, research limited to pilot projects or donor reporting, external factors such as cultural traditions and treatment durations, unclear role of government and absence of regulations, policy and local organizational capacity, and limited local technical capacity and support.

In addition, some of the other barriers to m-Health implementation in the developing world can be summarized as follows:

- Balance between the monetization of m-Health services from competing interests and misalignment of market priorities to detailed considerations of clinical efficacy and impact.
- Lack of global standards and regulatory procedures, with economic evaluation procedures for mobile phone interventions (usually cited as m-Health) and the necessary impact measurements.
- Absence of user engagement elements in the design and development process of existing global m-Health solutions.
- Scaling-up of existing pilots, with economic plans tailored for sustainable long-term benefits.
- Shortage of well-trained rural mobile health workers and m-Health specialists in primary care settings.
- Resistance to change from current healthcare practices.
- Economic factors, bureaucracy, absence of m-Health policy advocates, and m-Health leadership.
- Lack of pivotal clinical evidence and large-scale m-Health evaluation strategies.

The notion that mobile phone solutions are the main technological determinants for global m-Health services has been shown in most cases to result in limited and debatable outcomes. Moreover, most of these services are either project-led or pilots whose solutions are characterized by lack of clinical evidence, long-term usage, scalability, and other challenges that remain largely unanswered. These issues warrant revision of the current status and implementation, with better alternative solutions to address these challenges and barriers.

The Fragmented Categorization of Global m-Health Themes and Services   The fragmentation of the most common m-Health themes and taxonomies is another challenge. The lack of proper categorization impacts on the successful development and implementation plans for sustainable and globally adaptable m-Health systems. The existing fragmentation reflects on both the complexity and diversity of global m-Health initiatives and on the barriers listed above. We highlighted in Chapter 5 some of the categorization options from the global perspective, with examples that include the m-Health classification from the reproductive, maternal, newborn, and child health foci (Labrique et al., 2013). The UN Foundation report on m-Health cited another six service elements that provided a good fit for the developing world (UN Foundation–Vodafone Foundation Partnership, 2009). These are education and awareness, remote monitoring, data collection, peer-to-peer communication among healthcare workers, disease and epidemic outbreak tracking, and diagnostic and treatment support.

Later, the WHO survey on m-Health provided comprehensive examples from 114 countries worldwide and outlined examples of services and applications under the following categories (WHO, 2011; Malvey and Slovensky, 2014; Mechael and Solninsky, 2008):

- Communications between individuals and health services (health call centers, emergency toll-free services).
- Communications between health services, individual appointment reminders, treatment compliance and raising awareness, and community mobilization and health promotions.
- Consultation between healthcare professionals.
- Intersectoral communications in emergencies (managing emergencies and disasters),
- Health monitoring and surveillance (patient monitoring, surveillance, health surveys, and data collection).
- Access to health information for health professionals at the point of care (patient records, information access, and decision support systems)

Figure 6.4 shows a summary of the level of adoption of these health initiatives. These were classified into four levels: established, pilots, informal, and not given, with the majority of these in the pilots category (WHO, 2011).

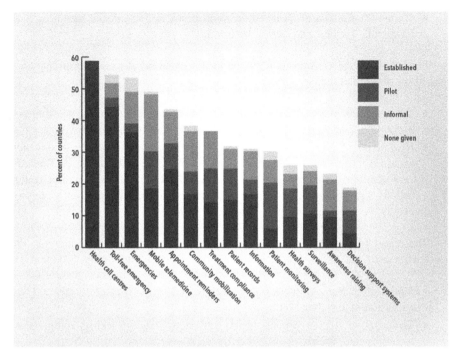

FIG. 6.4 Adoption of m-Health initiatives and phases globally. (Adapted from WHO m-Health Survey (WHO, 2011).)

The USAID m-Health program listed more than 215 m-Health projects in a compendium compiled for Africa, and highlighted the following five broad themes (Mendoza et al., 2013): behavior change communications, data collection, finance, logistics, and service delivery.

The online depository of the Global m-Health Initiative at Johns Hopkins University listed more than 137 global m-Health projects under the following categories (Johns Hopkins Global m-Health Initiative, 2014): maternal health, HIV/AIDS treatments, mental health, telemedicine, education and m-learning, tools and information systems, population surveillance, noncommunicable diseases and tobacco, and clinical care.

Another review highlighted different m-Health applications classified in three communication categories, reflecting the communications link used in each service (Källander et al., 2013):

- *One-Way Communication*, which includes m-Health applications that use SMS, sending data, push data, and call services.
- *Two-Way Communications*, which include applications that use sending data and receiving feedback, games, remote consultation and training, hotlines, and text lines.

- *Multi-Way Communications,* with applications that use social media such as Facebook, Twitter, and other platforms.

Finally, the GSMA proposed a mobile health services categorization framework based on two broad themes (GSMA, 2012):

- *Solutions across the Patient Pathway:* These include monitoring of wellness, prevention, diagnosis, and treatment that entails direct touch points with patients.
- *Healthcare System Strengthening Solutions:* These include emergency response, healthcare practitioner support, healthcare surveillance, and healthcare administration, along with applications that do not involve direct interactions with patients, but are primarily aimed at improving the efficiency of healthcare providers in delivering patient care.

The report by Boston Consulting Group categorized m-Health solutions in the following main categories and subthemes (Boston Consulting Group, 2012):

- *Information:* Public information/education; healthcare worker information and education; and public wellness.
- *Record/Data Access:* Public health and surveillance tracking; remote data recording and access.
- *Medical Services:* Diagnostic and treatment support; patient monitoring and compliance management.

In summary, this fragmented categorization and the examples summarized in Table 6.5 indicate a need for global agreement of common themes and solutions, as suggested in Table 6.1. This division also impacts the ongoing global standardization and regulatory efforts. The accurate framing and agreement of prioritized global m-Health themes and the translation of these into a unified categorization is vital for a new global m-Health framework embracing the technology, business, and clinical spheres.

Business, Sustainability, and Policy Planning Factors   The other important factors, misalignment of business and market interests, lack of proper sustainability, and policy planning, are also important elements in the functioning of the current framework.

From the sustainability aspects, the following issues and options need further considerations:

- The current global m-Health services need to use, for example, existing theoretical mechanisms of mobile phone adoption and appropriation models as applied for healthcare systems (Chib, 2013; Wirth et al., 2008).
- Development of alternative cost-effective and sustainable m-Health ecosystem (not necessarily exclusive to the cellular telecommunications model) that are best tailored for each country's healthcare demands, impact potentials, population demographics, and economies.

TABLE 6.5 Comparative Summary of Different Global m-Health Categorizations and Service Taxonomies

m-Health Themes and Services	UN–Vodafone Foundation (2009)	WHO (2011)	GSMA/PwC (2012)	Boston Consulting Group (2012)	RMNCH (Labrique et al., 2013)	USAID (Mendoza et al., 2013)
1	Education and awareness	Communications between health services and individuals (e.g., awareness)	Wellness, prevention (solutions across patient pathway)	Public information and education	Client education and behavioral change communication	Behavioral change and communications
2	Diagnostic and treatment support	Communications between individuals and health services	Diagnostics (solutions across patient pathway)	Diagnostics and treatments	Sensors and point-of-care diagnostics	
3	Remote monitoring	Health monitoring and surveillance (e.g., patient monitoring)	Monitoring (solutions across patient pathway)	Patient monitoring	Registries and vital events tracking	Service delivery
4	Data collection, disease, and epidemic tracking	Intersectoral communications in emergencies	Emergency response, healthcare surveillance (healthcare systems strengthening)	Remote data recording, public health, and surveillance	Data collection and reporting	Data collection
5					Electronic health records	
6		Access to health information for health professionals at point of care (e.g., DSS)			Electronic decision support systems	

(continued)

TABLE 6.5 (*Continued*)

m-Health Themes and Services	UN–Vodafone Foundation (2009)	WHO (2011)	GSMA/PwC (2012)	Boston Consulting Group (2012)	RMNCH (Labrique et al., 2013)	USAID (Mendoza et al., 2013)
7	Health worker communications	Consultation between healthcare professionals (e.g., mobile telemedicine)			Provider-to-provider communication	
8					Provider work planning and scheduling	
9			Healthcare practitioner support (healthcare systems strengthening)	Health worker education	Provider training and education	
10			Healthcare administration (healthcare systems strengthening)		Human resource management	
11					Supply chain management	Logistics
12					Financial transactions and incentives	Finance

From the business perspective, it is well known that most of the current global m-Health initiatives are predominantly led or funded by one or more of the following stakeholders:

- Charitable and philanthropic organizations and NGOs
- Universities and research institutions
- UN, WHO, and international development agencies
- Global telecommunications providers, mobile phone and medical devices industries
- Industry, health, IT companies, and multinational telecommunications ventures
- Local governmental health ministries and regional health organizations

The dominant role of telecommunications operators is evident in this process. With the likelihood that the same stakeholders will continue delivering the same m-Health services for the foreseeable future, this process warrants specific consideration with the mobile network operators (MNO). Typically, most of the MNOs plan their m-Health policy and operational strategies based on the monetization process of services through enhanced mobile phone use, consequently bringing the value-added service and profits to their networks by relevant business measures such as payments and incentives. The requirements and demands of the MNOs for scale, cost recovery, and risk, as opposed to the healthcare sector's requirements of clinical evidence, cost-effectiveness, and patient needs are still fuelling the gap. This lack of understanding mostly relies on attracting mobile phone users (patients, doctors, and healthcare workers) to the specific m-Health services being marketed, without detailed considerations of the clinical and health impact. This implicit marginalization of health evidence and long-term outcomes has been the main cause of the increasing gap and asymmetry between the monetary aspects and the clinical benefits of existing m-Health services. Recognizing this gap and the role of the MNOs in reducing it, the WHO listed the four common value drivers, as identified by the GSMA (WHO, 2015):

- To impact health outcomes
- To achieve competitive differentiation (i.e. to stand apart from competitors)
- To leverage the broader ICT and corporate capabilities
- To develop alternative revenue streams to traditional voice and data services

These common drivers are linked to the key "proof points" of the MNOs and their associated m-Health services. This linkage process is shown in Table 6.6 (WHO, 2015). These drivers and points of proof linkages are essential in the appropriate functioning of the m-Health framework's business sphere presented earlier.

Finally, from the policy and regulatory perspectives, there are several examples that address these issues in the m-Health context. One example identified the

TABLE 6.6 Summary of Value Drivers for MNOs and Proof Points of Demonstration

Value Driver	How the MNO Benefits	Proof Points of Demonstration
Impact of healthcare outcomes	Enhanced consumer brand Improved government/regulatory relations Improved employee satisfaction	Scale of the health problem Burden of disease Visibility of the health problem. Directness of the service's impact
Competitive differentiation	A unique value proposition created for the customer Being selected (or retained) as the mobile provider of choice Reduced "customer churn" (This refers to the number of customers who leave the operator or otherwise become inactive over a specified period of time) Improved usage of current or traditional operator services	Scope for competitive differentiation for the MNO: Competitive landscape Impact on core MNO's customer metrics: market share, subscriber base, customer churn
Effective use of existing MNO capabilities	Improved returns on existing investments	Potential to effectively use existing ICT capability
Diversification of revenue streams	New sources of revenue outside the traditional MNO business models	Potential for generating new business Reusability of existing technology or service Source of innovation/ capability

Source: Adapted from WHO (2015).

following set of themes and principles applied for m-Health (GSMA and PA Consulting Group, 2012):

- *Policy Issues*

    Patient Empowerment: The need to develop policies that promote user autonomy that will push for more m-Health adoption.

    Reimbursement: Development of schemes that reward positive healthcare outcomes.

- *Regulatory Issues*

    Medical Devices: Development of a proportionate approach to define what constitutes a medical device and how these are classified.

    Systems and Interfaces: The introduction of modularity that can encourage innovation and competition.

These suggested sets of principles and themes are largely proposed by and defined for the policymakers in the developed world. Their interpretation in the developing world remains challenging and difficult to implement. In most of the developing world and LMICs, neither public nor private insurance plans cover m-Health services, and clinicians are often not reimbursed for e-mail or phone consultations, with reimbursement policy skewed in favor of face-to-face medical treatment over digital or mobile applications (West, 2012). This emphasizes the need to develop separate sets of m-Health policy and regulatory themes better tailored for developing world environments. The privacy and security concerns are also important factors in future considerations of any regulatory framework, especially for advocates of more free mobile Internet-based m-Health services to be widely available in the developing world.

The need for developing new m-Health strategies to tackle the increasing impact of global health security is an important challenge that has been overlooked. Such strategy and relevant m-Health solutions need to address, for example, how to reduce the vulnerability of people around the world to new, acute, or rapidly spreading risks to health, particularly those threatening to cross international borders (Kickbusch et al., 2015).

6.4.2 Examples for Planning and Evaluating Global m-Health Frameworks

There are complex factors that need careful consideration in planning and evaluating m-Health frameworks in a global context. The rigorous planning for any m-Health evaluation strategy needs to be correlated with different policies and other management factors that can match the characteristics of the specific m-Health aspect being evaluated (e.g., patient education, remote monitoring, and data collection). Some of the common factors involved in any planning of successful m-Health evaluations include the following:

- Identification of specific objectives and setting selected priorities for the specific m-Health intervention to be evaluated.
- Adequacy of the available funding options and the selection of the specific m-Health intervention, with the relevant timing and outcomes of interest.
- Defining the specific relationship between the selected interventions and outcomes.
- Developing the relevant evaluation strategy and m-Health ecosystem that includes practical research, design, and analysis planning that is fit for purpose.

For establishing healthcare evaluation objectives, a valid entry might be to consider the recommendations published by the US Institute of Medicine, in its widely cited report entitled "Crossing the Quality Chasm" (Institute of Medicine, 2001). This report listed six general aims for healthcare system improvement:

- *Safe:* A system to be designed and developed that avoids errors and injuries to patients from the care that is intended to help them.

- *Effective:* Providing effective services based on scientific knowledge to all who could benefit.
- *Patient-Centered:* Providing care that is respectful of and responsive to individual patient preferences, needs, and values.
- *Timely:* Reducing the unnecessary and sometimes harmful delay for both those who receive and those who give care.
- *Efficient:* Avoiding waste, including waste of equipment, supplies, ideas, and energy.
- *Equitable:* Providing care that does not vary in quality because of personal characteristics, such as sex, ethnicity, geographic location, and socioeconomic status.

These can determine the possible healthcare objectives, but matching these to the particular m-Health evaluation framework may not be a completely straightforward process as there are other interests that need to be balanced and traded-off. In setting selected priorities, several factors need to be framed to address many of the societal, cultural, and policy issues addressed by the stakeholders, sponsors, and funders. These include, for example, the following:

- The significance of the m-Health intervention in tackling the targeted health problem, burden of disease, and impact on the patient population.
- The likelihood of the resultant outcomes and successful adoption of the m-Health solution into existing healthcare delivery practices.
- Sustainability of the intervention and long-term cost of resources and program management.

In practical terms, some of the existing global m-Health evaluations follow healthcare priorities that reflect specific programs or capacities that were set up to follow prioritized targets, without necessarily having the anticipated larger global impact and potential benefits.

Selection of the elements to be considered in planning and reporting mechanisms of the global m-Health evaluation process involves continued adaption and reconfiguration of these elements. Their practical applications need continued reassessing and revision, as seen fit to match a specific environment and intervention. From the m-Health perspective, these elements can be listed as follows:

- Defining the specific m-Health intervention and the basic healthcare problem.
- Design and development of the clinical and strategic objectives of the intervention.
- Planning of the project management with procurement and business issues for the appropriate development of the ecosystem.
- Relevant research design and analysis planning.
- Publication of the outcomes and wider dissemination of the results.

To illustrate these issues, an m-Health evaluation framework example that incorporates three basic stages is presented as follows (K4Health, 2014):

- *Concept Design and Initial Development:* This phase is focused on needs and the initial development stages of the system. The process includes, among other issues, understanding the context, consideration of stakeholders' engagement, initial design template, identification of case scenarios, software considerations, identifications of testing methodologies, design of interoperability, and security issues.
- *Solution Design and Testing:* This phase focuses on the solution and encompasses key technology decisions, content development and testing, prototyping, and usability testing with end users and target beneficiaries.
- *Implementation and Sustainability:* The main focus of the implementation phase is project management, partnership development, preparation for launch, monitoring, and evaluation of scale-up.

The process components are the nuts and bolts of the proposed implementation strategy and need to be considered alongside solutions for design and testing. For the sustainability phase these include financing and governance models, outcomes measurements, and return of investments planning. Details of this framework are given elsewhere (K4Health, 2014; Kumar et al., 2013).

Another example cited in this context is the WHO m-Health framework developed for reproductive, maternal, newborn, and child health (RMNCH) applications (Mehl et al., 2015). This framework is based on five key operational questions, with seven healthcare constraint elements, as shown in Table 6.7. This framework incorporated some of the value drivers and MNO's "proof points" shown in Table 6.6.

A further example is a proposed high-level m-Health framework based on four interacting components: patients, applications, healthcare providers, and information technologies (Varshney, 2014). However, this model illustrated a generic representation of the proposed framework, with uncertainty as to its adaptability for global health requirements and its applicability in the complex interacting mechanisms of the specific global m-Health stakeholders' requirements and constraints.

Regarding the technology component in any framework development process, it is also worth considering the rapid advances in future mobile and Internet communication technologies, as explained in Chapter 4. They include, for example, the concept of 5G, Internet-of-Things and Machine-to-Machine communications, and other technological advances. These will have a profound impact on the effectiveness of the role of mobile technology on future global m-Health services, ecosystems, and business strategies. Moreover, the emerging M2M and IoT communicating devices and their respective ecosystems will play a different, if not critical, role in future global m-Health applications and service models. However, the role of the MNO will continue in the foreseeable future, albeit with diminished impact, with perhaps more markets of noncellular or more economically friendly modes of communication appearing on the scene for these services. There is a likelihood that global

TABLE 6.7   Key Component of the WHO m-Health Framework for RMNCH and Applications

Framework Questions	Illustrative Options	Examples
WHEN: Details the time period (e.g., pregnancy) along the RMNCH continuum in which an m-Health solution is focused	Adolescence Pregnancy Birth Childhood	During infancy
WHAT: Details the specific health interventions for which m-Health serves as a catalyst improving quality and/or coverage	Malaria treatment PMTCT, breastfeeding supplementation Tobacco cessation	Postpartum care
HEALTH CONSTRAINTS: Refer to the challenges and barriers that impede optimal health promotion, diagnosis, and care, at which m-Health strategies are targeted. These include the following: *Information:* Client data health and vital events *Availability*: Right type of care available to those who would need it, appropriate type of service providers and materials *Cost:* Direct and indirect costs of treatment *Efficiency:* Timely and efficient access to the appropriate health services *Quality:* Technical ability of health services *Acceptability:* Alignment of health services with individual, social, and cultural norms *Utilization:* Impeding an individual's or group's use of a particular health service or treatment	Geographic inaccessibility Poor demand for service	Low demand of services
HOW: Describes the specific m-Health strategies, mobile technology functions and usage are employed to improve health	Client education and behavioral change Sensor and point of care diagnostics	SMS reminders about upcoming vaccination
WHERE: Details the information flow deployed in the m-Health implementation, visualizing the interactions ("touchpoints") between factors at different levels of the health system through use of the m-Health applications	Home, PHC, district facility, client, provider, laboratory, national health information system	SMS reminders sent to client's phone

Source: Adapted from Mehl et al. (2015) and WHO (2015).

telecommunications and network operators are contemplating such developments in the foreseeable future. It is thus timely for global health organizations and telecommunications bodies interested in m-Health to fully discuss and revise their current strategies and framework development plans, in anticipation and probable eventuality of the introduction of these new technological advances.

There are also barriers to knowledge mobilization (KM) in the global m-Health context. These have not been addressed or studied extensively so far. They are usually attributed to the complexity and dynamics of KM between the m-Health knowledge creators, healthcare policymakers, and patients. To date, there is no clear KM framework that can define the most appropriate m-Health research knowledge benefits to organizations. The challenges to be addressed include appropriate m-Health interventions, knowledge, or evidence; context, culture, and organization; and the process of facilitation, monitoring, and evaluation (Ferlie et al., 2015).

In summary, we can conclude that for the successful design, development, and evaluation process of new global m-Health frameworks, the following points must be considered:

- Detailed understanding of the complex dynamics, traits, barriers and constraints associated with existing frameworks, as described above.
- Embedding future advancements of the main m-Health building blocks in the new framework structure.
- Development of architectures and assessment tools required for the framework's planning and evaluation process.
- Understanding the specific cultural, political, and societal constraints and requirements in the areas where the framework is being developed and deployed.
- Considerations of the localized m-Health stakeholders from specific health services and social and business enterprise status and conditions.

Finally, translating these considerations to new and workable global m-Health frameworks applicable to the developing world, with its major healthcare challenges, requires vision, leadership, specialist human resources and efforts, combined with extensive interdisciplinary planning and a multistep collaborative approach from all the stakeholders interested in the long-term viability of m-Health.

6.5 SUMMARY

Most of the developing world and particularly LMICs face a plethora of challenges to their healthcare delivery. These countries face severe lack of specialist healthcare resources, as well as having some of the largest burdens of disease and extreme poverty, compounded by high population growth rates. Additionally, healthcare access to all reaches of society is generally low in these countries.

In this chapter, we have presented and discussed the existing status of m-Health in its current form and format from the global perspective. We have presented in detail the current landscape of numerous m-Health initiatives and projects conducted in the last decade in different continental settings. We have also discussed some of the common global m-Health themes and analyzed their current barriers and impact. Further, we have discussed the increasing incidence of global conflicts, especially in Africa and the Middle East. The countries in these war-torn regions face increasing healthcare problems, with massive burdens of disease, poverty, and a large-scale refugee crisis. These demand further action from governments and international health institutions to adapt to the crises with new healthcare services, including m-Health. We have highlighted the urgency of these scenarios and the need for effective global m-Health initiatives in areas such as conflict zones and natural disaster areas.

Most of the global m-Health initiatives and programs seem to lack the necessary clinical rigor and suitable measures for their long-term health impact and efficacy. The proliferation of m-Health "pilotitis" (small, technically driven pilots) in these countries as opposed to large-scale and sustainability plans was and will remain the challenge for the different stakeholders. We have suggested a need to develop new workable m-Health systems and related frameworks to overcome the barriers associated with current systems and models. It is more likely that in the absence of vast injections of funding and expertise into developing countries, the current m-Health models and scenarios are likely to continue in the foreseeable future, including more pilots with limited larger scale adoption due to the availability of mobile phones and the perceptions by users and industry that this is what m-Health is. The dominant role of the Mobile Network Operators (MNOs) into the m-Health arena will continue for the foreseeable future in these countries. However, from what has been achieved so far, a necessary change of direction is required for the current interpretation of m-Health. The key to successful adoption of future global m-Health strategies is that the governments in these countries should take the lead to create the essential large-scale mobile health programs. These need to be best suited for their local healthcare problems and they also need to be more effective and sustainable for larger populations of patients.

Finally, the future ethical, security, and privacy challenges of existing and future m-Health systems, particularly for the developing world and LMICs, have not been widely studied. These important themes warrant further work, particularly on the privacy and security complexities that will emerge from the anticipated use of future m-Health systems and devices and not only the mobile phone-based ones. These will tend to generate massive amounts of big health data, with dilemmas on how best to use and interpret it to tackle the problems of public health. In the absence of clear global regulations or standards in this context, particularly from the ethical and legal governance, there are risks of creating additional barriers rather than finding solutions.

It is also worth mentioning that the BRICS countries (Brazil, the Russian Federation, India, China, and South Africa) will emerge in the next few years as a potentially powerful force for advocating new m-Health strategies that can benefit healthcare systems in these and other countries globally. The powerful economic role and impact of these countries in shaping the future of global m-Health is inevitable. In particular, noncommunicable diseases represent the biggest health problem facing BRICS, with increasing prevalence in these countries. For example, the Russian Federation has one of the world's highest rates of cardiovascular disease, with China and India having two of the highest diabetes burdens. These major healthcare concerns warrant more collective and effective efforts for coordinated and actionable "BRICS m-Health" programs to alleviate the burden of these diseases.

It is imperative that the interpretation of the *m* in any future global m-Health initiatives especially in LMIC be more focused on the *medical* and *mobile* benefits and less aligned to the *monetary* benefits. Global m-Health is continuing to evolve and mature, with heightened expectations of more tangible and effective benefits, especially to the low- and middle-income countries. However, many barriers and challenges remain before achieving very low-cost, sustainable and accessible m-Health services to the populations most in need.

REFERENCES

Ahmed T, Lucas H, Khan AS, Islam R, Bhuiya A, and Iqbal M (2014) eHealth and mHealth initiatives in Bangladesh: a scoping study. *BMC Health Services Research* 14: 260.

Al Hilfi TK, Lafta R, and Burnham G (2013) Health services in Iraq. *The Lancet* 381: 939–948.

Alhyas L, McKay M, and Majeed A (2012) Prevalence of type 2 diabetes in the States of the Co-Operation Council for the Arab States of the Gulf: a systematic review. *PLoS One* 7(8):e40948.

Ali AA, Abdulsalam I, and Hasan AM (2011) *Iraq health information system review and assessment*, Ministry of Health/World Health Organization, Baghdad, Iraq. Available at http://applications.emro.who.int/dsaf/libcat/EMROPD_110.pdf (accessed October 2014).

Aranda-Jan BI, Mohutsiwa-Dibe N, and Loukanova S (2014) Systematic review on what works, what does not work and why implementation of mobile health (mHealth) projects in Africa. *BMC Public Health* 14: 188.

Banks K (2011) *Safe motherhood: mobile healthcare in the Philippines, National Geographic*. Available at voices.nationalgeographic.com/2011/05/10/safe-motherhood-mobile-healthcare-in-the-philippines/ (accessed October 2014).

Barrington J, Wereko-Brobby O, Ward P, Mwafongo W, and Kungulwe S (2010) SMS for life: a pilot project to improve anti-malarial drug supply management in rural Tanzania using standard technology. *Malaria Journal* 2010(9):298.

Bashshur R, Shannon G, Krupinski E, and Grigsby J (2011) Policy: the taxonomy of telemedicine. *Telemedicine and e-Health* 17(6):484–494.

Bassett IV, Govindasamy D, Erlwanger AS, Hyle EP, Kranzer K, Van Schaik N, Noubary F, Paltiel DA, Wood R, Walensky RP, Losina E, Bekker LG, and Freedberg KA (2014) Mobile HIV screening in Cape Town, South Africa: clinical impact, cost and cost-effectiveness. *PLoS One* 9(1):e85197.

Bates M (2015) *IEEE Pulse*. Available at pulse.embs.org/november-2015/the-present-and-future-of-low-cost-diagnostics (accessed January 2016).

Benjamin P (2013) mHealth: hope or hype. In: Donner J and Mechael P (Eds.), *m-Health in Practice: Mobile Technology for Health Promotion in the Developing World*. London: Bloomsbury, pp. 64–73.

Beratarrechea A, Lee AG, Willner JM, Jahangir E, Ciapponi A, and Rubinstein A (2014) The impact of mobile health interventions on chronic disease outcomes in developing countries: a systematic review. *Journal of Telemedicine and e-Health* 20(1):75–82.

Boston Consulting Group (2012) *The socio-economic impact of mobile health*, Report by BSC and Telenor Group. Available at http://www.telenor.com/wp-content/uploads/2012/05/BCG-Telenor-Mobile-Health-Report-May-20121.pdf (accessed October 2014).

Broughton E, Nunez D, and Moreno I (2014) Cost-effectiveness of improving healthcare to people with HIV in Nicaragua. *Nursing Research and Practice*. doi: http://dx.doi.org/10.1155/2014/232046.

Bustreo F, Genovese E, Omobono E, Axelsson H, and Bannon I (2005) *Improving child health in post-conflict countries: can the World Bank contribute?* The World Bank, Children and Youth Framework for Action Report, June 2005, pp. 1–65. Available at siteresources.worldbank.org/INTCPR/Resources/ImprovingChildHealthInPost-ConflictCountries.pdf (accessed October 2014).

Callaway DW, Peabody CR, Hoffman A, Cote E, Moulton S, Baez AA, and Nathanson L (2012) Disaster mobile health technology: lessons from Haiti. *Prehospital and Disaster Medicine* 27(2):148–152.

Cargo M (2012) *GSM m-Health initiative*. Available at www.gsma.com/mobilefordevelopment/mhealth-in-kenya-great-potential-but-many-challenges-to-address (accessed October 2014).

Cargo M (2013) *Pan-African m-Health initiative: addressing four key barriers in m-Health*, GSM m-Health Initiative. Available at http://www.gsma.com/mobilefordevelopment/pan-african-mhealth-initiative-addressing-4-key-barriers-in-mhealth (accessed October 2014).

Chib A, Law FB, Ahmad MN, and Ismail NM (2012) Moving mountains with mobiles: spatiotemporal perspectives on m-Health in Nepal. *Journal of Media and Communication Research* 28(52):100–114.

Chib A (2013) The promise and peril of m-Health in developing countries. *Mobile Media and Communication* 1(1):69–75.

Chigona W, Mudenda MN, and Metfula SA (2013) A review on m-Health research in developing countries. *Journal of Community Informatics* 9(2):1712–4441. Available at http://ci-journal.net/index.php/ciej/article/view/941/1011 (accessed October 2014).

Claudia B, Claudia A, Lange I, and Iñigo M (2012) SIGSAC software: a tool for the management of chronic disease and telecare. *Nurse Informatics* 2012: 56.

Compendium of m-Health projects, Compendium List and Summaries of 51 Global m-Health Projects. Available at www.globalproblems-globalsolutions-files.org/unf_website/assets/publications/technology/mhealth/mHealth_compendium_full.pdf (accessed October 2014).

Corker J (2010) Ligne Verte, toll-free hotline: using cell phones to increase access to family planning information in the Democratic Republic of Congo, *Cases in Public Health Communication & Marketing*, Vol. 4, pp. 23–37. Available at www.casesjournal.org/volume4 (accessed October 2014).

Curioso WH, Karras BT, Campos PE, Buendia C, Holmes KK, and Kimball AM (2005) Design and implementation of Cell-PREVEN: a real-time surveillance system for adverse events using cell phones in Peru. *Archives of AMIA Annul Symposium Proceedings*, pp. 176–180.

Dass R (2015) *m-Health has great potential in India because of growing mobile penetration in the country*, Health Cursor. Available at www.healthcursor.com/mhealth-has-great-potential-in-india-because-of-growing-mobile-penetration-in-the-country (accessed April, 2015).

Davey S and Davey A (2013) m-Health: can it improve Indian public health system? *National Journal of Community Medicine* 4(3):545–549. Available at http://njcmindia.org/uploads/4-3_545-549.pdf (accessed October 2014).

DeSouza SI, Rashmi MR, Vasanthi AP, Maria JS, and Rodrigues R (2014) Mobile phones: the next step towards healthcare delivery in rural India? *PLoS One* 9(8):e104895.

Donner J (2008) Research approaches to mobile use in the developing world: a review of literature. *The Information Society* 24(3):140–159.

Donner J and Mechael P (2013) *m-Health in Practice: Mobile Technology for Health Promotion in the Developing World*. London: Bloomsbury.

Economist Intelligence Unit (2012) The future of healthcare in Africa. *The Economist*. Available at www.economistinsights.com/sites/default/files/downloads/EIU- Janssen_HealthcareAfrica_Report_Web.pdf (accessed October 2014).

Farrell C and Bora G (2012) *Mobile application reinforces frontline health workers' knowledge, confidence, and credibility*, Frontline Health Workers' Coalition. Available at http://frontlinehealthworkers.org/mobile-application-reinforces-frontline-health-workers-knowledge-confidence-and-credibility/ (accessed October 2014).

Feder J.L (2010) Cell phone medicine brings care to patients in developing nations. *Health Affairs* 29(2):259–263.

Ferguson J (2014) *A digital health revolution in the making*, Global Agenda Council on Digital HealthWorld Economic Forum. Available at http://agenda.weforum.org/2014/01/digital-health-revolution-making (accessed May 2015).

Ferlie E, Crilly T, Jashapara A, Trenholm S, Peckham A, and Currie G (2015) Knowledge mobilization in healthcare organizations: a view from the resource-based view of the firm. *International Journal of Health Policy and Management* 4(3):127–130.

Folaranmi T (2014) *m-Health in Africa: challenges and opportunities*, Blavatnik School of Government, University of Oxford. Available at blogs.bsg.ox.ac.uk/2014/01/30/mhealth-in-africa-challenges-and-opportunities (accessed October 2014).

Foundation for Sustainable Development (FSD) (2014) *Child and maternal health issues in Nicaragua*. Available at www.fsdinternational.org/country/nicaragua/healthissues (accessed October 2014).

Francis T (2011) *Perspectives on healthcare in Latin America*, McKinsey & Company. Available at www.mckinsey.com.br/LatAm4/Data/Perspectives_on_Healthcare_in_Latin_America.pdf (accessed October 2014).

Free C, Phillips G, Watson L, Galli L, Felix L, Edwards P, Patel V, and Haines A (2013) The effectiveness of mobile health technologies to improve healthcare service delivery processes: a systematic review and meta-analysis. *PLoS One Medicine* 10(1):1–26.

Gatherer A (2015) *5G and the next billion mobile users: a view from India*. Available at http://www.comsoc.org/ctn/5g-and-next-billion-mobile-users-view-india?utm_source=Real%20Magnet&utm_medium=Email&utm_campaign=83166720 (accessed January 2016).

Goldberg R (2012) *Inequality and healthcare in Peru*, MedLife. Available at www.medlifeweb.org/blog/item/147-inequality-and-healthcare-in-peru.html (accessed October 2014).

Gomez-Marquez J (2013) Adhere.IO. In: Donner J and Mechael P (Eds.), *M-Health in Practice: Mobile Technology for Health Promotion in the Developing World*. London: Bloomsbury, pp. 162–172.

Grameen Foundation (2014) Breaking the Cycle: Grameen Foundation Annual Report, 2014–2014. Available at http://www.grameenfoundation.org/sites/grameenfoundation.org (accessed January 2015).

Grand View Research (2014) m-Health Market Analysis and Segment Forecasts to 2020. Report published in February 2014. Available at www.grandviewresearch.com/industry-analysis/mhealth-market/request (accessed October 2014).

GSMA (2012) *m-Health and the EU regulatory framework for medical devices.* Available at www.gsma.com/connectedliving/wp-content/uploads/2012/03/mHealth_Regulatory_medicaldevices_10_12.pdf (accessed October 2014).

GSMA Development Fund (2013) *Women & mobile: a global opportunity—a study on the mobile phone gender gap in low- and middle-income countries.* Available at www.gsma.com/connectedwomen/wp-content/uploads/2013/01/GSMA_Women_and_Mobile-A_Global_Opportunity.pdf (accessed October 2014).

GSMA Intelligence (2013) *Mobile economy Latin America 2013: a report from GSMA on Latin America's mobile market.* Available at usmediaconsulting.com/img/uploads/pdf/GSMA_ME_LatAm_Report_2013.pdf (accessed October 2014).

GSMA Intelligence (2015) *The mobile economy 2015, London: GSMA Intelligence.* Available at www.gsmamobileeconomy.com (accessed April 2015).

GSMA and PA Consulting Group (2012) *Policy and regulation for innovation in mobile health.* Available at www.gsma.com/mobilefordevelopment/wp-content/uploads/2012/04/policyandregulationforinnovationinmobilehealth.pdf. (accessed October 2014).

GSMA South Africa m-Health Initiative (2013) Pan-African m-Health Initiative South Africa m-Health, Feasibility Report, 1–36. Available at www.gsma.com/mobilefordevelopment/wp-content/uploads/2013/09/South-Africa-mHealth-Feasibility-Report-2013.pdf (accessed October 2014).

Haar RJ and Rubenstein LS (2012) Health in fragile and post-conflict states: a review of current understanding and challenges ahead. *Medicine, Conflict, and Survival* 4(28):289–316.

Haddad NS, Istepanian RSH, Philip N, Khazaal FAK, Hamdan AT, Pickles T, Amso N, and Gregory JW (2014) A feasibility study of mobile phone text messaging to support education and management of type 2 diabetes in Iraq, *Diabetes Technology & Therapeutics* 16(6):454–459.

Hall CS, Fottrell E, Wilkinson S, and Byass P (2014) Assessing the impact of m-health interventions in low- and middle-income countries: what has been shown to work? *Global Health Action* 7: 25606.

Hayder AA and Bachani AM (2014) *JHU-IKU mobile health for child injury prevention in Malaysia (M-CHILD),* Grant abstract, National Institute of Health (NIH). Available at grantome.com/grant/NIH/R21-TW009930-01 (accessed October 2014).

Health eVillages (2014) *Make a difference: heal the villages.* Available at http://www.healthevillages.org/wp-content/themes/foundation/images/HeV_Brochure.pdf (accessed October 2014).

Howitt P, Darzi A, Guang-Zhong Y, Hutan A, Rifat A, James B, Alex B, Anthony MJB, Josip C, Lesong C, Graham SC, Nathan F, Simon A J G, Karen K, Dominic K, Myutan K, Robert AM, Azeem M, Stephen M, Robert M, Hugh AP, Steven DR, Peter CS, Molly MS, Michael RT, Charles V, and Elizabeth W (2012) Technologies for global health. *The Lancet* 380: 507–535.

IBM (2006) *Healthcare in China toward greater access, efficiency and quality,* IBMBusiness Consulting Services. Available at http://www-935.ibm.com/services/us/imc/pdf/g510-6268-healthcare-china.pdf (accessed January 2016).

Committee on Quality Health Care in America and Institute of Medicine (2001) *Crossing the Quality Chasm: A New Health System for the 21st Century,* Washington, DC: National Academies Press.

Iribarren SJ, Beck SL, Pearce PF, Chirico C, Etchevarria M, and Rubinstein F (2014) m-Health intervention development to support patients with active tuberculosis. *Journal of Mobile Technology in Medicine* 3(2):16–27.

Istepanian RSH and Zhang YT (2012) 4G health: the long-term evolution of m-health (Guest editorial). *IEEE Transactions on Information Technology in Biomedicine* 16(1):1–5.

Istepanian RSH, Jovanov E, and Zhang YT (2004) m-Health: beyond seamless mobility for global wireless healthcare connectivity (Editorial). *IEEE Transactions on Information Technology in Biomedicine* 8(4):405–414.

Istepanian RSH, Laxminarayan S, and Pattichis C (Eds.) (2006) *M-Health: Emerging Mobile Health Systems*, London: Springer.

Istepanian RSH, Zitouni K, Harry D, Moutosammy N, Sungoor A, Tang B, and Earle K (2009) Evaluation of a mobile phone telemonitoring system for the intensification of glycaemic control in patients with complicated diabetes mellitus. *Journal of Telemedicine and Telecare* 15: 125–128.

Istepanian RSH, Mousa A, Haddad N, Sungoor A, Hammadan T, Soran H, and Al-Anzi T (2014) The potential of m-health systems for diabetes management in post-conflict regions: a case study from Iraq. *Conference Proceedings of the IEEE Engineering in Medicine and Biology Society* 2014: 3650–3653.

ITU–T (International Telecommunication Union)-T (2014) Overview of Disaster Relief Systems, Network Resilience and Recovery, ITU-T Focus Group Technical Report on Disaster Relief Systems, Network Resilience and Recovery, Version 1.0, pp. 1–20. Available at www.itu.int/dms_pub/itu-t/opb/fg/T-FG-DRNRR-2014-PDF-E.pdf (accessed May 2015).

Iwaya LH, Gomes MA, Simplício MA, Carvalho TC, Dominicini CK, Sakuragui RR, Rebelo MS, Gutierrez MA, Näslund M, and Håkansson P (2013) Mobile health in emerging countries: a survey of research initiatives in Brazil. *International Journal of Medical Informatics* 82(5):283–298.

Jabir M, Abdul-Salam I, Suheil DM, Al-Hilli W, Abul-Hassan S, Al-Zuheiri A, Al-Ba'aj R, Dekan A, Tunçalp O, and Souza JP (2013) Maternal near miss and quality of maternal healthcare in Baghdad, Iraq. *BMC Pregnancy and Childbirth* 13: 11.

Johns Hopkins Global m-Health Initiative (2014) Johns Hopkins University. Available at http://www.jhumhealth.org/ (accessed October 2014).

K4Health (2014) *The m-health planning guide: key considerations for integrating mobile technology into health programs*, Knowledge for Health. Available at www.k4health.org (accessed January 2015).

Källander K, Tibenderana JK, Akpogheneta OJ, Strachan DL, Hill Z, ten Asbroek AH, Conteh L, Kirkwood BR, and Meek SR (2013) Mobile health (m-health) approaches and lessons for increased performance and retention of community health workers in low- and middle-income countries: a review. *Journal of Medical Internet Research* 15(1):e17.

Kappal R, Mehndiratta A, Anandaraj P, and Tsanas A (2014) Current impact, future prospects and implications of mobile healthcare in India. *Central Asian Journal of Global Health* 3(1). doi: 10.5195/cajgh.2014.116.

Kickbusch I, Orbinski J, Winkler T, and Schnabel A (2015) We need a sustainable development goal 18 on global health security. *The Lancet* 385: 1069.

Kulendran M, Leff D R, Kerr, K, Tekkis P P, Athanasiou T, and Darzi A (2013) Global cancer burden and sustainable health development. *The Lancet* 381(9865):427–429.

Kumar S, Nilsen W J, Abernethy A, Atienza A, Patrick K, Pavel M, Riley WT, Shar A, Spring B, Spruijt-Metz D, Hedeker D, Honavar V, Kravitz R, Lefebvre RC, Mohr DC, Murphy SA, Quinn C, Shusterman V, and Swendeman D (2013) Mobile health technology evaluation: the mHealth evidence workshop. *American Journal of Preventive Medicine* 45(2):228–236.

Kwan A, Mechael P, and Kaonga NN (2013) State of behaviour change initiatives and how mobile phones are transforming it. In: Donner J and Mechael P (Eds.), *mHealth in Practice: Mobile Technology for Health Promotion in the Developing World*. London: Bloomsbury, pp. 15–31.

Labrique AB, Vasudevan L, Kochi E, Fabricant R, and Mehl G (2013) m-health innovations as health system strengthening tools: twelve common applications and a visual framework. *Global Health: Science and Practice* 1(2):160–171.

Lange I (2013) Tele-self-management support for type 2 diabetes care: working through public primary care centers in Santiago, Chile. In: Donner J and Mechael P (Eds.), *m-Health in Practice: Mobile Technology for Health Promotion in the Developing World*, London: Bloomsbury, pp. 74–86.

Lange I, Campos S, Urrutia M, Bustamante C, Alcayaga C, Tellez A, Pérez JC, Villarroel L, Chamorro G, and O'Connor A, Piette J (2010) Effect of a tele-care model on self-management and metabolic control among patients with type 2 diabetes in primary care centers in Santiago, Chile. *Revista Medica de Chile*, 138(6):729–737 (in Spanish; abstract in English available at www.ncbi.nlm.nih.gov/pubmed/20919483 (accessed October 2013).

Lester RT, Ritvo P, Mills EJ, Kariri A, Karanja S, and Chung MH (2010) Effects of a mobile phone message service on anti-retroviral treatment adherence in Kenya, WelTel Kenya 1: a randomised trial. *The Lancet* 376(9755):1838–1845.

Levy D (2012) *Emerging m-health: paths for growth*. Available at www.pwc.com/mhealth (accessed October 2014).

Li H, Zhang T, Chi H, Chen Y, Lid Y, and Wang J (2014) Mobile health in China: current status and future development. *Asian Journal of Psychiatry* 10: 101–104.

Lomazzi M, Borisch B, and Laaser U (2014) The millennium development goals: experiences, achievements and what's next. *Global Health Action* 2014: 7.

Lunde S (2013) *The mHealth case in India: Telco-led transformation of healthcare service delivery in India*, Wipro Council for Industry Research. Available at http://www.wipro.com/documents/the-mHealth-case-in-India.pdf (accessed January 2016).

Malvey D and Slovensky DJ (2014) *m-Health: Transforming Healthcare*. New York: Springer.

MAMA Aponjon project (2013) Formative Research Report. Available at www. mobilemamaalliance.org (accessed October 2014).

McQueen S, Konopka S, Palmer N, Morgan G, Bitrus S, and Okoko Y (2012) *m-Health Compendium*, 1st edn. Arlington, VA: African Strategies for Health Project, pp. 1–68. Available at www.cap-tb.org/sites/default/files/documents/MHealth.USAID.Compendium. pdf (accessed October 2014).

Mechael P (2009) The case for m-health in developing countries. *Innovations* 4(1):103–118.

Mechael PN and Solninsky D (2008) *Towards the Development of m-Health Strategy: A Literature Review*, WHO and Millennium Villages Project. Geneva: WHO.

Mechael P, Batavia H, Kaonga N, Searle S, Kwan A, Goldberger A, Fu L, and Ossman J (2010) *Barriers and gaps affecting m-health in low- and middle-income countries*, Policy White Paper. New York: Columbia University Center for Global Health and Economic

Development Earth Institute. Available at http://www.globalproblems-globalsolutionsfiles. org/pdfs/mHealth_Barriers_White_Paper.pdf (accessed October 2014).

Medhanyie AA, Moser A, Spigt M, Yebyo H, Little A, Dinant G, and Blanco R (2015) Mobile health data collection at primary healthcare in Ethiopia: a feasible challenge. *Journal of Clinical Epidemiology* 68(1):80–86.

Mehl G, Vasudevan L, Gonsalves L, Berg M, Seimon T, Temmerman M, and Labrique A (2015) Harnessing m-health in low-resource settings to overcome health system constraints and achieve universal access to health. In: Marsch LS, Lord S, and Dallery J (Eds.), *Behavioral Healthcare and Technology: Using Science-Based Innovations to Transform Practice*, Oxford University Press, pp. 239–264.

Mendoza G, Okoko L, Morgan G, and Konopka S (2013) *m-Health Compendium*, Vol. 2, Arlington, VA: African Strategies for Health Project, Management Sciences for Health, pp. 1–85.

Merry P (2014) *The case for nutritional health delivered over mobile in Tanzania, GSMA m-Health*. Available at www.gsma.com/mobilefordevelopment/the-case-for-nutritional-health-delivered-over-mobile-in-tanzania (accessed October 2014).

m-Health Alliance (2012a) *Baseline evaluation of the m-health ecosystem and the performance of the mHealth Alliance*, pp. 1–24. Available at www.mhealthknowledge.org/sites/default/ files/14_baseline_evaulation_report2013.pdf (accessed October 2014).

m-Health Alliance (2012b) *Leveraging mobile technologies to promote maternal and newborn health: the current landscape and opportunities for advancement in low-resource settings*, The Center for Innovation & Technology in Public Health Institute, Oakland, CA, pp. 1–36. Available at www.mhealthknowledge.org/sites/default/files/17_leveraging_mobile_ technologies_to_promote_maternal_newborn_health.pdf (accessed October 2014).

m-Health Knowledge (2014) Available at http://www.mhealthknowledge.org/ (accessed October 2014).

Ministry of Health, Iraq and World Health Organization (2006) *Chronic non-communicable disease risk factors survey in Iraq*. Available at www.who.int/chp/steps/IraqSTEPSReport2006 .pdf (accessed October 2014).

Mohammed O (2015) *Google is bringing cheap smartphones to Africa, but it has a problem: China got there first*, Quartz Africa, August 19. Available at http://qz.com/482807/google-is-bringing-cheap-smartphones-to-africa-but-it-has-a-problem-china-got-there-first/ (accessed January 2016).

Morrison J, Shrestha NR, Hayes B, and Zimmerman M (2013) Mobile phone support for rural health workers in Nepal through "celemedicine." *Journal of Nepal Medical Association* 52 (191):538–542.

MoTeCH (2014) *Mobile technology for community health in Ghana*. Available at ghsmotech. org (accessed October 2014).

m-Powering Frontline Health Workers (2015) *Opportunities to improve maternal, neonatal and child health in India through smartphones and 3G connectivity solutions*, Partnership Report, m-Powering Frontline Health Workers in collaborations with the United Nations Foundation and Qualcomm Wireless Reach, pp. 1–21. Available at: https://www .qualcomm.com/.../opportunities-to-improve-maternal-neonal (accessed January 2015).

Mulvaney DJ, Woodward B, Datta S, Harvey PD, Vyas AL, Thakkar B, Farooq O, and Istepanian RSH (2012a) Monitoring heart disease and diabetes with mobile Internet communications. *International Journal of Telemedicine and Applications*. doi: 10.1155/ 2012/195970.

Mulvaney DJ, Woodward B, Datta S, Philip N, and Istepanian RSH (2012b) Development of m-health monitoring systems in India and Iraq. *33rd Annual International Conference of the IEEE Engineering in Medicine and Biology Society*, San Diego, CA, pp. 288–291.

Navicharern R, Aungsuroch Y, and Thanasilp S (2009) Effects of multifaceted nurse-coaching intervention on diabetic complications and satisfaction of persons with type 2 diabetes. *Journal of the Medical Association of Thailand* 92(8):1102–1112.

New Cities Foundation (2013) *An urban e-Health project in Rio.* Available at www.newcities foundation.org/wp-content/uploads/PDF/Research/New-Cities-Foundation-E-HealthFull-Report .pdf (accessed October 2014).

Ni Z, Wu B, Samples CJ, and Shaw RJ (2014) Mobile technology for health care in rural China. *International Journal of Nursing Science* 1(323):e324.

Novartis Philippines (2014) *DoH, Novartis, partners launch country's first mobile phone-based leprosy referral system.* Available at www.novartis.com.ph/newsroom/2014/news_2014-02-05_001.html (accessed October 2014).

O'Donovan J and Bersin A (2015) Controlling ebola through m-Health strategies. *The Lancet. Global Health* 3(1):e22.

Oleaga M (2014) Digital healthcare in Latin America: Chileans prefer seeking online healthcare info via search engines than mobile apps, *Latin Post*. Available at www.latinpost.com/articles/14422/20140609/digital-healthcare-popular-latin-america-chileans-finding-online-information-valuable.htm (accessed October 2014).

Osborn J (2013) MOTECH. In: Donner J and Mechael P (Eds.), *m-Health in Practice: Mobile Technology for Health Promotion in the Developing World.* London: Bloomsbury, pp. 100–118.

Pak J (2013) Malaysia's mobile clinics provide home care for elderly. *BBC News*. Available at http://www.bbc.co.uk/news/business-24516288 (accessed October 2014).

Paradis G and Chiolero A (2011) The cardiovascular and chronic disease epidemic in low- and middle-income countries: a global health challenge. *Journal of the American College of Cardiology* 57(17):1775–1777.

Partners in Health (2013) *Finding patients in Mexico with home visits and m-Health technology.* Available at www.pih.org/blog/finding-patients-in-mexico-with-home-visits-and-mhealth-technology1 (accessed October 2014).

Philbreck, WC (2013) *m-Health and MNCH: State of the Evidence, Trends, Gaps, Stakeholder Needs, and Opportunities for Future Research on the Use of Mobile Technology to Improve Maternal, Newborn, and Child Health.* m-Health Alliance, pp. 1–52. Available at www. mhealthknowledge.org/sites/default/files/15_un_007_evidencegapreport_digital_aaa.pdf (accessed October 2014).

Philippines Department of Health (2013) *Philippines eHealth strategic framework and plan (2013–2017).* Available at www.doh.gov.ph/sites/default/files/Philippines_eHealthStrategic FrameworkPlan_February02_2014_Release02.pdf (accessed October 2014).

Pillai P (2012) m-Health: the future of health is mobile. *Middle East Hospital*, pp. 4–12. Available at http://middleeasthospital.com/Jan2012lo.pdf (accessed October 2014).

PwC (2013) *Socio-economic impact of m-health: an assessment report for Brazil and Mexico*, PricewaterhouseCoopers analysis report. Available at www.pwc.com/mx/es/industrias/archivo/2013-06-socio-economic-impact-of-mhealth-brazil-and-mexico.pdf (accessed October 2014).

Quraishi S and Quraishi H (2013) Freedom HIV/AIDS: mobile phone games for health communication and behavioural change. In: Donner J and Mechael P (Eds.), *m-Health in*

Practice: Mobile Technology for Health Promotion in the Developing World. London: Bloomsbury, pp. 146–161.

Ramachandran D (2013) Mobile persuasive messages for rural maternal health. In: Donner J and Mechael P (Eds.), *m-Health in Practice: Mobile Technology for Health Promotion in the Developing World.* London: Bloomsbury, pp. 87–99.

Research2Guidance (2014) *m-Health app developer economics 2014.* Available at mhealtheconomics.com/mhealth-developer-economics-report/ (accessed October 2014).

Roberts B, Patelb P, and McKeea M (2012) Non-communicable diseases and post-conflict countries: Editorial. *Bulletin of the World Health Organisation* 90: 2–2A.

Roome E, Raven J, and Martineau T (2014) Human resource management in post-conflict health systems: review of research and knowledge gaps. *Conflict and Health* 8: 1–18.

Royal Tropical Institute (2011) *m-Health for maternal and newborn health in resource-poor and health system settings, Sierra Leone*, Technical Brief: Feasibility Study, Final Report. Available at r4d.dfid.gov.uk/PDF/Outputs/Misc_MaternalHealth/mHealth-Sierra-Leone-Phase-1-Final-research-report-for-DFID-08Sep11.pdf (accessed October 2014).

Royston G, Hagar C, Long LA, McMahon D, Pakenham-Walsh N, and Wadhwani N (2015) Mobile health-care information for all: a global challenge. *The Lancet. Global Health* 3: e356–e357.

Save the Children Federation (2014) *State of the world's mothers 2014: saving mothers and children in humanitarian crises.* Available at www.savethechildren.org/ISBN1-888393-28-9 (accessed October 2014).

Spiegel PB, Checchi F, Colombo S, and Paik E (2010) Health-care needs of people affected by conflict: future trends and changing frameworks, *The Lancet* 341: 45.

Sultan SM and Mohan P (2009) MediNet: a mobile healthcare management system for the Caribbean region. *6th Annual International Conference on Mobile and Ubiquitous Systems: Networking & Services*, MobiQuitous, Toronto, Canada, July 13–16, 2016, pp. 1–2.

Sultan S and Mohan P (2013) Experiences from the MediNet project: the programmer's perspective. In: Donner J and Mechael P (Eds.), *m-Health in Practice: Mobile Technology for Health Promotion in the Developing World.* London: Bloomsbury, pp. 119–134.

Takenga C, Berndt RD, Musongya O, Kitero J, Katoke R, Molo K, Kazingufu B, Meni M, Vikandy M, and Takenga H (2014) An ICT-based diabetes management system tested for healthcare delivery in the African context. *International Journal of Telemedicine and Applications* 2014, 1–10.

Text Me! Flash Me! (2014) Centre for Health Market Innovations. Available at healthmarketinnovations.org/program/text-me-flash-me-helpline (accessed October 2014).

Uganda Bureau of Statistics (2013) Statistical Abstract-2013. Available at http://www.ubos. org/onlinefiles/uploads/ubos/pdf%20documents/abstracts/Statistical%20Abstract% 202013.pdf (accessed October 2014).

UN Foundation–Vodafone Foundation Partnership (2009) *m-Health for development: the opportunity of mobile technology for healthcare in the developing world*, Washington, DC.

UNHCR (United Nations High Commissioner for Refugees) (2015) *UNHCR global appeal update: population of concern.* Available at www.unhcr.org/pages/49c3646c11.html (accessed June 2015).

United Nations (2014) The Millennium Development Goals Report. Available at www.un.org/ millenniumgoals/2014.pdf (accessed February 2015).

Van Beijm H and Hoefman B (2013) Text to change: pioneers in using mobile phones as persuasive technology on health in Africa. In: Donner J and Mechael P (Eds.), *m-Health in*

Practice: Mobile Technology for Health Promotion in the Developing World. London: Bloomsbury, pp. 135–145.

Varshney U (2014) Mobile health: four emerging themes of research. *Decision Support Systems* 66: 20–35.

Vital Wave Consulting (2011) *m-Health in Ethiopia: strategies for a new framework.* Available at www.vitalwaveconsulting.com/pdf/2011.pdf (accessed October 2014).

Walsh B and Sifferlin A (2016) Zika's toll: a virus with links to birth defects sends fear through the Americas. *Time Magazine*, February 15, 32–27.

Wang H, Xu J, and Xu T (2007) Factors contributing to high costs and inequality in China's health care system. *Journal of the American Medical Association* 298(16):1928–1930.

WelTel (2014) *An m-Health patient engagement service for healthcare providers.* Available at www.weltel.org (accessed October 2014).

West DM (2012) How mobile devices are transforming healthcare. *Issues in Technology Innovation*, Vol. 18, Center for Technology Innovation at Brookings. Available at www.brookings.edu/research/papers/2012/05/22-mobile-health-west (accessed October 2014).

West DM and Valentini E (2013) *How mobile devices are transforming disaster relief and public safety issues in technology innovations*, Centre for Technology Innovations at Brookings. Available at www.insidepolitics.org/brookingsreports/Disaster%20Relief.pdf (accessed October 2014).

World Health Organization (2009) *A basic health services package for Iraq*, A Joint Ministry of Health-Iraq and WHO Report—2009. Available at http://applications.emro.who.int/dsaf/libcat/EMROPD_2009_109.pdf (accessed October 2014).

World Health Organization (2010) *Monitoring the Building Blocks of Health Systems: A Handbook of Indicators and Their Measurement Strategies.* Geneva: WHO Press, pp. 1–93.

World Health Organization (2011) *m-Health: new horizon for health through mobile technologies*, Global Observatory for e-Health Services, WHO, Geneva, Switzerland. Available at http://www.who.int/goe/publications/goe_mhealth_web.pdf (accessed October 2014).

World Health Organization (2012a) *Information and communication technologies for women's and child health: a planning workbook*, The Partnership for Maternal, Newborn & Child Health. Available at www.who.int/pmnch/knowledge/publications/ict_mhealth.pdf (accessed October 2014).

World Health Organization (2012b) *Prevention and control of non-communicable diseases: guidelines for primary health care in low resource settings.* Available at http://apps.who.int/iris/bitstream/10665/76173/1/9789241548397_eng.pdf (accessed October 2014).

World Health Organization (2012c) *mHealth for NCDs*, WHO-ITU Joint Work Plan. Available at www.who.int/nmh/events/2012/mhealth_background.pdf (accessed October 2014).

World Health Organization (2014) *Interim infection prevention and control guidance for care of patients with suspected or confirmed filovirus haemorrhagic fever in health-care settings, with focus on Ebola*, WHO, December. Available at http://apps.who.int/iris/bitstream/10665/130596/1/WHO_HIS_SDS_2014.4_eng.pdf?ua=1&ua=1&ua=1 (accessed May 2015).

World Health Organization (2015) Report of the Ebola Interim Assessment Panel. Available at http://www.who.int/csr/resources/publications/ebola/report-by-panel.pdf?ua=1 (accessed July 2015).

Wicklund E (2015) *Lessons learned in the fight against ebola.* Available at mhealthnews.com, http://www.mhealthnews.com/news/lessons-learned-fight-against-ebola (accessed April 2015).

Wirth W, Von Pape T, and Karnowski V (2008) An integrative model of mobile phone appropriation. *Journal of Computer-Mediated Communication* 13: 593–617.

Witter, S (2012) Health financing in fragile and post-conflict states: what do we know and what are the gaps? *Social Science and Medicine* 75(12):2370–2377.

Wood R and Richardson ET (2013) Prioritizing healthcare delivery in a conflict zone: comment on TB/HIV co-infection care in conflict-affected settings—a mapping of health facilities in the Goma area, Democratic Republic of Congo. *International Journal of Health Policy and Management* 1: 231–232.

World Bank (2014a) *The Economic Impact of the 2014 Ebola Epidemic: Short- and Medium-Term Estimates for West Africa*, October, pp. 1–71. Available at www.forbes.com/sites/peteguest/2014/10/23/counting-the-economic-cost-of-ebola/ (accessed December 2014).

World Bank (2014b) *Poverty overview*. Available at www.worldbank.org/en/topic/poverty/overview (accessed October 2014).

World Health Organization and United Nation Foundation (2015) *A practical guide for engaging with mobile network operators in m-Health*, WHO. Available at http://apps.who.int/iris/bitstream/10665/170275/1/9789241508766_eng.pdf?ua=1 (accessed April 2015).

Xiaohui Y, Han H, Jiadong D, Liurong W, Cheng L, Xueli Z, Haihua L, Ying H, Ke S, Na L, West D, Bleiberg J (2013) *mHealth in China and the United States: how mobile technology is transforming healthcare in the world's two largest economies*, Centre of Technology Innovations at Brookings. Available at http://www.brookings.edu/~/media/research/files/reports/2014/03/12-mhealth-china-united-states-health-care/mhealth_finalx.pdf (accessed January 2016).

Zurovac D, Sudoi RK, Akhwale WS, Ndiritu MH, Hamer DHK, Rowe AK, and Snow RW (2011) The effect of mobile phone text message reminders on Kenyan health workers' adherence to malaria treatment guidelines: a cluster-randomised trial. *The Lancet* 378: 795–803.

7

m-HEALTH ECOSYSTEMS, INTEROPERABILITY STANDARDS, AND MARKETS

I think that wireless has the opportunity to solve a whole bunch of problems, including, I believe, world poverty. So the two areas that I talk a lot about and really believe are important are the wireless impact on medical technology and on social networking. Both of those two things are going to be revolutionary.

Marty Cooper, who led the team at Motorola that invented the mobile phone

7.1 INTRODUCTION

A decade ago at the inception of the m-Health concept, early systems were developed for basic applications, such as mobile remote monitoring, mobile e-Health access to electronic personal records (EPR) and electronic health records (EHR), and mobile real-time imaging diagnostics. These and other applications were all designed and developed on an *ad hoc* basis with no tailored interoperability or recognized standards. Neither were any detailed proposals for mobile health ecosystems or policy frameworks presented during these early stages. A few years later, particularly after the concept drew the attention of major mobile phone and telecommunications industries, consistent efforts for developing different ecosystems, business models, and interoperability standards began in earnest. Today, with an estimated global m-Health market expected to reach tens of billions of U.S. dollars in the next few years, it is imperative to highlight current business and market issues impacting this growth and the potential future scenarios in this context.

m-Health: Fundamentals and Applications, First Edition. Robert S. H. Istepanian and Bryan Woodward.
© 2017 by The Institute of Electrical and Electronics Engineers, Inc. Published 2017 by John Wiley & Sons, Inc.

Stakeholders with a vested interest in m-Health include healthcare providers, health IT vendors, mobile telecommunications companies, medical device industries, pharmaceuticals conglomerates, various governmental and global regulatory bodies, universities, and research institutions. These are driving the current business and market aspects of m-Health and the dynamics behind them. Accordingly, to achieve the anticipated potential of any m-Health product or service, specific "ecosystems" and "m-architectures" are designed to fulfill these objectives. The majority of those proposed have evolved around the "mobile phone-centric" model and have been attributed to the best models that can use different mobile devices to support the practice of medicine and public health.

So far the full potential and promise of these models remain debatable, with yet-to-be-seen successful and effective scaling-up processes of this format of m-Health in many important clinical and healthcare delivery areas. The differentiation between "market- or consumer-driven" and "health outcome-driven" m-Health systems is being discussed extensively. The apparent form of such divergence, and sometimes collision, between the two approaches is clear in the current debate on the approach of m-Health ecosystems and the opportunities and challenges they pose. This view is also compounded by a multitude of reimbursement models, Return of Investment (ROI) schemes, business frameworks, and market analysis studies that are being continuously revised and updated. These can be attributed to the complex commercial factors and the impact of global market volatility and demands on the m-Health businesses that consequently affect market performance.

Most of the popular m-Health market models rely on an understanding of m-Health as a disruptive mechanism for changing the *healthcare business* to *health business*. This view is usually conceptualized by the need to shift healthcare approaches from reactive to proactive care, embracing different commercially available m-Health solutions, such as health tracking and health monitoring, with early predictions and wellness m-Health programs and products. This can be a narrow and risk-averse commercial view that only forms part of the whole picture, as it only focuses on the view from the developed world's consumer m-Health markets that has so far yielded debatable and lukewarm support from clinicians. It also collides with the well-established aims and objectives of m-Health in the developing world with its nonestablished markets. These conditions differ in their objectives and global priorities. The details of some of these issues were discussed in Chapter 6.

Further, many commercial and market proposals typically support healthcare systems in the developed world, which may in part enhance market shares as well as diversify and improve existing services. Most of these proposals are either incompatible or unworkable from the developing world's economic market perspective. The increasing realization that there is no global "one-size-fits-all" m-Health model or reference architecture necessitates a realignment of these assumptions to reflect the current and future status of m-Health. This process is necessary for two reasons. One is the current paradoxical status of m-Health from high investment levels and market hype. The other is the limited clinical acceptability and caution, compounded with slow scaling-up, both in established markets and also in other global healthcare economies. Although the entrepreneurial power and enthusiasm of m-Health is increasing globally, a number of important m-Health policy issues and investment strategies remain unclear, unanswered, and at best debatable.

One of the other issues that is important and being discussed extensively for the m-Health scaling-up process is standards and interoperability. In general, interoperability broadly refers to the ability of "entities" within a system to interact, while standards define those entities and their interactions. These terms are applicable to e-Health, health informatics, and m-Health.

In this chapter, we discuss some of these important issues and consider the challenges that are considered vital for the future of m-Health. It is true to say that we are in a veritable mobile health revolution (Cortez, 2014). The identification of m-Health stakeholders is followed by an overview of current m-Health ecosystems and different interoperability and standardization considerations. Business markets and business models are also presented, followed by some suggestions and recommendations for a new era of m-Health.

7.2   m-HEALTH STAKEHOLDERS AND ECOSYSTEMS

In this section we discuss the views that exist on m-Health ecosystems and their benefits. Since any m-Health business only exists with its stakeholders, it is vital first to identify the key stakeholders who constitute the basic players in these ecosystems, and their prospective roles in the management of their associated business models and organizations.

7.2.1   m-Health Stakeholders

It is widely understood from the business perspective that m-Health is not a "vertical domain" by itself, but a horizontal framework that cuts across and impacts all of its applicable health and social care areas. This view is evident in the different classifications of m-Health stakeholders so far. These classifications depend on how the proponents understand mobile health. Most of the main stakeholders are identified from the commercial viewpoint and are mainly attributed to the m-Health model that is based on mobile phones as the central technology element in providing an improved healthcare service delivery, cost-effectiveness, and enhanced patient expectations and experience. We highlight some of these classifications for clarification of the discussion.

The different m-Health stakeholders can be identified as (Malvey and Slovensky, 2014) "consumers" (patients); "providers" (health systems); medical staff (physicians and nurses); payers (insurers and employers); private sector investors; technology companies (e.g., application developers and startups); telecommunications and network service provider industries; big pharmaceutical and biotechnology companies; and the military.

This classification argues that m-Health stakeholders can be identified at the organizational level and at the industry level. They include those who can be influential in bringing a product to market, such as investors and regulators, and those who can promote technical adoptions such as physicians and advocacy groups (Malvey and Slovensky, 2014).

The GSMA has classified stakeholders associated with their m-Health ecosystem and architecture reference model as revolving around (GSMA, 2011) patients, clinicians, healthcare providers, mobile health service providers, mobile platforms, and mobile health devices or sensors.

This classification largely reflects the commercial influence of the mobile operators, telecommunications, medical devices, and other industries. This commercial model is based on developing mobile health services and products with the aim of maximizing the mobile network capabilities and providing multiple m-Health services. These include the areas we discussed earlier, such as mobile access of EPRs and EHRs, wellness monitoring, disease management, and other applications. Moreover, this also reflects the differentiation between the "user-centric" and "provider-centric" m-Health systems. This assumption is based largely on the mobile phone acting as the central service-enabling platform to provide services and functionalities (e.g., monitoring, education, and compliance).

Similar stakeholders were also identified by a McKinsey & Company study, for their remote health monitoring (RHM) m-Health ecosystem as patients, clinicians, device manufacturers, mobile network operators (MNOs), software platform providers, healthcare providers, payers, and regulators (McKinsey & Company, 2010). This classification is compatible with the GSMA categorization shown earlier, with the specific focus on m-Health monitoring.

A study by PricewaterhouseCoopers (PwC) identified stakeholders associated with their patient-centric value chain argument as physicians, hospitals, health insurers, pharmaceuticals, medical device companies, and government (PwC, 2014).

These studies illustrate that the most common m-Health stakeholders may be categorized as follows:

- User and consumers (patients, citizens, families)
- Medical and healthcare personnel (specialist clinicians, physicians and general practitioners, nurses, social and healthcare workers)
- Healthcare providers (hospitals, local clinics, social care providers—private and public)
- Business and finance enterprises (banks, health insurers and payers, NGOs, donors, governments)
- Technology and innovations (mobile phones and telecommunication industries, mobile network providers, medical devices manufacturers and vendors, pharmaceuticals, research and development (R&D) institutions, small-to-medium-sized enterprises (SMEs), health IT and Internet service and platform providers, universities, and noncivilian research centers)
- Policy and regulatory (legislative and legal institutions, standardization bodies, regulatory bodies, healthcare policy makers, and government health departments)

This categorization is illustrated in Fig. 7.1.

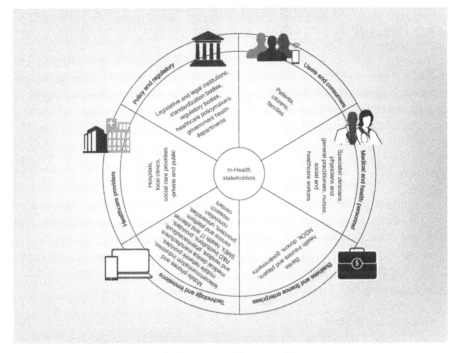

FIG. 7.1   m-Health stakeholders.

Each of these stakeholder groups and their constituent elements shown in Fig. 7.1 provide supporting roles in various ecosystem models that are developed and tailored for delivering m-Health services. The management of these stakeholders depends on the best approach being adopted to translate their associated business model into sustainable and successful m-Health services or products. This process can be tailored for better accessibility and integration for their specific healthcare target or organizational requirement. For example, a variety of large- and small-scale m-Health businesses support different m-Health applications (Apps) and cloud connectivity mechanisms for preventative disease management and wellness applications. Others use models for epidemic disease outbreaks and infectious diseases. All these solutions aim to provide accelerated information access and web services based on smart management processes between the different stakeholders involved in each ecosystem developed around specific technology and m-Health solutions. The important gap to bridge in the current understanding of m-Health by stakeholders is the need to identify the global business requirements, not only from the dominant view of the consumer-based perspective.

7.2.2   m-Health Ecosystems and Architectures

Surprisingly, there is no clear consensus as to what an m-Health ecosystem is all about. This can be attributed to several factors, including disparities between the market views of m-Health and the clinical notion of the concept, the nonuniformity of stakeholders' priorities in different global settings, and the knowledge gap of understanding m-Health as a scientific concept.

Typically, different m-Health ecosystems are being continuously proposed and developed, with varying levels of success, by businesses and institutions to provide multisector partnerships between stakeholders to accelerate the adoption of services and applications. In addition, it also facilitates the route to successful product commercialization and services from pilot stages to real-world healthcare environments; it also allows the adoption of any resulting innovations. The development of an m-Health ecosystem also allows a route to reduce barriers and accelerate the pace of innovation. These include the following:

- Facilitating collaborations and partnerships in the development of innovative m-Health products and solutions that can effectively meet patient expectations and clinical needs.
- Successful adoption of these solutions by reducing clinical barriers and facilitating the involvement of patients and clinicians in the ecosystem model.
- Enhancing ROI levels with the appropriate shared risk models.

One of the important issues that influence this process is an understanding of the key differences (and similarities) between healthcare challenges in the developed and the developing world. These include, for example, rising costs, access and choice options, healthcare expectations, barriers to reform, and resources. The individual roles of mobile telecommunication providers, SMEs, the medical device industry, healthcare service providers, and IT service vendors all differ in delivering well-being and preventative monitoring services. These range from the roles of the stakeholders in delivering more focused and specialist healthcare services, such as high-risk patient management or postsurgery monitoring. For example, the m-Health reimbursement service models advocated widely by the telecommunications industry, as linked to the source of value principles, can only be applied in certain m-Health applications and in certain settings.

The resulting imbalances, if they are not examined carefully, can generate detrimental variations in m-Health business plan projections, resulting in significant changes to their operational risk models. These can impact on the scaling-up and sustainability of services. In recent years, many m-Health ecosystem models have been introduced by interested organizations and mobile telecommunications and other businesses. Some are discussed extensively elsewhere (McKinsey & Company, 2010; GSMA, 2011; Krohn and Metcalf, 2012). We next present illustrative examples of some of these m-Health ecosystems for completeness, citing their advantages and limitations.

The World Bank (Dalberg) Ecosystem   This high level m-Health ecosystem is shown in Fig. 7.2 and was initially presented in the World Bank report on mobile applications for the health sector (World Bank, 2011). This model is based on

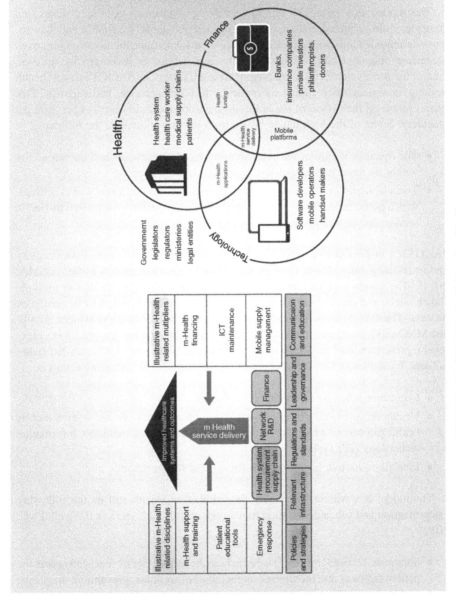

FIG. 7.2 The World Bank (Dalberg) m-Health ecosystem.

overlapping dynamic spheres of influence: healthcare, technology, and finance, where encompassing all of them is the influence of government, whose power to set regulations, policies, and strategies can affect the development and use of m-Health interventions.

The model also proposed different enabling platforms, such as policies and strategies, related infrastructures, regulations and standards, leadership and governance, communications, and education. These inputs determine the levels of improved healthcare outcomes, such as cost, quality, and access, as leveraged by different "multiplier access elements" (multipliers); these are factors such as ICT literacy, health training, complementary capital investments, ICT maintenance, and capacity. This model proposed the notion of variability in the implementation process for different resources. Some of the highlighted issues from this model include the following:

- The dynamic nature of the ecosystem development process and the variability within its input and output layers and relevant spheres.
- The adaptability of the model to healthcare systems in different global settings, with a potential matching to developing world healthcare systems and m-Health service delivery requirements.

The GSMA m-Health Ecosystem   This ecosystem is based on mobile connectivity and interfacing links between stakeholders of the models discussed earlier (GSMA, 2011). The essence of this model is in capturing the key roles of the multiple stakeholders, and playing out their individual roles and involvement in this complex process. The stakeholders include patients, physicians, healthcare providers, suppliers, MNO services, device manufacturers, payers, and regulators. Figure 7.3 is a high-level representation of this ecosystem, with interfaces between the associated stakeholders. Examples of this reference architecture include the following scenarios:

- Consumer purchase of mobile health services.
- Healthcare provider prescribes either a mobile health service with a mobile health gateway or a mobile health service connected to a healthcare information technology (IT) system.
- Prescribed mobile health service for disease management.

The model is a widely based MNO's vision of m-Health and on the following categorization and subcategorization framework of m-Health services (GSMA–PwC, 2012):

- *Solutions Across Pathway:* These include the subcategories that span across the patient pathway and monitoring applications of wellness, prevention, diagnosis, and treatment.
- *Healthcare System Strengthening:* This includes the subcategories of services of emergency response, healthcare practitioner support, healthcare surveillance, and administration.

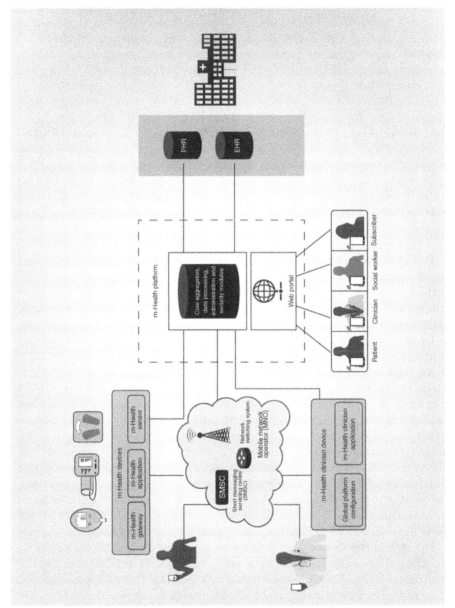

FIG. 7.3 Architecture of the GSMA m-Health ecosystem (Adapted from GSMA (2011).).

331

Although this model represents a feasible approach and view albeit from the MNO's perspective for some of the m-Health services and categories outlined earlier, the successful application and scaling-up of this model remains unclear. Developments of this model must include the rapid technological changes that ultimately necessitate revisions of the architecture. The applicability of this ecosystem in addressing the global m-Health requirements and constraints discussed in Chapter 6 are debatable and remain open for further work. Moreover, the validated authenticity of the model remains largely within the "m-Health pilotitis" sphere, with many open challenges remaining unaddressed, notably large-scale supporting clinical evidence studies.

Open m-Health and Qualcomm Life "2net" Ecosystem   This more recent architecture of the "open m-Health" platform is shown in Fig. 7.4; it illustrates the data flow and how it is accessed from third party Apps and devices (Open m-Health, 2016). This system catalyzes the establishment of a decentralized, innovative community with the aim of developing sharable m-Health tools with an open application programming interface (API), which constitutes a set of routines and protocols for building smart health applications. This allows independently developed software components to be mixed and matched, swapped and shared, similar to Lego blocks (Chen et al., 2012).

This architecture allows developers to standardize and store data, integrate or bring in data from another provider, and share or process the data to uncover more insight and better understanding. This open platform architecture represents a new ecosystem that provides interactive collaborative mechanisms between m-Health stakeholders, particularly for App developers, healthcare service providers, and other relevant businesses, to benefit and share their data and protocols to accelerate and optimize their business targets and deployment models.

Although this model provides many beneficial and collaborative structures for accelerating business adoption, it is prone to serious security and privacy risks. This is particularly true with recent developments in modern cybersecurity threats and hacking. Instead, the model can be largely used for m-Health research and business-related protocols for sharing the appropriate access and collaborative benefits, including a better understanding of the role of "big health data" generated from different mobile devices and Apps. Moreover, open platforms for health generally pose some medical ethics dilemmas, such as the rights of medical data sharing, processing, and ownership. Regardless of these issues, open m-Health architecture can be useful for accelerating the development of innovative solutions in the developing world, where these security restrictions and ethical challenges are less onerous.

The other m-Health ecosystem is the model developed by Qualcomm Life (2016). It consists of two main platforms representing the gateways, the 2net Gateway and Qualcomm Life Middleware, connecting the data sources with the Apps and services. These platforms allow the cooperation of participating businesses across the healthcare spectrum, as follows:

- Third party developers and companies building medical devices that have embedded wireless cellular technology, or medical devices wirelessly enabled

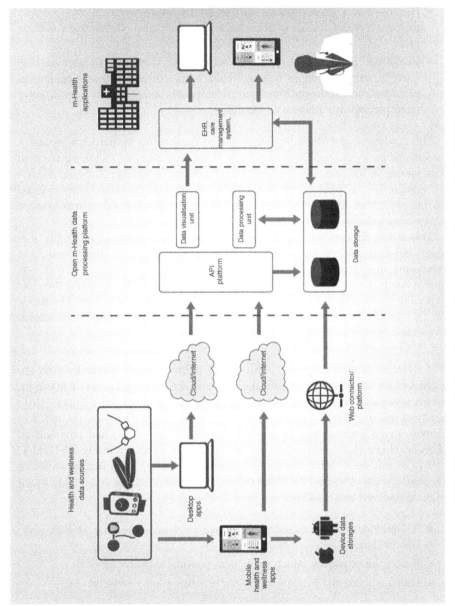

FIG. 7.4 Open m-Health architecture. (Adapted from Open m-Health (2016).).

through a connection with an external device, that is, the 2net Hub or 2net Mobile on a mobile phone or tablet, and application development companies that want to leverage smartphones to distribute data.

- Healthcare services companies, such as for disease management, which supply multiple medical devices to help patients and their caregivers to better manage care.

- Healthcare providers, payers, and accountable care organizations seeking scalable infrastructure to manage interactions across a patient's continuum of care to improve the quality of care, reduce costs, and enable providers to serve more patients at a fraction of the cost.

This ecosystem provides a flexible approach that benefits the participating stakeholders and business organizations using it. It has been reported that this ecosystem hosts more than 500 customers and technology partners, including collaborations with the "open m-Health" platform presented earlier, through the 2net Hub platform, to aid the integration of monitoring devices to improve type 1 diabetes management (Qualcomm Life, 2016).

This model remains within the corporate school of thinking on m-Health. It is largely developed for healthcare systems that follow consumer m-Health markets in monitoring, management, and wellness tracking applications. Furthermore, it reflects current trends of m-Health markets, particularly in the United States, where there are increasing levels of healthcare costs and insurance premiums, compounded by trends for increased consumerization of m-Health, with massive support of capital investment.

ITU e-Health M2M Ecosystem   This model follows the likely trends of an all-Internet connectivity that will form the core and access mechanisms of future m-Health applications. With recent developments of the Internet of Things (IoT) paradigm and machine-to-machine (M2M) communications systems presented in Chapter 4, the International Telecommunications Union (ITU) had proposed an M2M-enabled e-Health ecosystem for personal health, as shown in Fig 7.5 (ITU-T, 2014). In this ecosystem, there is an applicable variety of different m-Health monitoring areas of interest by global mobile telecommunications and MNOs, based on three principal stakeholders:

- ICT providers, which facilitate the storage, retrieval, processing, and transmission of data.
- Healthcare providers, which implement personal e-Health services.
- Users, who are the patients and caregivers using these services.

The sharp increase of different M2M m-Health solutions is commensurate with massive increases in investment and projected healthcare markets. These applications constitute potentially the most likely successful market sector in the IoT after the domestic home connectivity market (Frost and Sullivan, 2013). This emerging market

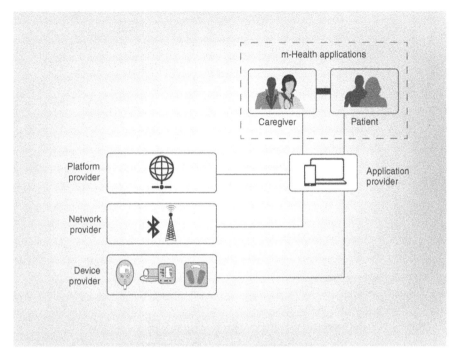

FIG. 7.5 ITU-T M2M-enabled e-Health (personal health) ecosystem. (Adapted from ITU-T (2014).)

will constitute the ultimate realization of "connected health" solutions in the near future, with a potential active role in the healthcare transformation process (Deloitte, 2015a).

This ITU-T (M2M) model also reflects the need to consider different "operational layers," such as devices, systems, and the data layers that are usually involved in the operational modalities of any proposed architecture. Although this is a draft model, it highlights the importance of the technology evolution in any m-Health ecosystem development. Further, future M2M-based m-Health (m-IoT) systems with their direct, low-cost mobility are associated with global Internet access. These will facilitate alternative services, with the option of lowering the associated care costs by enabling affordable realizations of efficient care solutions and improvement instead of any unnecessary reductions in healthcare quality to control cost. Overall, these and other examples represent snapshots of the various m-Health ecosystems and architectures that are continually being developed.

7.2.3 The m-Health Ecosystem Coordination Pyramid

It is evident from these examples that the development of successful m-Health ecosystems requires the combined aspects of healthcare quality, cost, and access to care; evolution of m-Health technologies; and the market and business drivers, to

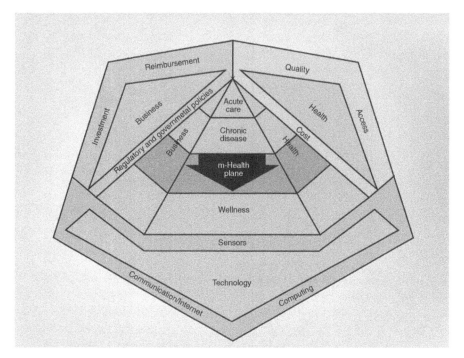

FIG. 7.6   The m-Health ecosystem coordination pyramid model.

achieve the required affordability, sustainability, and long-term global usability of these systems. This dynamic multidimensional principle can be translated into a proposed "ecosystem coordination pyramid" model shown in Fig. 7.6. In this model, any m-Health system, or indeed the future m-Health 2.0 system, can be envisaged as an adaptable sliding layer within the core of a multidimensional pyramid; the structure is representative of the whole layered ecosystem model and its associated dynamics. This m-Health layer optimizes the design process by balancing the constraints bounded by the different stakeholders' requirements to implement a "best effort" m-Health architecture. As shown in Fig. 7.6, these dimensions can be represented by three foundation layers consisting of healthcare dynamics and a patient pathway layer; technology and innovation layer, for example, sensors, computing, mobile communications; and business and regulatory layer, for example, reimbursement, investments, regulatory, and governance.

This model conforms to the notions presented in earlier models, particularly the dynamics and the spheres presented by the World Bank (Dalberg) model, and encapsulating the impact of the m-Health evolution. The development process of m-Health ecosystem architecture is a dynamic process that requires a scientific, economic, and mathematical approach to better understand it. These variables are affected by layers such as market volatility, technology developments, and patients care pathways. One approach is to represent different input and output variables as signals emulating the signal processing theory representation, similar to the model applied for the analysis of the financial data (Drakakis, 2009). The application of this new digital signal processing approach for the analysis and modeling of new dynamical m-Health ecosystem models in this pyramid representation is a new area of research that warrants further investigation.

7.3 m-HEALTH INTEROPERABILITY AND STANDARDIZATION

In general, to clarify the notion of interoperability for the lay reader, we are all mindful of the lack of interoperability between mobile phone companies, which have always frustrated customers by not agreeing to standardize their charging cables! Interoperability simply means that it is in everyone's interest to conform to certain agreed-upon standards of equipment, mobile devices, connectivity protocols, and so on, for the ability of a system or a product to work with other systems or products without special effort on the part of the customer. For example, there are numerous blood glucose monitoring and management systems on the market today, as presented in Chapter 5. These are available to patients and consumers to use with their smartphone Apps. However, the ability to connect the captured glucose levels and daily diets to their clinicians or health centers is not assured, as different healthcare providers adopt different IT systems that can work with some medical devices but not with others. This is particularly evident in the wide-ranging m-Health devices market for which there is increasing demand for interoperability and data integration. Hence, the necessity for global interoperability standards for m-Health systems is vital for their successful adoption.

Interoperability from the healthcare perspective can be defined as "the ability of different information technology systems and software applications to communicate, exchange data and use the information that has been exchanged" (HIMSS, 2013). Work on interoperability in m-Health-related issues as part of e-Health standards has been reported elsewhere (European Commission – CALLIOPE Network, 2010; ITU-T, 2012; Deloitte-European Commission, 2013; Stroetmann, 2014). In particular, the European Interoperability concept and framework has gained international recognition (Stroetmann, 2014). The European e-Health stakeholder group defined interoperability for e-Health as "the ability of two or more e-Health systems to use and exchange both computer interpretable data and human understandable data and knowledge" (European Commission e-Health Stakeholder Group, 2014). This study group divided interoperability into the following four layers:

- *Legal Interoperability:* This refers to the legal framework that supports the exchange of e-Health systems and the relevant interpretable data across different geographical boundaries and settings.
- *Organizational Interoperability:* This refers to the broader environment of policies, procedures, and bilateral cooperation needed to allow the seamless exchange of information between different organizations, regions, and countries. Human factors can be considered within this layer.

- *Semantic Interoperability:* This refers to the need to ensure that the precise meaning of exchanged information is interpretable by any other system or application that is not initially developed for this purpose.
- *Technical Interoperability:* This means the ability of two or more ICT applications to accept data from each other and perform a given task in an appropriate and satisfactory manner, without the need for extra operator intervention

From the health informatics perspective, interoperability, as defined by Health Level 7, is classified into three broad categories: technical, semantic, and process interoperability (HL7, 2016).

There is a clear gap in m-Health interoperability, and ongoing efforts remain largely "work in progress," partly because of the blurring boundaries between e-Health and m-Health. Here, we focus on the barriers relating to developing these standards.

As yet, there is no accepted definition of m-Health interoperability *per se*; however, it can be defined as "the ability to create mobile end-to-end solutions by allowing the structural interconnection between different m-Health components and devices from multiple vendors and services for the seamless exchange and interpretation of mobile data within formulated mobile health systems or networks."

The efforts required for the development of global m-Health interoperability standards remain a key challenge for the success and scaling-up of applications and their wider clinical acceptability and adoption. Today, a conundrum exists within the global interoperability standardization community for the development of such a framework; this is due to a wide spectrum of complex factors.

In recent years, there have been extensive efforts from international standards development organizations (SDO) and global industry alliances that are involved in developing, certifying, and marketing global interoperability standards for "personal health devices" and services. These can be considered relevant for m-Health application purposes. They include, for example, the Institute of Electrical and Electronics Engineers 11073 Personal Health Device Workgroup (IEEE 11073-PHD WG), Bluetooth Special Interest Group (Bluetooth SIG), Continua Health Alliance (CHA), and others, as shown in Table 7.1.

However, most of these SDO efforts are focused on emerging mobile personal devices and services that are tailored for use in high-income countries. This has been mainly due to a greater readiness of the commercial world to adopt these mobile devices and their technologies, while supported by substantial government and private industry investment.

Other e-Health interoperability standards originate under the Joint Initiative Council (JIC) from the following:

- International Organization for Standardization (ISO/IEEE-11073, 2010)
- Health Level Seven International (HL7, 2016)
- Digital Imaging and Communications in Medicine (DICOM, 2016) and
- Integrating the Healthcare Enterprise (IHE, 2016).

TABLE 7.1 Summary of Some of the Global Organizations Involved in e-Health Interoperability Standardization

Standardization Organization/Alliance	Relevant Link/Details
European Committee for Standardization (CEN/TC 251: Health Informatics)	www.cen.eu
Continua Health Alliance	www.continuaalliance.org/
epSOS (European Patients Smart Open Services)	www.espsos.eu
GS1 Healthcare	www.gs1.org/healthcare
DICOM (Digital Imaging and Communications in Medicine)	www.medical.nema.org
HL7: Electronic Health Information Systems	www.hl7.org/
ISO/TC 215: Health Informatics	www.iso.org/iso/iso_technical_committee?commid=54960
ISO/IEEE-11073: Medical/Health Device Communication Standards	http://www.iso.org
ITU-T Q28/16: Multimedia Framework for e-Health Applications	http://itu.int/ITU-T/studygroups/com16/sg16-q28.html
ITU-T FG M2M: Machine-to-Machine Service Layer	http://itu.int/en/ITU-T/focusgroups/m2m/
ITU-T IoT-GSI: Internet of Things Global Standards Initiative	http://itu.int/en/ITU-T/gsi/iot/
WHO e-Health Standardization and Coordination Group	www.who.int/ehscg/en/
WHO Global Observatory for eHealth	www.who.int/goe/en/

Source: Adapted from ITU-T (2012).

There are also global organizations outside the JIC umbrella, such as the WHO and the ITU, with their normative and informational coordination roles for the development and dissemination of standards.

From the health informatics standardization perspective, the JIC was established from seven SDO organizations: HL7, ISO Technical Committee 215 (ISO TC 215), European Committee for Normalization (CEN) TC 251, Clinical Data Interchange Standards Consortium (CDISC), GS-1, International Health Terminology Standardization Organization (IHTSDO), and IHE. Their remit was to facilitate coordination and harmonization of standards development of global health informatics (JIC, 2014). There is now a "JIC Standards Set – Patient Summary," which declared that it would focus its efforts by contributing to "better global patient health outcomes by providing strategic leadership in the specification of sets of implementable standards for health information sharing" (JIC, 2014).

Most of this global standardization work is based on the interpretation of m-Health as a subset of e-Health, and as such, many of the relevant issues surrounding the scaling-up and harmonization of e-Health are applicable to m-Health; for example, the need for national e-Health strategies and stronger e-Health governance frameworks.

These can be particularly applicable for LMICs, with different benefits that can be gained from these services for their healthcare systems (Payne, 2013).

Table 7.1 is a summary of some of these global organizations, SDOs, and other alliances involved in "e-Health" interoperability standardization (ITU-T, 2012).

From the m-Health perspective, there is a gap with a "disruptive diversity" effect on global m-Health interoperability and standardization. This gap can be attributed to the following issues:

- Fragmentation in current standardization efforts due to the overlapping and blurring nature between the concepts of "e-Health" and "m-Health," whereas m-Health should not be considered as part of e-Health, as illustrated throughout this book. Furthermore, most of the current e-Health and health informatics standards are based on the notion of m-Health as "mobile phone-centric" services for personal health solutions, without deeper consideration of the wider aspects of the concept.

- The unclear difference between clinical demands and healthcare benefits, and patient requirements of m-Health systems and devices in high-income countries, as opposed to the same demands and benefits in other global healthcare settings (e.g., LMICs).

- The conflicting barriers to existing mobile health markets between standardization organizations, vendors, telecommunications providers, and regulatory authorities. These barriers are reflected by the different m-Health ecosystem, market, and business strategies adopted by stakeholders, and they can lead to fragmented paths and fragile standards compliance levels.

- Limited access and engagement with international organizations of m-Health standardization issues, especially from LMICs. In addition to the absence of a clear understanding of the benefits of these standards from many LMIC governments, there is a lack of awareness of the necessity of m-Health standards within their healthcare systems and infrastructures.

7.3.1 European m-Health Interoperability

Interoperability from the European Commission's perspective can be summarized from the outcomes on the European m-Health Green Paper (European Commission, 2014). The Commission published the responses by 211 respondents from public authorities, healthcare providers, patients' organizations, and web entrepreneurs inside and outside the European Union. The following outcomes summarize the EU m-Health interoperability issues (European Commission, 2015):

- A majority (110) of the respondents supported the actions proposed in the e-Health Action Plan (European Commission, 2012). In particular, the need to foster the use of international standards was quoted by at least 18 respondents. The main standard development organizations named were Continua/ITU-T, IHE profiles, HL7, IEEE, and SNOMED CT. A series of additional actions was also put forward, such as promoting the establishment of open standards for interoperability.

- A specific approach to procurers, especially public procurers, was also supported. This recommended the need to identify *de facto* interoperability specifications through the ICT standards multistakeholders platform.
- There was a consensus on the need to develop an EU e-Health Interoperability framework. A few respondents emphasized the importance of continuing or reinforcing international cooperation in the field (EU–US MoU Japan WHO).
- There was also a clear consensus among the respondents that EU and national actions should seek to ensure interoperability of m-Health solutions with EHRs as this would be beneficial for enhancing continuity of care, patient empowerment, and research.

These issues reflect the different arguments relating to the current status of m-Health standards and interoperability, with several barriers that hinder the scale-up process and wider clinical acceptability of m-Health. The absence of standards that mandate interoperability between m-Health solutions and devices impedes the innovation toward wider economies of scale. This also prevents m-Health investments from being used well and limits the scalability of such solutions (European Commission, 2014). For example, most of the global SDO efforts have focused so far on personal health device markets as part of e-Health and health informatics standardization. These efforts have resulted in the publication of more than 20 international standards and more than 90 CHA-certified products available for global health device markets (Zhong et al., 2013).

7.3.2 Barriers to m-Health Interoperability and Standards

Although these standards can be considered an important prerequisite for the successful adoption and scaling-up of m-Health solutions, there are several barriers to such proliferation; these are mostly attributable to existing gaps between the current regulations, health policies, and market realities. These barriers include the following:

Market Adoption Barriers for Current Interoperability Standards (Zhong et al., 2013)

- *Granularity of the Regulation Policy:* This barrier is about the correlation between the degree of granularity to be provided by SDO regulators about the "intended usage" of particular personal health devices or services, and their subsequent tailoring of the interoperability standards to better satisfy the technical, business, and regulatory requirements.
- *Command and Control Issues of Devices:* This barrier concerns the lack of different personal health device vendors to adapt specific command and control functionalities, as specified by these standards, due to the lack of regulatory guidance and the precautionary approach taken by the vendors.

- *The Selective Connection Policies by Vendors Between Peer Devices:* This barrier is based on the lack of a universal "plug and play" standard for personal health devices, due to the choice of some vendors in selecting which of their products will follow the required interoperable standards and guidelines and which will not follow them. This is due to the market competitiveness between different vendors or service providers, over which the SDO has no control.

Barriers Attributed to the Content of Standards (Zhong et al., 2013)

- *Identifiers of Device and User Issues:* These relate to device IDs (DID) and user IDs (UID) of an end-to-end information system for personal health services, which often uses multiple health devices and different types of users. Current regulation policies toward ID mapping between DIDs and UIDs (e.g., a mobile phone connected to a weighing scale) have not been explicitly clarified yet, which becomes a potential barrier for market entry. Both vendors and SDOs expect clear guidance about this issue from the relevant regulation authority.
- *Duplicate Data Issues:* "Duplicate data" occurs when a specific "peer" device receives multiple copies of a single data set, either by duplicated transmission via the same channel or by overlapped transmission via different channels. This phenomenon affects many personal health-related applications, and leads some device vendors to decide that the peer device (the receiving data device) should decide whether to merge the duplicate data. Another group of vendors decides that the data merging process may aggravate the regulatory burden of the peer device, and may complicate and delay the data processing. Consequently, vendors and SDOs are expected to have clear guidance from the appropriate regulation authorities.

Interoperability Barriers and Financial Support Issues of m-Health in LMIC Settings   The interoperability barriers and financial issues related to m-Health in LMICs have not yet been addressed adequately. In an extensive literature review and gap analysis conducted by the m-Health Alliance, these issues have been addressed to achieve better use of interoperable m-Health scaled systems, particularly in LMICs (Payne, 2013):

- There was a lack of clarity about the actual standards and interoperability issues that are peculiar to m-Health. Thus, steps should be taken to determine how to approach the broader e-Health ecosystem through the lens of m-Health. This was regarded as a "critical" or "central" issue by several informants of this study.
- The opportunities for the use of standards to achieve interoperable m-Health at scale can be categorized as building capacity and increasing standards access; filling in standards gaps for LMICs; LMIC engagement and mediation; promoting coordination and alignment; and strengthening m-Health and e-Health governance.

- Several cross-cutting recommendations were given to frame future coordination procedures, especially for LMICs: shifting the market dynamics to incentivize interoperability; alignment with national e-Health strategies; and the establishment of regional e-Health standardization collaboratives.

The m-Health market fragmentation and lack of adequate funding and investment in different LMICs, where healthcare delivery is often funded through public resources, is another challenge that needs strategic standards solutions. The issues of proper procurement processes and standards formulations also need addressing.

Clinical Aspects of Standards and Interoperability   Several studies have addressed clinical barriers and the importance of standards and interoperability on the scaling-up process of m-Health (Tomlinson et al., 2013):

- Existing standards for research should be reconsidered to provide guidance as to when scale-up is appropriate.
- Establishing an open m-Health architecture should be based on a robust platform with standards for App development, which would facilitate scalable and sustainable health information systems.
- Scale-up of m-Health in LMICs should be preceded by efficacy and effectiveness.
- Trials, so that they are founded on an appropriate evidence base.
- Governments, funders, and industry must cooperate in setting standards to create a self-governing, commercially viable ecosystem for innovation.

Security and Privacy Challenges   This is a complex and widely discussed topic with m-Health standardization bodies, and many unanswered challenges remain pending. The purpose here is to illustrate the magnitude of these challenges and the complexity of the associated tasks, particularly in the standards context.

We cite an example of the security concerns of the widely known ISO/IEEE-11073 standard, especially those parts that relate to personal health device communications. In particular, we discuss security measures in this standard, and focus on those parts that relate to wireless glucose meters, continuous glucose monitoring (CGM), and insulin pumps. These measures are applied for different mobile diabetes management and wireless monitoring devices used for many successful m-Health commercial products and market applications. Table 7.2 is a summary of the current status of the ISO/IEEE-11073 standard associated with CGM and artificial pancreas systems.

The standard defines the formats for information sent wirelessly between specific diabetes sensors and the diabetes data managers used for collecting the glucose data acquired by these devices and for the mutual exchange of information.

As explained in Chapter 5, these diabetes devices and sensors are off-the-shelf continuous glucose monitors, fingerstick blood glucose devices, and insulin pumps, usually referred to as "PHD" in the standards documentation. They collect the glucose data and then transmit it to a gateway, such as a smartphone, tablet, or personal computer, initially for the purpose of data collection and display. The gateway then

TABLE 7.2   IEEE 11073 Standards Used for Wireless Glucose Meters, CGM, and Insulin Pumps

Devices	Standard	Status
Glucose meters	IEEE-11073-10417	Approved
CGM	IEEE-11073-10425	Approved
Insulin pump	IEEE-11073-10419	Under development

Source: Adapted from O'Keeffe et al. (2015) and ISO/EEE-11073 (2010).

transmits the data for additional analysis to a remote specialist diabetes center for further observation and feedback advice by a specialist or diabetes nurse. These standards control the communication paths between a PHD and the gateway, assumed to be a point-to-point connection.

Typically, the PHD communicates with a single gateway at a specific point. The gateways can also communicate with a plurality of PHDs simultaneously, using separate point-to-point connections (ISO/IEEE-11073, 2010). However, recent research highlighted the security vulnerabilities in these standards and the need for better understanding the incorporation of robust security measures to counter these threats (O'Keeffe et al., 2015). For example, cybersecurity attacks can threaten the wireless link between a CGM and a microcontroller unit in the wireless management system; and if the link is not encrypted properly, it may be possible to introduce erroneous data to the target device. Deliberately incorrect low glucose data sent to an unprotected mobile computing platform may cause the algorithm to deliver excessive insulin, whereas incorrect low glucose values could cause it to deliver too little (O'Keeffe et al., 2015).

These and other scenarios pose a serious threat to standardized component connectivity in m-Health diabetes management systems. Furthermore, recent developments of IoT and M2M communications in healthcare markets and wireless disease management sectors make these challenging issues more urgent for further consideration by the global standardization and regulatory bodies.

As an overall conclusion from this analysis of the current landscape on m-Health interoperability and standards, the following issues need further study or development:

- There is a need to develop an m-Health interoperability framework and relevant standards. This development is vital for the future and sustainability of m-Health services and applications. However, the development is a complex process that requires global efforts and expert coordination between stakeholders and standardization bodies.
- m-Health interoperability can be interpreted as a multilayered process (e.g., semantic, technical, organizational, and legal), with each layer having its own architectural structure that requires further detailed multifaceted work.
- There is blurring and uncertainty between e-Health and m-Health interoperability requirements. This is due to the existing knowledge gap of the full spectrum of m-Health as a separate concept but also as a complement to

e-Health in some areas of healthcare services. In this context, there is increasing demand, at least from the European Union, for m-Health interoperability with EHRs, as this would be beneficial for enhancing continuity of care, patient empowerment, and research. This topic is an important but overlooked issue subject to further discussions and research.

- There is a need to address the correlation between m-Health standards and clinical requirements for LMICs that match those for developed countries. These issues cannot be developed separately and without such a correlation process. This synchronization is vital for the global usability of these m-Health systems in LMIC care settings. Furthermore, m-Health standards must meet those currently at the development stage, such as IoT, cloud, and 5G networks.
- Interoperable m-Health services cannot be fully integrated and operational without underpinning the necessary legal and different ethical uncertainties inherent in these systems and services. The same principle is applicable to the security and privacy issues.

The importance of m-Health standards to patients, physicians, and businesses remains vital. The development process of m-Health interoperability, with relevant standards, is basically the development of common m-Health platforms for consensus among the major global stakeholders. It is vital to have global coordinated efforts in the planning and development of workable m-Health interoperability standards that are applicable to wider markets.

7.4 m-HEALTH MARKETS AND BUSINESS MODELS

The m-Health markets and business models are controversial topics that have been continuously debated and revised for more than a decade. In an article entitled "m-Health: Health and Appiness," *The Economist* summarized the m-Health market by saying "those pouring money into health-related gadgets and Apps believe they can work the miracle of making healthcare better and cheaper" (*The Economist*, 2014). This statement sums up what has been the essence of the m-Health market and the current paradoxical status of the massive increases in healthcare expenditure, in both the developed and the developing worlds. This includes the growing m-Health investment and the disproportionately low level of m-Health adoption in healthcare services. One report has cited an assessment of five common m-Health business models from the global MNO perspective, based on key commercial capabilities that include a reliable healthcare service provider, strong consumer brand, large customer base, and broad retail distribution channels (Little, 2011). These are the "bit pipe" provider model, the "enabler-connectivity provider" model, the "joint partner" model, the "white-label" partner, and the "lead-label" partner.

This report also advocated that MNOs should not develop new m-Health ecosystems or leverage existing ecosystems in the developed market. For the emerging markets, it recommended the focus on simpler m-Health services such as SMS and leading the development of new ecosystems.

Many recent market analysis reports show that healthcare providers the world over are facing a formidable challenge to manage rapidly increasing costs. For example, healthcare spending in North America is anticipated to grow on an average of 4.9% annually in 2014–2018, driven partly by the expansion of insurance coverage under the U.S. implementation of the 2010 Patient Protection and Affordable Care Act (ACA). U.S. healthcare spending, already the highest in the world, is likely to reach 17.9% of GDP by 2018 (Deloitte, 2015b).

In the United Kingdom, the net expenditure (resource plus capital, minus depreciation) of the National Health Service (NHS) has increased from £64.173 billion in 2003–2004 to £113.3 billion in 2014–2015, with a planned expenditure increase for 2015–2016 at £116.574 billion (NHS Confederation, 2015). These increasing expenditure figures are leading to estimates of a funding gap in NHS England of £30 billion by 2020–2021 (Appleby, 2014).

These unsustainable increasing costs are likely to drive fast adoption of new m-Health applications in different healthcare services. These will act as transformative digital tools in the way physicians, payers, patients, and other healthcare stakeholders interact to tackle the increases. Many market research studies indicate increasing levels of investments in digital and mobile health technologies, with some estimates reaching in excess of $4.1 billion in 2014 (Sullivan, 2015). The leading market investment areas in 2015 were wellness, big data analytics, population health, and navigating the care system (StartUp Health 2015). However, there is still timid adoption of m-Health into mainstream healthcare services, excluding the personal fitness and health tracking markets.

This conundrum of lack of m-Health uptake commensurate with increasing healthcare costs is illustrated in Fig. 7.7. This shows the m-Health "hype cycle" since its inception. It is well known that the two general components of any emerging technology are the hype cycle, represented by the bell curve, and the backlash phases, represented as expectations that the technology soars then quickly subsides. The second is the S-shaped "business maturity and adoption" level that follows the contour, and is superimposed on the first level, representing the move of that emerging technology to wider acceptance (Moore, 2014).

In the latest Gartner hype cycle representation of emerging technologies, "mobile health monitoring" (not m-Health) is shown within the edge of the bell, or the "trough of disillusionment" phase (Gartner Inc., 2014). The main m-Health chasm was crossed with the introduction of the smartphone in 2007, although there was an earlier chasm shortly after the innovation trigger in 2003. However, its current predicted position coincides with the market projection of the probable option of majority acceptance in the next decade or so, provided that m-Health can finally cross the "trough of disillusionment" by its wider acceptance from clinicians, patients, and healthcare service providers. The main barrier to this crossing is not technological, but medical. This is a pending challenge, which awaits the formulation of acceptable and workable business models tailored to the wider m-Health markets.

The major point about mobile health is that it introduces attractive business solutions and alternatives for both patients and clinicians in offering less expensive

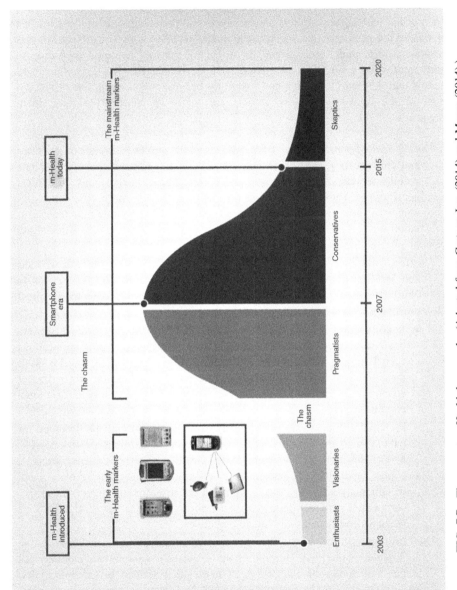

FIG. 7.7 The current m-Health hype cycle. (Adapted from Gartner, Inc. (2014) and Moore (2014).)

347

healthcare solutions, new ways to manage care, empower patients, and potentially to provide better health outcomes using powerful and widely used technologies.

As explained in earlier sections, many existing m-Health business models developed in recent years were complex, precarious, or difficult to implement on a larger scale, with costly implications to payers and patients. Some of these models advocated the principle that specialist IT healthcare service providers and payers (insurance companies) are charged for m-Health services rendered according to their nature and scale. The perceived wisdom is that new m-Health business models will continue to evolve and hopefully become more acceptable clinically. Most of the relevant business models are likely to conform to three main categories:

The Clinically Focused Business Model: This model provides efficient transactions and services for the different stakeholders, including health service providers, mobile networks, hospital costs, and pharmacies, with the focus on clinical applications. These are designed primarily to provide more efficient use of clinicians' time for different healthcare settings. They include mobile consultation services and remote access to patient health data through the efficient use of smartphones and tablets.

The Consumer-Driven Business Model: This model is based on m-Health products and services that are widely perceived as consumer based, as opposed to clinically based. This is one of the most widely used and accepted business models, with the following aspects:

- It embraces m-Health Apps business models with the medical sensing devices industry. These are usually used for health monitoring applications such as systems used for wellness and disease management. The wider use of these solutions is understood to provide reductions in hospital care and better economic outcomes of their increasing costs, especially for chronic conditions.

- Its popularity is consistent with increasing m-Health markets designed for patients who are more disciplined or concerned about their health and well-being. This includes a wide range of m-Health products designed to monitor, collect, and interpret medical and behavioral data, including diet, sleep, weight, daily activities, and so on. A major barrier to this model is the longer compliance and continuity of using these solutions for straightforward wellness and behavioral change objectives.

The Continuity of Connected Care Business Model: This model is based on the deployment of m-Health solutions in the hospital and continuity of care services arena. It includes the integration of m-Health solutions in existing EHRs, hospital emergency care, patient tracking, and other specialist care services, using different mobile infrastructure configurations. Different m-Health cloud, smart data mining, and analytic services are considered part of this model. Security and privacy, especially their vulnerability to outside security attacks and breaches of confidentiality, are some of the challenges for this model.

These models are largely advocated by stakeholders that are outside the health domain, mainly from healthcare ICT technology and innovation companies, SMEs, mobile telecommunication providers, and/or medical devices industries. However, since the impact of m-Health is still evolving, and is relatively modest compared to the size of healthcare economies, there are virtually unlimited opportunities for these businesses, which should be mindful that their end users will be mainly patients and clinicians.

To reflect on this note, a recent analysis on m-Health business issues found that solutions have begun the trend of embracing the following six points (PwC, 2014):

- *Interoperability:* Interoperable with sensors and other mobile or non-mobile devices to share vast amounts of data with other applications, such as EHRs and existing healthcare plans.
- *Integration:* Integrated into existing activities and workflows of providers and patients to provide the support needed for new behaviors.
- *Intelligence:* Offer problem-solving ability to provide real-time, qualitative solutions based on existing data in order to realize productivity gains.
- *Socialization:* Act as a hub by sharing information across a broad community to provide support, coaching, recommendations, and other forms of assistance.
- *Outcomes:* Provide a return on investment in terms of cost, access, and quality of care based on healthcare objectives.
- *Engagement:* Enable patient involvement and the provision of ubiquitous and instant feedback in order to realize new behaviors and/or sustain desired performance.

In another survey, 27 informants were asked to determine the most important issues facing the implementation of m-Health from the viewpoint of those in the U.S. health system and those working in the U.S. m-Health markets. Their responses were summarized as the following five common aspects (Whittaker, 2012):

- *Policy and Regulatory:* These included privacy and security; clear FDA m-Health regulations; prescribing regulations relevant to m-Health; and mobile spectrum availability.
- *Wireless Networks:* These included compatibility across multiple networks, multiple platforms, and proprietary systems; cost to the public or end user; and coverage in remote areas.
- *Health Systems:* These included lack of sustainable business models, reimbursement and understanding of the value m-Health may provide; clinical roles accountability and integration into clinical practice; integration into EHRs and health information systems; competing health IT priorities; and broader opportunity cost issues.
- *m-Health Practices:* These included lack of knowledge of how to do it well; wrong focus on the technology or on advantaged populations (those who do not

need it); governance of m-Health and publicly available applications not evaluated and without basis in theory or evidence; and stand-alone or siloed initiatives due to existing platforms or proprietary systems.

- *Research:* This included a need for more high-quality research; a need to demonstrate efficacy and cost-effectiveness; a mismatch in pace and flexibility between research and technology development; and measurement of reach or access for the underserved.

Although the above concerns reflect the views of U.S. stakeholders, it could be argued that some of them are also applicable to other m-Health markets and businesses, especially in the developed world.

Alleviating these aspects can be considered a legitimate starting point for discussions on the current status of the m-Health market by the stakeholders. This would allow m-Health to move forward from the era of "pilotitis" to large scale-up adoption for demanding healthcare delivery. However, these aspects remain predictions, so extensive work and global coordination are still needed to enable m-Health markets and technological solutions to evolve and mature. In LMICs, there are separate issues that face m-Health businesses and markets that stem from different factors and barriers:

- *Limited Investment and Donor Contributions:* These investment levels continue, but they usually allocate insufficient resources and time to generate robust evidence that can provide scaling-up of m-Health services. Investment is mainly focused on mobile technology and cost-effective innovations, rather than on business approaches that aim to build large-scale or national m-Health services.
- *Absence or Lack of Established National m-Health Strategies:* Although there are many pilots in these countries, very few exhibit national m-Health strategy and government leadership, so promising pilots have little chance of being scaled up.
- *Clash of Existing Public Health Culture and Technology Innovations:* The decision makers and public health experts often see the application of m-Health technologies as a distraction or hindering factor in their health programs. Their inexperience and lack of knowledge and their hesitation to engage with new technologies usually hinder the development of large-scale m-Health initiatives.
- *Ambiguity of Business Models:* There is a lack of understanding between the mobile communications sectors, healthcare service providers, and governmental sectors. The business models advocated by the mobile network operators are usually either financially not viable or unworkable for appropriate scale-up.
- *Lack of Standardization, Regulatory, and Security Strategies:* These should be tailored to different m-Health applications and services.

Some of these challenges are also connected to other aspects listed earlier. Their future interconnections are inevitable for securing successful global m-Health business strategies. Although increasing m-Health initiatives are being deployed in numerous LMICs, they remain mainly in the "pilotitis" sphere, with sporadic scaling-up business models. What is needed are new planning strategies for m-Health services to move forward to initiate national healthcare delivery plans that can benefit the wider population who are in need of much improved services.

7.5 SUMMARY

The business of m-Health is one of the most dynamic and successful healthcare innovations and market successes. Billions of U.S. dollars are invested annually and with an increasing trend. Unprecedented investment levels and associated business dynamics have not yet crystallized with tangible patient benefits, nor have they led to the much hyped transformative benefits for healthcare delivery services. However, this picture might change in the next few years, with the influx of technology innovations related to m-Health. These will probably change the current landscape by reducing the cost of medical devices and enabling access to alternative mobile connectivity with smarter computing paradigms. These constitute the basic building elements of the new m-Health era presented in this book. With these developments on the horizon, m-Health stakeholders and businesses are poised to embrace a step change to develop alternative understandings of their existing market models. However, there remain many serious challenges that we have highlighted in this chapter. There are barriers that require global collaboration and better understanding of the concept, notably the vital transformation from "consumer gadgetization" to the "science of m-Health."

The lack of consensus in understanding and developing a successful m-Health ecosystem remains a challenge for businesses. Some of the technology and economic issues that need consideration in this context are as follows:

- The economics of "m-Health gadgetization" and its reversal, by lowering the cost and affordability of m-Health devices and technologies for everyone to use.
- The adoption of new technological evolutionary models by differentiating between the current Apps-centric m-Health model and the evolving m-Health technologies model.
- Understanding the dynamics of stakeholders in each care setting and the adaptability of the m-Health models for different healthcare segments.
- Better use of m-Health for "people empowerment," with accessible and acceptable m-Health systems connected seamlessly to their daily routines and activities.

Most of the business-related studies and models assume the narrow market definitions and consumer overview of m-Health that we have discussed in earlier chapters. There is no comprehensive and systematic m-Health economic study to date that can provide a scientific analysis of global investments in m-Health. This research is important, if not vital, to quantify the contributions of the global industries, businesses, donors, and other research funding agencies allocated for m-Health applications, compared to similar funding for other healthcare delivery sectors.

This will allow policymakers, regulators, and other decision-making bodies to depict a more comprehensive picture of the current landscape of m-Health, such as lack of clear clinical evidence and impact on global health.

REFERENCES

Appleby J (2014) *NHS funding: past and future*, The King's Fund, http://www.kingsfund.org.uk/blog/2014/10/nhs-funding-past-and-future (accessed November 2014).

Chen C, Haddad D, Selsky J, Hoffman JE, Kravitz RL, Estrin DE, and Sim I (2012) Making sense of mobile health data: an open architecture to improve individual- and population-level health. *Journal of Medical Internet Research* 14(4):e112.

Cortez N (2014) The mobile health revolution? *UC Davis Law Review* 47(4):1173.

Deloitte (2015a) *Connected health: how digital technology is transforming health and social care*, Deloitte Centre for Health Solutions. Available at http://www2.deloitte.com/content/dam/Deloitte/uk/Documents/life-sciences-health-care/deloitte-uk-connected-health.pdf (accessed June 2015).

Deloitte (2015b) *2015 Global healthcare outlook: common goals, competing priorities*. Available at http://www2.deloitte.com/content/dam/Deloitte/global/Documents/Life-Sciences-Health-Care/gx-lshc-2015-health-care-outlook-global.pdf (accessed June 2015).

Deloitte (European Commission) (2013) e-Health EIF e-Health European Interoperability Framework, European Commission, ISA Work Programme. Available at http://www.uems.eu/__data/assets/pdf_file/0010/1504/eHealth_European_Interoperability_Framework_Study_report_-_July_2013.pdf (accessed October 2014).

DICOM (Digital Imaging and Communications in Medicine) (2016) Available at www.medical.nema.org (accessed January 2016).

Drakakis K (2009) Application of signal processing to the analysis of financial data. *IEEE Signal Processing Magazine*, September, pp. 156–158.

The Economist (2014) *m-Health: health and happiness*. Available at http://www.economist.com/news/business/21595461-those-pouring-money-health-related-mobile-gadgets-and-apps-believe-they-can-work (accessed October 2014).

European Commission (2012) *eHealth Action Plan 2012–2020: Innovative Healthcare for the 21st Century*, Communication from the Commissions to the European Parliament, the Council, the European Economic and Social Committee and Committee of the Regions {SWD, 2012 413, 414 Final}, pp. 1–14. Available at http://ec.europa.eu/health/ehealth/docs/com_2012_736_en.pdf (accessed October 2014).

European Commission (2014) Green Paper on Mobile Health ('m-Health') {SWD (2014) 135 final}, Brussels. Available at ec.europa.eu/information_society/newsroom/cf/dae/document.cfm? (accessed December 2014).

European Commission (2015) Summary Report on the Public Consultation on the Green Paper on Mobile Health. Available at http://ec.europa.eu/digital-agenda/en/news/mhealth-europe-preparing-ground-consultation-results-published-today (accessed May 2015).

European Commission – CALLIOPE (CALL for InterOPErability) Network (2010) European eHealth Interoperability Roadmap, Final European Progress Report. Available at http://www.ehgi.eu/Download/European%20eHealth%20Interoperability%20Roadmap%20[CALLIOPE%20-%20published%20by%20DG%20INFSO].pdf (accessed October 2014).

European Commission e-Health Stakeholder Group (2014) eHealth Stakeholder Group Report-Perspectives and Recommendations on Interoperability, Final Report, March. Available at ec.europa.eu/information_society/newsroom/cf/dae/document.cfm (accessed October 2014).

Frost and Sullivan (2013) *Embracing M2M opportunities in healthcare: the case of Orange Business Services and Telefonica.* Available at http://www.reportlinker.com/p01889355-summary/Embracing-M2M-Opportunities-in-Healthcare-The-Case-of-Orange-Business-Services-and-Telefonica-.html (accessed October 2014).

Gartner Inc. (2014) *Gartner's 2014 hype cycle for emerging technologies maps the journey to digital business.* Available at http://www.gartner.com/newsroom/id/2819918 (accessed May 2015).

GSMA (2011) *Connected mobile health devices: a reference architecture V1.0,* January. Available at www.gsma.com/connectedliving/wp-content/uploads/2012/03/connectedmobile healthdevicesareferencearchitecture.pdf (accessed October 2014).

GSMA–PwC (2012) *Touching lives through mobile health: assessment of the global market opportunity.* Available at www.pwc.com (accessed October 2014).

HIMSS (Healthcare Information and Management Systems Society) (2013) *What is interoperability?* Available at http://www.himss.org/library/interoperability-standards/what-is-interoperability (accessed October 2013).

HL7 (Health Level Seven International) (2016) Available at www.hl7.org (accessed January 2016).

IHE (Integrating the Healthcare Enterprise) (2016) Available at www.ihe.net (accessed January 2016).

ISO/IEEE-11073 (2010) *Standard for health informatics: personal health device communication,* Institute of Electrical and Electronics Engineers. Available at https://en.wikipedia.org/wiki/ISO/IEEE_11073_Personal_Health_Data_(PHD)_Standards (accessed October 2014).

ITU-T (International Telecommunication Union) (2012) *e-Health standards and interoperability,* ITU-T Technology Watch Report. Available at https://www.itu.int/dms_pub/itu-t/oth/23/01/T23010000170001PDFE.pdf (accessed October 2014).

ITU-T (International Telecommunication Union) (2014) *M2M-enabled ecosystems: e-Health,* ITU-T Focus Group on M2M Service Layer, Focus Group Technical Report. Available at https://www.itu.int/dms_pub/itu-t/opb/fg/T-FG-M2M-2014-D0.2-PDF-E.pdf (accessed October 2014).

JIC (Joint Initiative Council) (2014) *Joint initiative on SDO global health informatics standardization.* Available at http://www.jointinitiativecouncil.org (accessed October 2014).

Krohn R and Metcalf D (2012) *m-Health: from smartphones to smart systems,* Healthcare Information and Management Systems Society (HIMSS), Annual Conference and Exhibition, file:///C:/Users/Bryan/Desktop/mHealth%20Book/material%20from%20Robert's%20e-mails/Slies-%20HIMSS%202012%20mHealth%20Krohn%20MetcalfFINAL%20(1).pdf (accessed January 2016).

Little AD (2011) *Capturing value in the mHealth oasis: an opportunity for mobile network operators.* Available at http://www.adlittle.co.uk/uploads/tx_extthoughtleadership/ADL_mHealth.pdf (accessed October 2014).

Malvey D and Slovensky DJ (2014) *m-Health: Transforming Healthcare.* New York: Springer Publishing Company.

McKinsey & Company (2010) *m-Health: a new vision for healthcare*, GSMA. Available at http://www.gsma.com/connectedliving/wp-content/uploads/2012/03/gsmamckinseym healthreport.pdf (accessed October 2014).

Moore G (2014) *Crossing the Chasm: Marketing and Selling Disruptive Products to Mainstream Customers (Collins Business Essentials)*, 3rd edn. Harper Collins.

NHS Confederation (2015) *Key statistics on the NHS*. Available at http://www.nhsconfed.org/resources/key-statistics-on-the-nhs (accessed July 2015).

O'Keeffe DT, Maraka S, Basu A, Keith-Hynes P, and Kudva YC (2015) Cybersecurity in artificial pancreas experiments. *Diabetes Technology & Therapeutics* 17(8):1–3.

Open m-Health (2016) Available at www.openmhealth.org (accessed January 2016).

Payne JD (2013) *The state of standards and interoperability for m-Health among low-and middle-income countries*, m-Health Alliance Report. Available at http://www.mhealthknowledge.org/sites/default/files/12_state_of_standards_report_2013.pdf (accessed October 2014).

PwC (PricewaterhouseCoopers) (2014) *Emerging m-Health: paths for growth, a global research study about the opportunities and challenges of mobile health from the perspective of patients, payers and providers*. Available at www.pwc.com/gx/en/healthcare/mhealth/assets/pwc-emerging-mhealth-full.pdf (accessed January 2015).

Qualcomm Life (2016) Available at www.qualcommlife.com/wireless-health (accessed January 2016).

StartUp Health (2015) *StartUp health insights: digital health funding rankings*, 2015 Midyear Report. Available at https://www.startuphealth.com/content/insights-2015q2 (accessed June 2015).

Stroetmann KA (2014) Health system efficiency and ehealth interoperability: How much interoperability do we need? In: Rocha A et al. (eds.), *New Perspectives in Information Systems and Technologies, Volume 2*, Advances in Intelligent Systems and Computing (Book 276), Switzerland: Springer International Publishing, pp. 395–406.

Sullivan M (2015) *2014 Digital health investment exceeded total of three previous years combined, Rock Health says*. Available at http://venturebeat.com/2015/01/02/2014-digital-health-investment-exceeded-total-of-three-previous-years-combined-rock-health-says (accessed July 2015).

Tomlinson M, Rotheram-Borus MJ, Swartz L, and Tsai AC (2013) Scaling up m-Health: where is the evidence? *PLoS Medicine* 10(2):e1001382.

Whittaker R (2012) Issues in m-Health: findings from key informant interviews. *Journal of Medical Internet Research* 14(5):e129.

Qiang CZ, Yamamichi M, Hausman V, and Altman D (2011) *Mobile application for the health sector*, ICT Sector, World Bank Report, December 2011. Available at http://www.cmamforum.org/Pool/Resources/Mobile-applications-for-health-report-WBank-2012.pdf (accessed October 2014).

Zhong D, Kirwan MJ, and Duan X (2013) Regulatory barriers blocking standardization of interoperability. *Journal of Medical Internet Research m-Health and u-Health* 1(2):e13.

8

THE FUTURE OF m-HEALTH: PROGRESS OR RETROGRESSION?

Infirmity doth still neglect all office whereto our health is bound; we are not ourselves when nature, being oppress'd, commands the mind to suffer with the body.

William Shakespeare, King Lear, Act 2, Scene 4

8.1 INTRODUCTION

This book started with a historical perspective of m-Health from its early beginnings more than a decade ago. We conclude by outlining some of the challenges, likely futuristic scenarios, and trends. A glance at the volume of the remarkable online m-Health resources available on the Internet reflects the massive global business, medical, technological, and research activities in various healthcare disciplines (Appendix). These resources also reflect the major strides that the concept has achieved from its early, humble academic beginnings to its status at this moment.

It is inevitable that m-Health will be associated with cellular mobile phone technology for the foreseeable future. Although this marriage of convenience is likely to continue, a radical change is required for its progress to avoid retrogression. The existing gap between the consumerism of m-Health and the underlying science that can transform many of our future healthcare delivery services will narrow only if new thinking and direction are brought to bear.

The future remains an open book and as the famous quote goes: "in order to predict the future you need to invent it." It is not difficult to envisage all manner of amazing

m-Health: Fundamentals and Applications, First Edition. Robert S. H. Istepanian and Bryan Woodward.
© 2017 by The Institute of Electrical and Electronics Engineers, Inc. Published 2017 by John Wiley & Sons, Inc.

futuristic scenarios in this context. There will be few technology limitations, but there will be limitations to what people want or will accept. For example, not everyone wants to be constantly wired up to have their health monitored, even if it is in their own interests (except perhaps for the proponents of the quantified self-enhancing self-knowledge through self-tracking). Will the future of m-Health be purely "wearable health" and well-being monitoring with smart mobile phones? Or will it be the compact and intelligent self-diagnosing and monitoring device similar to Start Trek's tricorder?

Realistic scenarios of m-Health, with a little imagination, can be much broader than these examples, with truly transformative healthcare outcomes and an overarching vision that goes beyond the narrow confines of its current manifestations.

We hope that in the future a new *m-healthology* science will be established for studying the various multidisciplinary areas of m-Health, uniting medicine, engineering, computing, social, business, and other disciplines that it already embraces. Only the future can reveal this, but the hope is that current consumer-led m-Health gizmos and products dominating the scene under the umbrella of mobile health will only be part of the bigger picture and not the whole one.

This future evolution will be dependent on several technological, clinical, and economic factors that will drive such processes. These include, for example, the following:

- The uncertainty and balance between health technology-driven market opportunities and the increasing demand for transformative healthcare delivery services, with clear evidence-based clinical outcomes. The narrowing of this chasm will not be achieved by the increasing gap that is tilting m-Health more toward consumerism rather than to clinically validated solutions.
- Future societal and cultural changes embracing large-scale mobile healthcare services.
- Future advancement in m-Health innovations and how best to exploit these effectively by introducing a new generation of affordable m-Health systems that can be adopted globally. For example, the development of cheap smartphones with cost-effective wireless connectivity to a variety of medical devices that are affordable to the poor can achieve major strides in this direction.
- More coordinated roles from governments and international health organizations toward better understanding of the future role of m-Health from a global perspective. These include the development of workable strategies that require, for example, better alignment of global standardization, interoperability, and regulatory efforts, supported by well-defined m-Health aims and objectives defined by clinical evidence targets.

For more than a decade, m-Health has been synonymous with the use of mobile phone and cellular technologies, and more recently with its health App phenomenon. This association has largely characterized it as a disruptive innovation, displacing prior ways of doing things at much lower costs and with continued high expectations for ways in which it can transform healthcare; however, only a few Apps have yet proved functional and sustainable (Malvey and Slovensky, 2014). Furthermore, this m-Health model and its associated strategies are more suited to the healthcare needs of the developed world of high-income countries, as opposed to the global health challenges of low-income and developing countries. These issues have led to the continuous ongoing debate associated with existing barriers that have hindered the large-scale adoption of m-Health so far. These include the following:

- Inadequate m-Health strategies by healthcare providers and payers, especially in high-income countries.
- Lack of widely successful and workable business models with return of investment frameworks adaptable to global markets.
- Existing uncertainty of the reliability, clinical efficacy, and large-scale evidence of these systems.
- Absence of global regulatory m-Health frameworks and fragmented m-Health policies.
- Security and privacy concerns and the lack of mitigation procedures, including any viable alternatives for secure Internet or open-architecture connectivity models.
- Lack of appropriate frugal m-Health solutions dedicated to global healthcare challenges.

Addressing these and other factors that are being continuously debated by global m-Health communities requires serious revisiting and questioning of the current understanding of m-Health, replacing it with a new vision that transcends the current status, leading to more successful future m-Health systems. In this chapter, we present some of these potential directions and speculate how m-Health can be reshaped so that it can evolve from its current interpretation to a truly unique and transformative healthcare innovation concept.

8.2   FUTURE TRENDS OF m-HEALTH

When the Wright brothers first took to the air in their flying machine in 1903, few people could have predicted jet fighter aircraft that could fly faster than the speed of sound, or achieve 1500 mph, or climb at 50,000 ft/min. Few could imagine commercial passenger aircraft that could fly halfway around the world nonstop, carrying 500 or more passengers. Nor could they have predicted the fantastical notion of space flight that in 1969 allowed men to walk on the moon, or for unmanned spacecraft to travel to the far reaches of our solar system, or for the immense laboratory of the International Space Station to orbit the Earth indefinitely, with crew and supplies coming and going by a regular rocket-propelled shuttle service.

Similarly, but rather less spectacularly, when the unknown Robert Plath, the Northwest Airlines pilot who invented travel luggage with wheels, few ever thought that his invention would radically change the way people travel with their suitcases.

So it is with healthcare. Medical advances have been equally impressive and just as difficult for people to predict in the last century when they looked up awestruck at the Wright brothers flying overhead like birds. It is certain that few could have predicted the use of sophisticated anesthetics that took the terror out of crude operations, antibiotics like penicillin that fought infection, organ transplants of the heart, kidneys, lungs, and even hands, keyhole surgery, hip replacement surgery, and a thousand other amazing achievements.

Most recent market statistics predict a massive growth in the m-Health market in the next few years. However, many confusing questions exist as to what direction this growth will take. Does it, for example, reflect the billions of U.S. dollars that are already invested, or will be invested in new m-Health market solutions, with the aim of delivering the much-hyped transformative quality, efficiency, and access to healthcare? Or will these be used to provide purposeful systems with clear healthcare targets, outcomes, and organizational change solutions? Or will they simply provide better global healthcare solutions to many underserved areas and poorer regions of the world? Can someone predict the future of the Apple "HealthKit"? Are there any lessons learnt from the much-hyped and now defunct "Google Health?" The answer to all these questions is uncertain and as yet unclear.

Regardless of the confusing fact that the global market growth of m-Health will not only lead to innovative solutions with diversifying functionalities, it will also cover many social, scientific, and engineering disciplines, targeting a range of new medical challenges. Yet this adds to the confusion because many of these questions will remain unanswered, or at best ambiguous. This is particularly true of the clinical angle, especially since many of the much-hyped m-Health solutions of today provide fragile evidence of the anticipated promises of better access, lower cost, and efficient healthcare delivery outcomes, combined with sketchy and uncertain clinical evidence.

The future of m-Health as it now stands is on a par with looking into a crystal ball and wondering what else will be available in 10, 20, 50 years from now. From the technological standpoint, m-Health can definitely lead to tangible and effective changes in current healthcare delivery services, provided that a new vision is adopted, resulting in generations of m-Health systems that are better designed and developed. These also need to be appropriately focused on the healthcare needs of patients, and associated with lower costs to providers. This is a difficult equation to balance, especially in an uncertain economic climate, and with existing competing, lucrative market potential. Future technological developments of m-Health can be crystallized as follows (Istepanian and Zhang, 2012):

- Future m-Health services and applications with more efficient and effective outcomes.
- Emerging mobile access networks, sensor connectivity, and computing systems better tailored for healthcare needs.

- Better business models, ecosystem development, and deployment as and when it is needed to deliver better and more efficient healthcare with lower costs.
- Personalization and prevention modes that are accessible and workable in different healthcare environments.

The successful realization of these elements will be governed by many competing factors, and the realignment of various stakeholders that we alluded to in earlier chapters. The evolution of m-Health technologies, increased patient expectations, targeted clinician demands, and global challenges are critical factors in this process.

We aim in the following sections to present some of the current trends and to address future challenges. We will also outline a future vision of m-Health, focusing on the technological advances and global healthcare challenges and requirements.

8.2.1 Mobile Communications, Internet, and Computing Challenges of m-Health

Most recent m-Health technological innovations were driven by three major technological developments:

- Massive developments in wearable and associated consumer sensing technologies and connected devices.
- Advances in smart mobile phone communications and their health Apps computing platforms.
- New developments in the World Wide Web, massive social networking, and mobile Internet access.

Although these developments represent technological advances of the main m-Health building blocks, they were largely propelled by increasing healthcare markets in the "m-Health monitoring" domain. Moreover, these developments have initiated successful global m-Health business opportunities in developed countries. As discussed in earlier chapters, this largely commercial drive consequently led to clinically debatable paths that contributed increasingly to the current paradoxical status of m-Health, with its relative success in some healthcare areas and failure in others. As we have seen, these costly health devices are out of reach for most potential patients in the developing world, and they lack the necessary mechanisms to achieve solutions to basic global healthcare needs; this has added further debate to this notion of m-Health.

Interestingly, these trends are continuing, with all their concomitant clinical uncertainties and speculative healthcare benefits. As we have also seen, the different financial and business models developed around these solutions provided a powerful economic argument for m-Health, but are as yet ineffective and frugal from a global health perspective. While m-Health businesses are essential to future systems, they should not necessarily continue to represent the exclusive driving force for m-Health development. This process leads to increased commoditization of consumer-based m-Health systems and is weakened by its global healthcare transformative capabilities and benefits; furthermore, it may lead to narrowing applicability, particularly for providing large-scale global healthcare delivery solutions.

As mentioned in earlier chapters, although the first forays into mobile health were introduced in the early years of this century, they were interpreted as advancements from the original definition of telemedicine as "medicine practiced at a distance." The clear vision was to reshape the structure of healthcare delivery globally around mobility rather than the traditional desktop platforms of telemedicine (Istepanian and Laxminarayan, 2000).

Later, when the term "m-Health" was first coined and defined as "mobile computing, medical sensors, and communications technologies for healthcare" (Istepanian et al., 2004), these trends remained the main catalyst, embracing many technological, engineering, and scientific disciplines that shaped m-Health into its current format. Nowadays, m-Health can bring about major improvements in people's lives, especially those who are elderly, disabled, or chronically ill (Silva et al., 2015). However, uncertainty still looms as to how best to implement these developments and to realign them to better and broader visions of m-Health that are acceptable to most people, not only to the few who can afford it. This new vision needs to form the basis of any future evolution that goes beyond the current nexus of smartphone-centric systems and their Apps. We will explain some of these future developments and observations next.

8.2.2   Challenges to Future Mobile Communications and Wearable Technologies

The future influence of wearable sensors and devices on mobile communication for m-Health will be important areas in the evolution of mobile health. Future trends in commercial consumer-based wearable technologies will be centered on the following aspects:

- Stylized and invisible systems, such as those embedded in watches and electronic fabrics.
- More personalized, reliable, and robust communications connectivity.
- Association with more cognitive, behavioral change, and social functionalities.

Examples of future developments include wearable health monitors with extremely low power sensors and embedded textiles, leading to personalized motivational health trackers (Misra et al., 2015; Ozturk et al., 2015). These devices will harness self-powered wearable sensors and although they are largely in their developmental and prototype stages, they could potentially trigger new cost-effective and energy-efficient monitoring solutions that go beyond today's relatively expensive wearable health trackers. Other developments include wearable "biostamp" sensors that represent thin, stretchy, and skin-like patches that are powered wirelessly, using near-field communications (NFC) to provide skin electronic sensing for body health monitoring (Perry 2015). Ongoing developments of new diabetes management solutions, particularly advances of "robo-pancreas" systems, include wearable glucose sensors that continuously sample the glucose concentration under the skin. These sensors wirelessly link to a compact insulin pump for administering the appropriate insulin levels via a pipette under the skin, using a computer algorithm housed in a compact wearable device (Ross 2015).

An important challenge for future wearable monitoring systems is their ability to be used longer term, particularly for accurately monitoring vital signs on a daily basis. Recent U.S. statistics indicate that one-third of American consumers who have owned a wearable product stopped using it within 6 months, while among the 1 in 10 American adults who own some form of activity tracker, half of them no longer use it (Ledger 2014). These statistics indicate a need for further work, particularly for proactive health monitoring and for self-coaching to support self-motivation by users of these devices (Schraefel and Churchill, 2014). The development of more economic and cost-effective versions of these devices beyond the current consumer markets is an important element for their clinical viability and wider use.

The security and privacy of future wearable devices and their fragile immunity toward sophisticated future cybersecurity threats such as medical data leaks, network security, and personally identifiable information also requires further work (Blum 2015). The concept of m-Health based or medical Internet of Services (m-IoS) will be another major research theme in the near future. This will provide intelligent new infrastructure of uniquely identifiable health-sensing devices capable of communicating wirelessly with each other and connecting seamlessly with healthcare providers and care services through the Internet and new communication channels.

From the communications perspective, the advent of new wireless standards and network connectivity with thrifty Internet access will take m-Health applications, especially for global health, to new dimensions. Remote healthcare developments in the foreseeable future will exploit the technology of the Fifth Generation (5G) of mobile communication systems, which will underpin the long-term evolution of m-Health and bring new services and consumer use models. Other mobile network technologies that constitute part of the 5G evolution can also be designed and developed to provide wireless broadband m-Health services on the scale of the metropolitan area network (MAN), including the IEEE 802.16e mobile WiMAX and the IEEE 802.16j multi-hop WiMAX systems (Alasti et al., 2010).

The competitive and effective usage of the constituent 5G technologies for m-Health applications will be a key commercial challenge in the future, with extensive work and debate. Alternatively, there are recent, albeit commercial, attempts to provide free global Internet access other than by traditional mobile and cellular connectivity links. For example, "Outernet" is a project that aims to provide free, worldwide Wi-Fi connectivity via hundreds of low-cost miniature satellites known as "cubesats" to be launched into Earth orbit (Yirka 2014). "Internet.org" is a similar initiative by Facebook collaborating with the telecommunications industries, with the aim of bringing the benefits of free global Internet access to everyone on the planet (Internet.org 2015). Regardless of the fact that these have commercial appeal and their long-term viability is yet to be seen, their potential benefits to mobile health are tremendous and transformative.

Another outside contender is the "Li-Fi Internet," which is claimed to be potentially a hundred times faster than present speeds (Taub 2015). In the health context, this could be termed "m-Li-Fi," a new medical light communications and Internet connectivity information medium for mobile health.

It is this juxtaposition of developments that has prompted the introduction of m-Health 2.0 that we defined earlier as the future of m-Health that could bring combined technological advances to provide better healthcare services at lower cost. This would be accomplished with more adaptable business models, ecosystem, and deployment scenarios, not only for high-income countries but most importantly also for poor countries and the developing world. We can envision some examples of future developments as follows:

- Personalized m-Health systems that are tailored to the functionalities of future 5G mobile communications standards. These systems, if designed and developed appropriately, will be able to deliver personalized multitasking healthcare services with multifunctional communications for universally accessible healthcare delivery that is compatible with different health environments and conditions.

- New "tactile m-Health 2.0 systems" with connected systems and objects. This new sphere will allow connected smart devices and sensors to be developed around the tactile Internet; this will provide more robust human reaction and interaction communications. However, these systems need to be designed and developed with cost-effective and affordable target formats, to be used globally in various clinical environments and compatible with standardized and interoperable functionalities.

- New smart m-Health 2.0 systems that can be exclusively tailored to hospital and home care services and applications. These include rehabilitation, elderly care, and assisted living.

- Development of new intelligent robotic-based m-Health systems that can communicate seamlessly with other smart devices and objects to support different healthcare delivery and disease management scenarios. These systems can be envisioned in various formats, from compact nanolevels of miniaturized robots to humanoid health companion formats. One example would be a miniaturized robotic system that can communicate wirelessly and deliver drugs directly to infected tumors in the body (Atakan and Akan, 2012; Krishnaswamy et al., 2012). Another example is a miniature mobile robot that can deliver minimally invasive therapy to the surface of the heart. A further example is a humanoid companion social robotic system programmed with multiple tasks required for long-term chronic disease monitoring and management or elderly care (Istepanian et al., 2014). However, the likelihood of the rise of the more automated and increasingly autonomous robotic technologies in many healthcare and medical surgery applications in the near future requires greater attention to ethical scrutiny and more medical robot ethical research.

- The development of new cybersecurity m-Health systems with resilient features that can give immunity to future security and privacy attacks. These include, for example, intelligent counterthreat systems to prevent or alert to any unauthorized hacking of medical data or personal information.

All of these developments are feasible in the near future and will constitute part of the new generation of m-Health that will go beyond its current confines.

8.2.3 m-Health Big Data, Social Networking, and Cognitive Apps

The science of big health data will be one of the main research themes in the future of m-Health. There is uncertainty associated with the massive amount of health data generated by different devices (e.g., fitness trackers, health wearables, and other monitoring devices) and social networks, and how this data can best be accessed and interpreted by clinical experts. The extraction of meaningful information will be determined by a combination of social networking and intelligent data sensing that is associated with big data analytic processing. The resultant technologies will be used increasingly to personalize future mobile health products and services, and to generate personalized healthcare outcomes at lower cost, with improved patient engagement and experience. Moreover, the appropriate integration of "persuasive technologies" with "intelligent data analytics," using seamless "wireless connectivity and cloud access" could be one of the anticipated future developments. These combined developments, termed as the Social Networking, Mobile, Analytics, and Cloud (SMAC) trends, will represent a powerful driving innovation path for future consumer-led m-Health solutions and services (PwC 2013).

The increasing role of social networking within future m-Health systems will be one of key development trends. The current examples of social networking platforms applied for different healthcare areas, as we discussed in Chapter 3, will evolve to their next levels of functionalities with future m-Health systems. These will include, for example, embedding "big health data analytics" for the prediction of disease epidemiology or for personalized health monitoring linked with individualized environmental and workplace conditions, especially in urban areas. One example is Big Data mapping of mobile tweets correlated with global WHO emotional data sets to better understand emotions and mental health issues and their relevant distributions (Larsen et al., 2015). Many challenges remain on how effectively these can be successfully integrated for viable long-term use, and how they can benefit all stakeholders, including healthcare providers and clinicians. All this remains open for future research.

Given the unprecedented popularity in the health Apps market, the promise of more clinically acceptable Apps for specific m-Health services, such as chronic disease management, is difficult to predict. Nonetheless, it is expected that the market power and popularity of mobile phone communications and medical devices will continue to push for new strategies. Developments will introduce more health-oriented Apps linked to smartphones, with intelligent wearable health devices (e.g., smart watches) increasingly becoming part of this cycle. New consumer health-related gizmos designed for mobile vital sign monitoring and management (e.g., cardiac, COPD, diabetes, and other chronic conditions), will dominate this sphere, together with compact "Star Trek Tricorder" look-alike intelligent systems for real-time mobile diagnostic and monitoring functions.

The 2015 launch of many "health vaults" by major mobile phone and medical industries, with their elegant push to the consumer health and wellness market, is their latest strategy in this hugely profitable market. Whether these will succeed where previous systems such as the now defunct "Google Health" have failed remains speculative. The recent launch of Apple's "HealthKit" is another attempt to revive this concept, albeit in more cautious and focused way.

Unless new m-Health strategies are adopted and lessons learnt from previous attempts, new m-Health services will remain constrained and limited at best. Apart from acceptability by healthcare professionals and patients, the cost of these market devices remains out of reach to the majority of patients, certainly in the developing world. Some possibilities might also exist for potential offshoot products from these developments, which can lead to clinically viable m-Health solutions that might address some of the uncertainties. These solutions would need to combine a workable pitch that will be both clinically acceptable and economically affordable, but this is as yet a future vision.

8.2.4   Open m-Health 2.0 Collaborative Registry

A robust and secure m-Health 2.0 open collaborative registry could be an important future development. It can be especially useful for collecting effective clinical evidence and relevant health data from different patient populations for various diseases. Any new models can benefit from the open architecture principle of shared m-Health data between different patients or populations, healthcare providers, and researchers, perhaps in similar ways that genomic data research has benefited in the past.

An earlier example has been proposed that advocates these principles, building on open-software architecture to address the many data sharing and "sense-making" bottlenecks that exist in current m-Health systems (Estrin and Sim 2010; Chen et al., 2012). However, open architecture proposals are based on the current mobile phone-centric model of m-Health with its smart App modalities. These have their own limitations and disadvantages, as discussed in earlier chapters. They also have serious inherent risks and fragility issues, particularly with their security, privacy, confidentiality, and data ownership, which can potentially formulate major risk factors and problematic challenges for patients and healthcare providers.

The development of m-Health 2.0 collaborative data registry architectures, with more robust and secure connectivity and immunity to these risks, is feasible. These intelligent open-network repositories of m-Health data, if set up appropriately, can provide a supportive geographical dashboard of the main thematic areas of m-Health data and identify activities, avoid duplication, and provide better use of global m-Health outcomes. These common themes will also facilitate the development of new knowledge bases as well as the reinforcement of new partnerships.

However, these structures need to be designed and developed with less suscepti-bility to current Internet security and privacy threats and coordination with global mobile telecommunications network operators and other m-Health vendors.

8.2.5 Rapprochement Between e-Health and m-Health

Developments that address the successful rapprochement between m-Health and its classical association with e-Health are vital for the future of the two concepts. For example, from the mobility aspect of e-Health, systems are being continually developed and marketed to achieve secure integration of electronic health records (EHRs) and personal health records (PHRs) using different global mobile access mechanisms. However, the robust design and development of such systems, with complete immunity from existing Internet privacy risks and cybersecurity attacks, remains a challenge. Where existing EHRs serve the information needs of healthcare professionals, many e-Health systems have PHRs that capture the health data of patients and are controlled by them. However, the robust design and development of new reconfigurable, clinically trustworthy mobile "tethering" mechanisms that can successfully link PHRs and EHRs for both organizational needs and patient require-ments in different global settings remains a future challenge. There is therefore a need to design and develop new "mobile e-Health" systems that can be more individualized for different clinical requirements; these would allow, for example, smarter medica-tion protocols that combine the delivery of specific drugs to treat individual patients based on their specific "genetic push" and "habitual pull" (Topol 2012). These systems also need to be implemented with workable standardized and interoperable m-Health platforms that can allow their seamless global access using future mobile communications systems. These and other rapprochement challenges between e-Health and m-Health will remain open topics for future work in the foreseeable future. A new global revision on the differentiation between the e-Health and m-Health concept is required to clear the current blurring views between the two concepts.

8.2.6 m-Health Globalization

The future of global m-Health is better illustrated by addressing the major global health-related challenges defined by the UN's Millennium Development Goals (United Nations 2014; Mechael and Solninsky 2008). More robust and sustainable m-Health solutions and technologies are required to address the more tailored and cost-effective healthcare connectivity configurations that complement (but are not exclusively based on) mobile phone-centric models. Details of the technology and other barriers to the greater use of current m-Health technologies for global health, particularly their effectiveness, are addressed elsewhere (Howitt et al., 2012; Free et al., 2013). These studies indicated poor clinical outcomes with no clear evidence of the benefits of m-Health for some healthcare service delivery applications (Free et al., 2013).

Whichever way we look at the rigor of these studies, the conclusion seems to be that "the jury is still out" on judging the benefits of current m-Health systems. The overall picture is disappointingly negative. Moreover, the economic and financial benefits of m-Health are also not clear from these studies and there is a need to conduct future studies to demonstrate objectively that current m-Health systems can save time and money, particularly for the developing world.

As a consequence of this status, it is imperative to develop new, alternative m-Health systems that are tailored toward global healthcare outcomes, especially in poor resource settings. These must embrace some frugal technology options of m-Health that can overcome some of the current barriers listed in the studies cited above. Some suggestions include the following:

- Development of alternative Internet-based m-Health access systems that will allow either free or very low-cost global communications. The m-Health Outernet system can potentially be used to provide free Internet access to health and medical care services as an economic alternative to cellular connectivity access. Such new approaches require global and philanthropic efforts, with appropriate funding to be successfully implemented.

- Development of low-cost and low-energy medical sensors that can be realized for different diseases and for remote patient monitoring, especially in poor settings.

- Design and development of new solar-powered communication devices using, for example, asynchronous transfer mode (ATM) for mobile transmission; these could be used as potential options for new m-Health communication systems, especially in remote and rural areas. This can be seen as an economic and viable option to the cellular mode of mobile communications.

- Development of new m-Health systems that exploit these evolving technologies to address some of the leading global diseases and causes of disability by 2030. These include unipolar depressive disorders, ischemic heart disease, cerebrovascular disease, COPD, and lower respiratory tract infections; the disability causes include road traffic accidents (WHO, 2008). So far there are few existing practical m-Health solutions for some of these diseases.

Future studies need to focus on new ecosystem and business models that address these new m-Health architectures and configurations. It is also important, due to the expected growth and importance of global health, to establish a UN-sponsored global m-Health organization to focus on the long-term planning and coordination of new strategies. These would include sustainable planning of different global m-Health initiatives that are tailored for the major long-term healthcare problems and economic conditions of the developing world.

8.3   CHALLENGES AND EXPECTATIONS: m-HEALTH "MARKET" VERSUS "SCIENCE"

For nearly a decade, most of the global institutions such as the WHO, UN, World Bank, ITU, and many other international and governmental bodies were and still are interested in embracing and supporting m-Health, albeit with varying results and outcomes. This support has not been crystallized into an agreed and coordinated approach to set up and establish a global entity dedicated to m-Health. The establishment of such an organization is both timely and vital considering its global importance. Assigned tasks could prioritize the benefits of m-Health and identify strategies to tackle major health problems, including addressing specific global health emergencies and disease epidemics. The absence of such a global entity can be attributed to a complex and conflicting range of factors from political to financial, organizational, and lack of leadership, among others.

The current unclear differentiation between the *for market* and *for health* m-Health systems is evidence of this lack of global coordination. It could be argued that there is no difference between the two. If this notion were true, then the hype of many years ago should have led to the large-scale, diverse, and successful clinical deployment of m-Health services. Instead, the reality is a disproportionate level of interest between the global markets and the business of m-Health from one end and the clinical uptake on the other. The rapprochement between the two ends is a complex task due to the push and pull requirements and complex driving factors that are difficult, although possible, to bridge. Furthermore, the business influence is increasing this gap toward more consumerism of m-Health products and services. However, a likely candidate for the "business-clinical" convergence to be successfully accomplished with proper m-Health strategies is disease prevention and prediction.

Most of current market-driven m-Health solutions are being developed for high-income countries, relying on their potential markets and user applicability. These solutions also need to address similar demands for low-income and developing world countries. Otherwise, further divergence of m-Health into very different paths will occur. This fact is evident by the need to develop more frugal m-Health technologies dedicated to wider global healthcare, and not limited to the healthcare demands in high-income countries. This is particularly the case with the development of 5G, which is seen as an urban system with little impact on rural areas (IEEE ComSoc Technology News, 2015). The barrier here is more commercial than technical: in rural areas, especially in developing countries, there is a "lack of potential revenue per square mile" (NGMN Alliance, 2015).

The recent demise of several high-profile market initiatives and products such as the "Google Glass" and "Google Health," which aimed to dominate health consumer markets, are examples of the "hyper-gadgetization" of m-Health with global brands and products, without careful consideration of healthcare necessities, user demands, and expectations. Similarly, the wearable and fitness tracking industries are increasingly pursuing rigorous marketing strategies to sell their products for the prevention and wellness market, with prohibitively expensive and out-of-reach products for most users in LMICs. These consumer-driven products, although thriving commercially, remain largely unknown territory in terms of their clinical validity, reliability, and efficacy. In addition to the serious concerns over data ownership, privacy, and security of the health-related data generated daily from these devices, this segregation gap is increasingly contributing to the uncertainty of the long-term benefits and viability of these m-Health products and services. Hence, there is a demand for a balancing vision for future m-Health systems to address this asymmetry, with more equilibrium and effective implementation strategy between their global healthcare benefits and commercial market potentials.

Another relevant factor is the popular interpretation of m-Health by clinicians. The widely held view is of cellular technology that enables healthcare delivery tasks. This is based on mobile phone-dependent solutions, such as mobile access and viewing of patient clinical data and information, or the use of mobile terminals for remote medical education and training, mobile access to health information, and clinical notifications. Others include mobile clinician or patient communications (e.g., text messaging and e-mail), systems for timely access to health information resources, devices for secure mobile access to hospital information systems (EHR, EMR, etc.), and clinical decision support systems. All these applications represent this popular "clinical traction" factor and a view of m-Health as simply a "mobile e-Health" concept. This view was naturally popularized by the wider use of mobile phones and smart terminals among clinicians and healthcare providers for these and many other applications.

It is vital to revise this view of communications, as it is important to notice that the two embodying communication terms of *mobile* (cellular) and *wireless* are diverging into different future m-Health paths. This separation within the future healthcare delivery roadmap will be based on the former's current association with "mobile phone-centric" m-Health compared to latter's future "wireless" connectivity paradigms. To illustrate this notion, we cite a few examples for clarification. The first is the future wireless connectivity of medical monitoring and diagnostic devices. These will be largely used in different m-Health applications, from medical sensing to diagnostic. The communications of these devices and sensors will eventually be designed and developed with alternative Internet connectivity links.

Another example is from the evolving areas of robotics in healthcare and surgery. The future vision is of humanoid robots that will act as health companions. These will interact and react emotionally and wirelessly with patients, analyze and process their daily activities, convey the necessary health advice and feedback, and inform healthcare providers of their medical progress. Such scenarios will also be implemented using M2M and IoT wireless connectivity paradigms. Many of these and other futuristic m-Health scenarios and applications are likely to be configured with similar connectivity links and will not necessarily be based on a smartphone cellular network as the main communication hub, but as an option. This future vision will not only alter the current clinical traction with an alternative view of m-Health but will also empower the notion of patient-centered m-Health. This notion includes future "wireless connectivity" systems that will be less dependent on "mobile phone" connectivity, as opposed to more economically viable Internet-based communication channels linking users, patients, and their healthcare providers.

8.3.1 The Business and Clinical Rapprochement of m-Health

m-Health will continue to be one of the key markets of different stakeholders, mainly mobile telecommunications operators, and mobile phone, pharmaceutical, and other medical device industries that will reach their products and services beyond their current markets. Emerging technologies and related products will be another influencing factor in the success of future m-Health solutions that are dependent on these systems. For example, adopting new reimbursement models for m-Health services will be subject to extensive debate in most of the developed countries in the future. This will be particularly true in the United States, with the new legislation of the U.S. "Affordable Care Act" that puts consumers back in charge of their healthcare, with more healthcare cost, coverage, and care benefits, using new forms of m-Health services. Similar discussions are also ongoing in many European countries regarding the most appropriate m-Health policies and reimbursement strategies. These require the development of new m-Health ecosystem models that reflect the expected market growth. As shown earlier, extensive work toward an ideal "m-Health ecosystem" has been reported in the past, but so far there is no accepted model that can be widely adopted successfully.

Most of the current m-Health models are market-driven or have specific areas of focus, so there is scope for rectifying this process and achieving new m-Health care models with wider clinical acceptability and implications for patients and benefits to healthcare providers. This can only be realized if all interested stakeholders adopt new, balanced approaches to adopt economically valid, but more clinically acceptable models. The earlier global initiatives between different organizations were not properly coordinated, leading to the current disparities and weaknesses of many m-Health initiatives. The globalization of decreasing healthcare disparities and inequality are paramount in achieving better health access and quality. The design and deployment of future mobile health technologies should aim to decrease healthcare disparities between countries and their populations, with lower profit margins to ensure better healthcare outcomes.

8.3.2 Multidisciplinary Aspects of m-Health

The future of m-Health will be based on developing new multidisciplinary collaboration aimed at tackling many of the current medical challenges. The key building pillars of m-Health will be central to such models, but cannot achieve the complete solutions alone.

As an example, new collaborative frameworks between diverse disciplines such as psychology, social sciences, design, and engineering will be vital for better understanding some of the current barriers and clinical constraints associated with existing m-Health solutions for chronic diseases. For example, there is increasing discussions about the role of m-Health as part of some health psychology disciplines (Dalku et al., 2015).

As presented in earlier chapters, many recent studies have illustrated this notion, with the necessity for more innovative smart, educational, and behavioral change systems to be designed with traditional clinical and m-Health interventions for many chronic diseases, such as diabetes and obesity. The successful correlation of the design and development of these m-Health systems, with targeted clinical outcomes, is a complex task and is not yet widely accepted or validated clinically.

There have also been extensive discussions on the preventative role of m-Health in tackling many of the urgent healthcare burdens of modern society, such as the association of certain lifestyle behaviors and diet with the onset of chronic conditions and cancer. However, many questions remain unanswered and more innovative m-Health research work is still required.

An important example in this direction is the global epidemic of childhood obesity. This major health problem will be a significant public health crisis, with urgent need for the injection of vital new and effective m-Health solutions for intervention design, delivery, and diffusion of treatment and prevention. Further examples are the early detection of prediabetes and some forms of cancer, such as cervical, colon, and bowel cancers. These innovations will require novel multi-disciplinary collaboration to exploit advances in medical sensors, biomaterials, engineering, genomics, and biology. Successful collaborative frameworks will open new horizons for m-Health in many important but as yet untapped areas. We will list some of these next. Other examples might include finding more innovative solutions to the prediction of cancer and other chronic conditions such as diabetes and cardiovascular diseases.

8.4 FUTURE m-HEALTH SCENARIOS

A new decade of affordable, viable, and clinically effective m-Health is important for its long-term continuation as a truly transformative healthcare innovation concept. This new era requires a vision of how the future of m-Health will look as it moves away from its current status (Sifferlin, 2015). This vision needs to be addressed, and must overcome many of the challenges and barriers that have dominated its progress in the past 10 years, especially in terms of affordability, wider clinical acceptability, and scaling-up. The future remains open as to how such a vision will be successfully realized, considering the level of these and other challenges from the economic, societal, and environmental arenas. Some of our modest suggestions and contributions toward the m-Health vision are presented next.

8.4.1 Frugal m-Health Systems

The development of frugal m-Health systems that are workable and usable by poor as well as rich patients is vital for the global success. The progressive role of m-Health as a transformative healthcare concept that can benefit larger patient populations worldwide has so far been largely belated in achieving these goals. Many influential factors, such as the twinning association of m-Health with mobile phones, the strong influence of global market forces, the high costs of sensing devices, and lack of global initiatives outside these circles have contributed to its current status. There is therefore an urgent need to seek new and more frugal communication and sensing systems associated with m-Health that are affordable and workable across different mobile standards and functional modalities (including but not exclusive to cellular networks). These new systems also need to harness the power of Internet connectivity. The development of open data analytical tools accessible to users and developers, tailored for the development of these systems, is vital for this vision.

The new m-Health era also needs to address the numerous challenges in current hospital care systems, not only in high-income countries but also in developing countries and poorly resourced ones. As seen in earlier chapters, there is increasing use of different smart mobile devices and terminals that are becoming the choice for access to EHRs and EPRs by clinicians in many leading hospitals in high-income countries. The recent uses of wearable bracelets and activity-monitoring trackers to monitor postsurgery patients and those with chronic conditions are good examples (Chiauzzi et al., 2015). However, most of these solutions are unlikely to be affordable in similar care settings in developing and poor-resource countries. Current wearable devices on the market are beyond the reach of healthcare providers in such countries. A good economic and business strategy for the design and development of affordable and reliable wearable monitoring devices that can be workable and applicable globally will be beneficial for scaled-up m-Health services. These and other m-Health examples are seen as a means of transcending cost boundaries and addressing the real healthcare disparities that can enable global m-Health continuum and success.

There is also a need for robust and effective "m-Health warning systems" for timely prediction of global epidemics; this requires more multidisciplinary collaborative work, as discussed in earlier chapters. The necessity of developing workable evidence-based m-Health frameworks to achieve effective and scalable m-Health objectives targeting urgent global healthcare challenges is also vital for the future vision of the new m-Health era. Any such framework in this context needs to address not only the key healthcare challenges but also the problematic issues associated with current m-Health systems and interventions.

The coordination and development of large-scale global randomized controlled trials with necessary clinical procedures can ultimately overcome most of the current poor evidence and limitations associated with pilots and localized studies. The development of these global pilots also needs to address the quality of intervention and how and when these m-Health implementations of scale can be achieved in various care environments. The need to develop new global m-Health interoperability standards is also vital in this process.

8.4.2 Nano m-Health and Precision Medicine

One of the transforming developments in future m-Health systems will be the successful rapprochement between m-Health and advances in the *omics*, *nano*, and *mem* areas. In recent years, the reality of a complete "digitization" of any person's genome can be achieved with fast and effective genomic sequencing technologies costing less than US$1000, and perhaps much less in the future. These innovations in genomic sequencing are opening up new research horizons that can contribute largely to the development of personalized medicine in the near future.

The benefits of personalized medicine are many and well known, and can be summarized as follows:

- Predicting the patient's susceptibility to specific diseases and treatments, improving disease detection, and preempting disease progression.
- Shifting the emphasis in medical care from reaction to prevention and better customizing disease prevention strategies for early prevention.
- Development of more effective drugs and avoiding those with predictable side effects.
- Reducing errors in new clinical trials and providing tangible benefits of time and cost.

Future developments in genomic sequencing, combined with m-Health developments, will contribute to accelerating research efforts and the realization of these goals.

Complementary research topics are "genomic signal processing" and "biomolecular communication." These methods combine advanced signal processing and digital communication theory with genomics to provide important clues for better understanding and developing future personalized treatments. Past and ongoing work is promising, especially with new applied methods that can potentially lead to major breakthroughs in future personalized diagnostic systems (Istepanian et al., 2011; Wang et al., 2003).

Recent advances in molecular diagnostics can lead to new research insights in the prognosis and treatment of many diseases. These developments can detect and measure the presence of genetic material or proteins associated with a specific health condition or disease, helping to uncover the underlying mechanisms of disease and enabling clinicians to tailor care at an individual level, thus facilitating new personalized medical treatments. Research in real-time molecular diagnostics that combine biomolecular processing embedded with real-time mobile sensing is ongoing. This "digital biology," if successfully developed and implemented in clinical settings, can provide major breakthroughs in many aspects of medicine. Some examples include the early detection of some forms of cancer that are difficult to diagnose, or the onset of diabetes and its prevention, especially in "prediabetics."

Another future research direction worth mentioning is "nano m-Health." Developments will aim, for example, to produce new smart nano M2M healthcare communication systems and sensing devices. These nano m-Health systems can potentially provide suitable mechanisms for molecular drug delivery in the body and transform smart medicine to new levels. These developments can play an important role in the future research of precision medicine and its different disciplines.

8.4.3 Green and "Mem-Health" Systems

There has been for many years increasing debate about the health concerns and environmental impacts of mobile phones and their network infrastructures.

These concerns include the amount of energy consumed, the effect of high mobile phone use on human health, and other risk implications (IARC, 2011; Hardell and Carlberg, 2015). They are compounded by the impact of global mobile use and its associated infrastructure on the effects of climate change. This impact has prompted increasing calls for more energy-efficient and environment-friendly communications. In particular, the need for more energy-efficient and green mobile communication systems will be potential research areas for new m-Health systems. For example, recent advances in developing new "memcomputers" (short for memory-crunching computers) that embrace new electronic energy principles (Di Ventra and Pershin 2013) will potentially become one such area of future research, with the aim of developing new intelligent "mem-Health" energy-efficient sensors and mobile communication systems.

Further, more work will be needed on the role of m-Health systems to better predict the health impact of increased urbanization, with concomitant traffic and industrial noise. Similar solutions might also be able to contribute to current environmental challenges, such as developing new geo-monitoring solutions in polluted areas and their effect on human health. Others include monitoring wildlife and animal sanctuaries, and better understanding their well-being and behaviors, especially in their natural habitats, as well as the impact of human intrusion and climate change on these habitats, especially for endangered species of animals and birds.

8.4.4 Cyber-Physical Connected m-Health Systems and Smart Hospitals

It is expected that future research in m-Health will cover multiple service sectors with which people interact and use on a daily basis to monitor their healthcare and well-being in various environments. As an example, there is an increasing link between m-Health and smart transport environments, where future m-Health systems could be used as potential solutions for the prevention of road traffic accidents by monitoring and alerting the driver's own behavior and fatigue (CTC & Associates 2015). Other examples include the development of m-Health systems embedded in intelligent and connected transportation systems (ITS), where smart vehicle manufacturing can potentially yield safer, efficient, and more reliable transport. Moreover, monitoring patients during air travel and in airports is another scenario, where risks to passengers before, during, and after long-haul flights or train journeys can be mitigated.

Most of these and other futuristic scenarios need large and complex sensing and processing structures, with spatially distributed systems that can interact and be controlled by a considerable number of distributed m-Health network computing and processing elements.

These will be at the core of new "trustworthy cyber-physical m-Health systems" that can be developed and tailored in the future for these and other health-related scenarios as part of the emerging paradigm of cyber-physical systems (CPS). CPS refers to a new generation of systems with integrated computational and physical capabilities that can interact with people through many new modalities (Baheti and Gill, 2011). This can also be viewed as one of the many alternatives to the IoT connectivity concept that we discussed earlier in Chapter 4. This emerging area will also bring major security and privacy challenges that will be the subject of future research, particularly in the diverse settings envisaged, from future digital hospitals to connected home, transportation, and car environments.

Finally, recent developments in the key m-Health components that we have discussed throughout this book are driving new approaches that aim to develop future generations of "smart hospitals." Such hospitals can achieve maximum collaborative communications, healthcare service coverage, resources management, and workflow consolidation by reducing cost and improving patient satisfaction. These future hospitals will be equipped with the latest infrastructure, including mobile, IT and other technological, data processing, and computing equipment to enable them to be patient-centered, thereby facilitating patient care, safety, and satisfaction; staff-centered, to improve productivity and efficiency; and also system- and innovation-centred to enable community integration, resource utilization, and improved economic performance. Today, most of the relevant mobile IT and networking enterprises and other IT for healthcare service providers are advocating different solutions to achieve these objectives. The smart hospital approach is still evolving but remains largely limited and at an early stage of development, with many challenges remaining open for further research and testing. These include, for example, the new 5G communications and M2M networking connections and their integration with hospital staff, workflow, and medical equipment, combined with cognitive computing for EHRs, applications of smart robotics in hospitals, multi-agents for data mining, and many others. The successful integration of future m-Health services within this vision is also an important challenge in this process.

8.5 SUMMARY

A decade ago, when the first m-Health book was published (Istepanian et al., 2006), most of the technical innovations, from intelligent wearable devices to wireless video streaming and mobile monitoring, were still in their early developmental stages. Today, these constitute major global markets, and there are m-Health products and devices available for a wide range of health and wellness applications. This rapid evolution and technical advancement is testament to the vitality and importance of m-Health in the healthcare technology and innovation sectors and to its potential role in transforming medical care in the digital world of the future. It is also noteworthy that there are around 30 international conferences and work-shops annually related to m-Health, one of the largest being mHealth Summit (2015).

In this final chapter, we have endeavored to look into a crystal ball, so to speak, to give some notion of future m-Health advances, developments, and challenges. We conclude with some final thoughts on this matter. There have been soul-searching discussions about where m-Health stands at this moment and where it is heading in the future. The reason for these continued agonized discussions is the clear lack of understanding of m-Health as a science, as opposed to the view of it as a technological and applied tool for healthcare delivery using mobile systems.

In the early days of the last decade when the notion of m-Health was first introduced, it not only caught the eye of academic and medical communities but, most important, also the attention of mobile phone operators and telecommunications industries. This was obviously based on the potential market opportunities that such a concept could bring to their businesses. This interest had a profound impact on shaping the vision of m-Health as a mobile healthcare delivery system. The massive commercial appeal and the proliferation of the m-Health Apps culture that we see today is a major part of this initial interest—and yet it is a misguided view, as we have also seen.

The public at large and patients in particular are increasingly accepting m-Health as a tool for improving their health or monitoring their wellness. Similarly, it has been favorably, albeit cautiously, embraced by a majority of interested healthcare providers and clinicians as a means of reducing cost and providing better healthcare services for their patients.

However, from what we have seen of the large evidence-based studies cited in this book, this narrow notion did not work at worst, or has worked to some degree for a few at best. The real picture is debatable and still not clear, but surely it was not a match to the original expectations and much-hyped perceived benefits. This misfiring eventually achieved less sustainability and more uncertainty as to the future of m-Health. Furthermore, the full potential and promise for its transformative characteristics and benefits for global health services, and tangible solutions to urgent health challenges, are not yet fully utilized nor understood, particularly for the poor and needy in this world.

m-Health is currently in flux and to a large extent a successful global business, but it is not yet a successful and clinically acceptable scientific domain. However, despite its immense success in the former and the relative skepticism within the latter, m-Health will continue to thrive in the coming years, with much promise in changing the future healthcare landscape.

Finally, m-Health ought to be like a great jazz record, where people listen to it repeatedly over the years. It has all the necessary success ingredients of improvisation and the spontaneity elements that users both crave and enjoy!

REFERENCES

Alasti M, Neekzad B, Hui J, and Vannithamby R (2010) Quality of service in WiMAX and LTE networks. *IEEE Communications Magazine* 48(5):104–111.

Atakan B and Akan OB (2012) Body area nanonetworks with molecular communications in nanomedicine. *IEEE Communications Magazine* 50(1):28–34.

Baheti R and Gill H (2011) Cyber-physical systems. In: *The Impact of Control Technology*, T. Samad and AM Annaswamy (Eds.). Available at http://ieeecss.org/sites/ieeecss.org/files/documents/IoCT-Part3-02CyberphysicalSystems.pdf (accessed October 2014).

Blum B (2015) *Are your wearables safe from cyber-security threats?* Accenture. Available at http://www.accenture.com/us-en/blogs/technology-blog/archive/2015/01/18/are-your-wearables-safe-from-cyber-security-threats.aspx (accessed May 2015).

Chen C, Haddad D, Selsky J, Hoffman JE, Kravitz RL, Estrin DE, and Sim I (2012) Making sense of mobile health data: an open architecture to improve individual- and population-level health. *Journal of Medical & Internet Research* 14(4):e112.

Chiauzzi E, Rodarte C, and DasMahapatra P (2015) Patient-centered activity monitoring in the self-management of chronic health conditions. *BMC Medicine* 13:77.

CTC & Associates (2015) *Monitoring motor vehicle driver fatigue*, Transport Research Synthesis Report, Minnesota Department of Transportation. Available at http://www.dot.state.mn.us/research/TRS/2015/TRS1501.pdf (accessed May 2015).

Dalku M, Nikopolu VA, and Efharis P. (2015) Why mHealth interventions are the new trend in health psychology? Effectiveness, applicability and critical points. *The European Health Psychologist* 17(3):129–136.

Di Ventra M and Pershin YV (2013) The parallel approach. *Nature Physics* 9:200–202.

Estrin D and Sim I (2010) Open m-Health architecture: an engine for healthcare innovation. *Science* 330/6005:759–760.

Free C, Philips G, Watson L, Galli L, Felix L, Edwards P, Patel V, and Haines A (2013) The effectiveness of mobile-health technologies to improve healthcare service delivery processes: a systematic review and meta-analysis. *PLoS Medicine* 10(1):e1001363.

Hardell R and Carlberg M (2015) Mobile phone and cordless phone use and the risk for glioma: analysis of pooled case–control studies in Sweden, 1997–2003 and 2007–2009. *Pathophysiology* 22:1–13.

Howitt P, Darzi A, Guang-Zhong Y, Hutan A, Rifat A, James B, Alex B, Anthony MJB, Josip C, Lesong C, Graham SC, Nathan F, Simon AJG, Karen K, Dominic K, Myutan K, Robert AM, Azeem M, Stephen M, Robert M, Hugh AP, Steven DR, Peter CS, Molly MS, Michael RT, Charles V, and Elizabeth W (2012) Technologies for global health. *The Lancet* 380: 507–535.

IARC (International Agency for Research on Cancer) (2011) *IARC classified radiofrequency electromagnetic fields as possibly ᐧ carcinogenic to humans*, IARC-WHO, Release No. 208. Available at http://www.iarc.fr/en/media-centre/pr/2011/pdfs/pr208_E.pdf (accessed October 2014).

IEEE ComSoc Technology News (2015) Available at http://www.comsoc.org/ctn (accessed January 2015).

Internet.org (2015) *Making the Internet affordable*. Available at https://internet.org/contact (accessed May 2015).

Istepanian RSH and Laxminarayan S (2000) Unwired e-med: the next generation of wireless and Internet telemedicine systems. *IEEE Transactions on Information Technology in Biomedicine* 4(3):189–194.

Istepanian RSH and Zhang YT (2012) Guest editorial: introduction to the special section: 4G health—the long-term evolution of m-Health. *IEEE Transactions on Information Technology in Biomedicine* 16(1):1–5.

Istepanian RSH, Jovanov E, and Zhang YT (2004) Guest editorial introduction to the special section on m-Health: beyond seamless mobility for global wireless healthcare connectivity. *IEEE Transactions on Information Technology in Biomedicine* 8(4): 405–412.

Istepanian RSH, Laxminarayan S, and Pattichis CS (2006) *M-Health: Emerging Mobile Health Systems*. London: Springer International Publishing.

Istepanian RSH, Sungoor A, and Nebel JC (2011) Comparative analysis of genomic signal processing for microarray data clustering, *IEEE Transactions on NanoBioScience* 10(4):225–238.

Istepanian RSH, Good A, and Philip N (2014) Smart social robotics for 4G health applications. In: L. Roa Romero (Ed.), *XIII Mediterranean Conference on Medical and Biological Engineering and Computing 2013*, IFMBE Proceedings, Vol. 41, London: Springer International Publishing, pp. 1919–1922.

Krishnaswamy D, Ramanathan R, and Qamar A (2012) Collaborative wireless nanobots for tumor discovery and drug delivery, *Proceedings of the 2012 IEEE International Conference on Communications (ICC)*, Ottawa, Canada, June 10–15, 2012, pp. 6203–6208.

Larsen ME, Boonstra TW, Batterham PJ, O'Dea B, Paris C, and Christensen H (2015) We feel: mapping emotion on Twitter. *IEEE Journal of Biomedical and Health Informatics* 19(4):1246–1252.

Ledger D (2014) *Inside wearables: Part 2: a look at the uncertain future of smart wearable devices and five industry developments that will be necessary for meaningful mass market adoption and sustained engagement*. White Paper, Endeavour Partners, pp. 1–18. Available at endeavourpartners.net/assets/Endeavour-Partners-Inside-Wearables-Part-2-July-2014. pdf (accessed May 2015).

Malvey D and Slovensky DJ (2014) *m-Health: Transforming Healthcare*. New York: Springer.

Mechael PN and Solninsky D (2008) *Towards the development of m-health strategy: a literature review*, WHO and Millennium Villages Project, WHO, Geneva. Available at http://www.who.int/goe/mobile_health/mHealthReview_Aug09.pdf (accessed October 2014).

mHealth Summit (2015) Available at http://www.mhealthsummit.org (accessed January 2016).

Misra A, Bozkurt A, Calhoun B, Jackson T, Jur J, Lach J, Bongmook L, Muth J, Oralkan O, and Health Insight (2015) Available at http://mhealthinsight.com/2015/11/20/mhealth-events-for-2016/ (accessed January 2016).

El Hattachi R and J. Erfanian (Eds.) (2015) *5G White Paper*, NGMN Alliance, February 17. Available at http://www.ngmn.org (accessed January 2016).

Ozturk M, Trolier-McKinstry S, Vashaee D, Wentzloff D, and Yong Z (2015) Flexible technologies for self-powered wearable health and environmental sensing. *IEEE Proceedings* 103(4):665–681.

Perry TS (2015) A temporary tattoo that senses through your skin: the biostamp can replace today's clunky biomedical sensors. *IEEE Spectrum*, June, 26–31.

PwC (2013) *Through the looking glass: emerging trends*, PricewaterhouseCoopers, 21st Convergence India 2013 Expo, January 16–18, Pragati Maidan, Delhi. Available at https://www.pwc.in/en_IN/in/assets/pdfs/publications/2013/convergence-event-report-pwc.pdf (accessed October 2014).

Ross PE (2015) Diabetes has a new enemy: robo-pancreas, sensors, actuators and algorithms can automatically control blood sugar. *IEEE Spectrum*, June, 32–35.

Schraefel MC and Churchill EF (2014) Wealth creation: using computer science to support proactive health. *IEEE Computer* 47(1):70–72.

Sifferlin A (2015) Why (almost) everyone is embracing the digital doctor. *Time Magazine*, November 9. Available at http://time.com/4092350/why-almost-everyone-is-embracing-the-digital-doctor/ (accessed January 2016).

Silva BMC, Rodrigues JJPC, de la Torre Diez I, Lopez-Coronado M, and Saleem K (2015) Mobile heath: a review of current state in 2015. *Journal of Biomedical Informatics* 56: 265–272.

Taub B (2015) *New Li-Fi Internet is 100 times faster than Wi-Fi.* Available at http://www.iflscience.com/technology/li-fi-internet-could-be-100-times-faster-wi-fi-0 (accessed January 2016).

Topol E (2012) *The Creative Destruction of Medicine.* New York: Basic Books.

United Nations (2014) *The Millennium Development Goals Report.* Available at www.un.org/millenniumgoals/2014.pdf (accessed February 2015).

Wang XH, Istepanian RSH, Song YH, and May EE (2003) Review of application of coding theory in genetic sequence analysis, *Proceedings of the 5th International Workshop on Enterprise Networking and Computing in Healthcare Industry (Healthcom 2003)*, June 6–7, 2003, Santa Monica, CA, pp. 5–9.

WHO (World Health Organization) (2008) *The Global Burden of Disease: 2004 Update.* Geneva: WHO.

Yirka B (2014) *Project Outernet, looking to bring free Internet to entire world*, TechXplore.com. Available at http://techxplore.com/news/2014-02-outernet-free-internet-entire-world.html (accessed May 2015).

APPENDIX

USEFUL m-HEALTH ONLINE RESOURCES

These lists represent some of the online resources available to readers interested to learn more about m-Health's recent news, developments, and related topics. This is not an exhaustive list but a representative snapshot of the large volume of links relating directly or indirectly to different areas and advancement activities of mobile health.

A Google and Google Scholar search in June 2016 revealed the following hits:

Google: Mobile health: 189,000,000; m-Health: 83,300,000; mHealth: 1,940,000; m-Health App: 31,400,000

Google Scholar: 4,650,000; m-Health: 61,000; mHealth: 37,200; m-Health Apps: 2420

Online News and Newsletter Sites

MobilealthNews: http://mobihealthnews.com

mHealth Watch: http://mhealthwatch.com

mHealth News: http://www.mhealthnews.com

Fierce Mobile Healthcare: http://www.fiercemobilehealthcare.com

mHealth Insight: http://mhealthinsight.com

m-Health: Fundamentals and Applications, First Edition. Robert S. H. Istepanian and Bryan Woodward.
© 2017 by The Institute of Electrical and Electronics Engineers, Inc. Published 2017 by John Wiley & Sons, Inc.

m-Health and m-Health-Related Alliances and Foundations

mHealthKnowledege: http://www.mhealthknowledge.org

United Nations Foundation-Mobile Health for Development: http://www.unfoundation.org

Groupe Spéciale Mobile Association (GSMA–m-Health): http://www.gsma.com/mobilefordevelopment/programmes/mhealth

Healthcare Information and Management Systems Society: http://www.himss.org

Continua Alliance: http://www.conti/alliance.org.

European Connected Health Alliance: http://www.echalliance.com.

Annual m-Health Summit, Conferences, and Related Meetings

Connected Health Conference (Previously mHealth Summit): http://www.pchaconference.org/

MobiHealth: http://mobihealth.name

Digital Health Summit: http://digitalhealthsummit.com.

Health 2.0: http://www.health2con.com

Annual International Conference of IEEE Engineering in Medicine and Biology Society: http://embc.embs.org/

m-Health and m-Health-Related Online and Academic Journals

The Journal of mHealth: http://www.thejournalofmhealth.com

m-Health: http://www.themhealth.org/

The Journal of Mobile Technology in Medicine: http://www.journalmtm.com

Journal of Medical Internet Research: mHealth and uHealth: http://mhealth.jmir.org/

IEEE Journal of Biomedical and Health Informatics (previously IEEE Transactions on Information Technology in Biomedicine (1997–2012): http://ieeexplore.ieee.org/xpl/RecentIssue.jsp?punumber=4233

All links accessed June 2016.

INDEX

m-Health: Fundamentals and Applications, First Edition. Robert S. H. Istepanian and Bryan Woodward.
© 2017 The Institute of Electrical and Electronics Engineers, Inc. Published 2017 by John Wiley & Sons, Inc.

IEEE Press Series in Biomedical Engineering

The focus of our series is to introduce current and emerging technologies to biomedical and electrical engineering practitioners, researchers, and students. This series seeks to foster interdisciplinary biomedical engineering education to satisfy the needs of the industrial and academic areas. This requires an innovative approach that overcomes the difficulties associated with the traditional textbooks and edited collections.

Series Editor: Metin Akay, University of Houston, Houston, Texas